ANDREA DAY   MELISSA PHILLIPS   LORRAINE NORDMANN

PROFESSIONAL
BEAUTY
THERAPY

WORKBOOK

FOURTH EDITION

Professional Beauty Therapy Workbook
4th Edition
Andrea Day
Melissa Phillips
Lorraine Nordmann

Portfolio lead: Fiona Hammond
Product manager: Sandy Jayadev, Sophie Kaliniecki
Content developer: Eleanor Yeoell, James Cole
Project editor: Alex Chambers
Cover designer: Watershed Art & Design (Leigh Ashforth)
Text designer: Ruth Comey
Project Designer: Nikita Bansal
Permissions/Photo researcher: Liz McShane
Indexer: Mei Yen Chua
Proofreader: Jennifer Bulter
Cover: Stocksy United/Alexandra Bergam
Typeset by KnowledgeWorks Global Ltd.

Any URLs contained in this publication were checked for currency during the production process. Note, however, that the publisher cannot vouch for the ongoing currency of URLs.

Fourth edition published in 2022.

Adaptation of Lorraine Nordmann's:

Beauty Basics Level 1, revised 3rd Edition (9781473710603)
Beauty Therapy Foundations Level 2, 7th Edition (ISBN 9781473734562)
Professional Beauty Therapy Level 3, 5th Edition (ISBN 9781473734777)

For product information and technology assistance,
in Australia call **1300 790 853**;
in New Zealand call **0800 449 725**

For permission to use material from this text or product, please email **aust.permissions@cengage.com**

National Library of Australia Cataloguing-in-Publication Data
ISBN: 9780170446266
A catalogue record for this book is available from the National Library of Australia.

Cengage Learning Australia
Level 7, 80 Dorcas Street
South Melbourne, Victoria Australia 3205

Cengage Learning New Zealand
Unit 4B Rosedale Office Park
331 Rosedale Road, Albany, North Shore 0632, NZ

For learning solutions, visit **cengage.com.au**

Printed in China by 1010 Printing International Limited.
1 2 3 4 5 6 7 26 25 24 23 22

BRIEF CONTENTS

PART 1: THE BUSINESS OF BEAUTY

SECTION 1
THE BEAUTY PROFESSIONAL — 2

1	Comply with organisational requirements within a personal services environment	3
2	Provide salon services to clients	19
3	Advise on beauty products and services	32
4	Sell to the retail customer	43
5	Research and apply beauty industry information	50
6	Conduct salon financial transactions	62

SECTION 2
HEALTH AND SAFETY — 71

| 7 | Apply safe hygiene, health and work practices | 72 |
| 8 | Maintain infection control standards | 102 |

PART 2: THE PRACTICE OF BEAUTY

SECTION 3
NAIL TREATMENTS — 127

| 9 | Manicure and pedicure services | 128 |
| 10 | Nail art and advanced nail art | 157 |

SECTION 4
HAIR TREATMENT AND REMOVAL — 168

11	Waxing services	169
12	Lash and brow services	184
13	Eyelash extensions	197

SECTION 5
MAKE-UP — 209

14	Design and apply make-up	210
15	Remedial camouflage make-up	226
16	Photographic make-up	235

SECTION 6
BODY TREATMENTS AND SERVICES — 243

17	Cosmetic tanning	244
18	Provide body massages	257
19	Aromatherapy massages	278
20	Aromatic plant oil blends	293

SECTION 7
FACIAL TREATMENTS — 306

| 21 | Provide facial treatments and skincare recommendations | 307 |
| 22 | Specialised facial treatments | 348 |

PART 3: THE SCIENCE OF BEAUTY

SECTION 8
SKIN AND BODY SCIENCE — 370

23	Promote healthy nutritional options in a beauty therapy context	371
24	Body structures and systems	387
25	Skin science	419
26	Incorporate knowledge of anatomy and physiology into beauty therapy	454

PART 4: ONLINE-ONLY CHAPTERS

SECTION 9
SPA SERVICES

On1	Work in a spa therapies framework
On2	Provide spa therapies
On3	Stone therapy massages
On4	Indian head massages for relaxation

SECTION 10
BODY SERVICES

On5	Reflexology
On6	Provide superficial lymph drainage massages
On7	Nail enhancements and electric file equipment

SECTION 11
HAIR REDUCTION

| On8 | Permanent hair reduction |
| On9 & On10 | Female and male intimate waxing |

SECTION 12
SKIN SERVICES

On11	Provide diathermy treatments
On12	Piercing ear lobes
On13	Provide micro-dermabrasion treatments

Cross-sector appendix: Using social media and online platforms 466

CONTENTS

Guide to the text viii
Guide to the online resources x
Preface xii
About the authors xiii
Acknowledgements xiv

PART 1 THE BUSINESS OF BEAUTY — 1

SECTION 1 THE BEAUTY PROFESSIONAL — 2

1 Comply with organisational requirements within a personal services environment 3
Introduction 3
Your employment rights and responsibilities 3
Working within organisational requirements 7
Supporting your work team 8
Maintaining personal presentation 15
Developing effective work habits 16

2 Provide salon services to clients 19
Introduction 19
Receiving clients 19
Develop rapport, anticipate contingencies, promote products and services, and process sales, returns and refunds 24
Scheduling appointments for clients 27
Responding to client complaints 28
Responding to clients with special needs 30

3 Advise on beauty products and services 32
Introduction 32
Developing product knowledge 32
Recommending salon services 36
Preparing a client for product demonstration 38
Demonstrating beauty care products 39

4 Sell to the retail customer 43
Introduction 43
Establishing customer needs 43
Providing advice on products and services 46
Facilitating the sale of products and services 48

5 Research and apply beauty industry information 50
Introduction 50
Sourcing and using information on the beauty industry 50
Researching legal and ethical issues 53
Updating your knowledge of the beauty industry and products 59

6 Conduct salon financial transactions 62
Introduction 62
Operating POS equipment 62
Completing POS transactions 64
Completing refunds 67
Removing takings from register or terminal 68
Reconciling takings 70

SECTION 2 HEALTH AND SAFETY — 71

7 Apply safe hygiene, health and work practices — 72
Introduction — 72
Protecting yourself against infection risks — 72
Applying salon safety procedures — 74
Using electricity safely — 84
Minimising infection risks in the salon environment — 89
Following infection control procedures — 90
Following procedures for emergency situations — 98
Cleaning the salon — 99

8 Maintain infection control standards — 102
Introduction — 102
Complying with infection control legal obligations — 102
Monitoring hygiene of premises — 105
Maintaining infection control for skin penetration treatments — 106
Sterilising equipment and maintaining steriliser — 121
Maintaining awareness of clinic design for control of infection risks — 124

PART 2 THE PRACTICE OF BEAUTY — 126

SECTION 3 NAIL TREATMENTS — 127

9 Manicure and pedicure services — 128
Introduction — 128
Establishing client priorities — 128
Preparing for the nail service — 141
Providing the nail service — 144
Reviewing the service — 152
Cleaning the treatment area — 155

10 Nail art and advanced nail art — 157
Introduction — 157
Establishing client priorities — 157
Preparing the service area — 161
Applying nail art — 163
Applying advanced nail art — 164
Reviewing the service — 165
Cleaning the service area — 166

SECTION 4 HAIR TREATMENT AND REMOVAL — 168

11 Waxing services — 169
Introduction — 169
Establishing client priorities — 169
Preparing for waxing service — 176
Applying and removing wax — 177
Reviewing waxing service and providing post-service advice — 179
Cleaning the treatment area — 182

12 Lash and brow services — 184
Introduction — 184
Establishing client priorities — 184
Preparing for a lash and brow service — 188
Chemically treating eyelashes and eyebrows — 190
Shaping eyebrows — 191
Reviewing the service and providing post-service advice — 192
Cleaning the treatment area — 195

13 Eyelash extensions — 197
Introduction — 197
Establishing client priorities — 197
Preparing the treatment area, yourself and the client — 200
Applying eyelash extensions — 203
Reviewing the service and providing post-service advice — 206
Cleaning the treatment area — 208

SECTION 5 MAKE-UP — 209

14 Design and apply make-up — 210
Introduction — 210
Establishing make-up requirements — 210
Designing the make-up plan and application techniques — 213
Providing make-up services — 222
Applying false eyelashes — 223
Providing post-service advice — 224
Cleaning the service area — 225

15 Remedial camouflage make-up — 226
Introduction — 226
Establishing remedial camouflage make-up requirements — 226
Designing the make-up plan — 230
Applying remedial camouflage make-up — 232
Reviewing the service — 233
Cleaning the treatment area — 234

16 Photographic make-up — 235
Introduction — 235
Analysing photography context — 235
Establishing make-up requirements — 237
Designing the make-up plan — 238
Applying make-up for photography — 240
Cleaning tools and equipment — 241

SECTION 6 BODY TREATMENTS AND SERVICES — 243

17 Cosmetic tanning — 244
Introduction — 244
Establishing client priorities — 244
Preparing to apply cosmetic tanning products — 246
Applying product with the spray gun — 250
Reviewing the service — 252
Cleaning the treatment area — 255

18 Provide body massages — 257
Introduction — 257
Establishing client priorities — 257
Designing the massage treatment — 266
Preparing for body massage treatments — 270
Providing body massages — 272
Reviewing massage and providing body care advice — 274
Cleaning the treatment area — 277

19 Aromatherapy massages — 278
Introduction — 278
Establishing client priorities — 278
Designing and recommending aromatherapy massage treatment — 281
Preparing for aromatherapy massage treatment — 286
Providing aromatherapy massage treatment — 287
Reviewing massage treatment and providing post-treatment advice — 288
Cleaning the treatment area — 290

20 Aromatic plant oil blends — 293
Introduction — 293
Establishing client priorities — 293
Designing oil blends — 295
Setting up for blending — 301
Preparing aromatic oil blends — 302
Providing treatment using aromatic oil blends — 302
Reviewing treatment and providing post-treatment advice — 303
Cleaning the treatment area — 305

SECTION 7 FACIAL TREATMENTS 306

21 Provide facial treatments and skincare recommendations 307
Introduction 307
Establishing client priorities 307
Designing and recommending facial treatments 311
Preparing for facial treatments 313
Cleansing and exfoliating the skin and performing extractions 314
Providing facial massage 316
Applying specialised products 317
Reviewing the facial treatment 343
Recommending post-treatment skincare regimen 345
Cleaning the treatment area 346

22 Specialised facial treatments 348
Introduction 348
Establishing client priorities 348
Designing and recommending specialised facials 351
Preparing for specialised facial treatments 353
Cleansing skin using ultrasonic or galvanic 355
Removing minor skin blemishes and infusing serums 360
Completing the treatment 363
Providing post-treatment advice 363
Cleaning the treatment area 368

PART 3 THE SCIENCE OF BEAUTY 369

SECTION 8 SKIN AND BODY SCIENCE 370

23 Promote healthy nutritional options in a beauty therapy context 371
Introduction 371
Principles of nutrition 371
Applying knowledge of body systems to beauty therapy treatments 382
Providing advice on dietary guidelines 384

24 Body structures and systems 387
Introduction 387
Structural levels of organisation 387
The musculoskeletal system 389
The nervous system 401
The circulatory system 403
The endocrine system 405
Organ systems and their relationship to a healthy body 408

25 Skin science 419
Introduction 419
Skin structure 419
Glands of the skin 431
Appearance of skin 433
Growth, development, ageing and healing of human skin 444
Scope of practice 450

26 Incorporate knowledge of anatomy and physiology into beauty therapy 454
Introduction 454
Using knowledge of body and skin structures and systems to determine client needs and design the treatment 454
Advising the client 458
Recording relevant data 463
Maintaining knowledge of anatomy and physiology 464

Cross-sector appendix: Using social media and online platforms 466
Introduction 466
Source information on the general impacts of social media 466
Comply with industry and organisational ethical and professional codes of conduct for online activities 473
Maintain personal online presence consistent with organisational standards 478
Protect customer privacy and maintain confidentiality of organisational information 480
Source information on copyright 485
Engage professionally with customers online 487

Index 493

Guide to the text

As you read this text you will find a number of features in every chapter to enhance your study of beauty therapy and help you understand how the theory is applied in the real world.

PART-OPENING FEATURES

A **Section list** outlines the section and chapters contained in each part for easy reference.

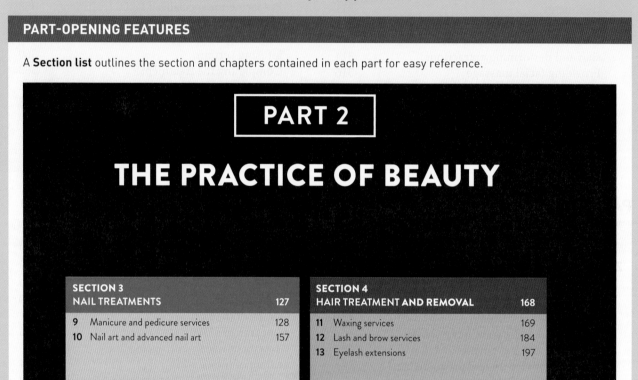

PART 2

THE PRACTICE OF BEAUTY

SECTION 3
NAIL TREATMENTS 127

9 Manicure and pedicure services 128
10 Nail art and advanced nail art 157

SECTION 4
HAIR TREATMENT AND REMOVAL 168

11 Waxing services 169
12 Lash and brow services 184
13 Eyelash extensions 197

SECTION-OPENING FEATURES

Refer to the **Chapter list** for an outline of the chapters in each section.

SECTION 6

BODY TREATMENTS AND SERVICES

17 Cosmetic tanning
18 Provide body massages
19 Aromatherapy massages
20 Aromatic plant oil blends

CHAPTER-OPENING FEATURES

Identify the key concepts you will engage with through the **Learning objectives** at the start of each chapter.

CHAPTER 17: COSMETIC TANNING

LEARNING OBJECTIVES

After completing this chapter, you should be able to:

LO17.1 establish client priorities with thorough consultation, skin analysis and patch testing

LO17.2 select and prepare tanning products and equipment, and prepare yourself, the treatment area and the client for comfort and according to WHS legislative requirements

LO17.3 use spray tanning equipment to apply cosmetic tan according to salon protocol and the treatment plan

LO17.4 review the service, providing aftercare and homecare recommendations and a future treatment plan

LO17.5 clean the treatment area in preparation for the next client.

INTRODUCTION

Cosmetic tanning products have been in Australia and New Zealand since 1960 when Coppertone® released its first cosmetic tanning product QT®, or 'Quick Tanning Lotion'. Cosmetic tanning is a temporary tan that gives the skin a healthy, bronzed appearance without the harmful effects of UV radiation.

The learning activities in this chapter allow you to demonstrate your knowledge and understanding of the unit of competency *SHBBBOS007 Apply cosmetic tanning products*, using the performance and knowledge evidences.

FEATURES WITHIN CHAPTERS

Throughout each chapter you will find icons that help you to understand how the activities in the workbook link to assessable material in your course and relate directly to the training package. The **PE icons** at the beginning of a section identify where the **performance evidence** is addressed in the chapter and can be applied to the knowledge across the entire section. The **KE icons** highlight when the **knowledge evidence** is referenced and indicate a specific learning.

Performance Evidence

Knowledge Evidence

LO1.1 YOUR EMPLOYMENT RIGHTS AND RESPONSIBILITIES

Performance Evidence

Knowing your rights and entitlements as a worker means that you will get what you deserve and that your employer can't treat you unfairly. Likewise, you need to understand your responsibilities so that you don't find yourself in the wrong, or making false accusations.

When an icon is found at a subheading throughout the section, this highlights a specific learning.

Industrial awards for hairdressing and beauty employees

Knowledge Evidence

There are aspects to the Hair and Beauty Industry Award that are relevant to your job role; for example, if you work only on nails and are qualified at Certificate II level, you can expect to be paid less than a person that is qualified at Diploma Level and is able to perform specialised facials and permanent hair removal.

Guide to the online resources

Cengage is pleased to provide you with a selection of resources
that will help you prepare your lectures and assessments.
Contact your Cengage learning consultant for more information.

MINDTAP

Premium online teaching and learning tools are available on the *MindTap* platform – the personalised eLearning solution.

MindTap is a flexible and easy-to-use platform that helps build student confidence and gives you a clear picture of their progress. We partner with you to ease the transition to digital – we're with you every step of the way.

The *Cengage Mobile App* puts your course directly into students' hands with course materials available on their smartphone or tablet. Students can read on the go, complete practice quizzes or participate in interactive real-time activities. MindTap is full of innovative resources to support critical thinking, and help your students move from memorisation to mastery!

MindTap is a premium purchasable eLearning tool. Contact your Cengage learning consultant to find out how MindTap can transform your course.

WORKBOOK SOLUTIONS MANUAL

The **Solutions manual** provides detailed answers to the questions in the Workbook.

MAPPING GRID

The **Mapping grid** is a simple grid that shows how the content of this book relates to the units of competency across qualifications in the Beauty Services Training Package.

INSTRUCTORS' CHOICE RESOURCE PACK

This optional, purchasable pack of premium resources provides additional teaching support, saving time and adding more depth to your classes. These resources cover additional content with an exclusive selection of engaging features aligned with the text. Contact your Cengage learning consultant to find out more.

ONLINE WORKBOOK CHAPTERS

Workbook chapters written to accompany the online-only chapters and enhance student understanding.

ONLINE CHAPTERS SOLUTIONS MANUAL

The **Solutions manual for online chapters** provides detailed answers to every question in the workbook.

ARTWORK FROM THE WORKBOOK

Add the **digital files** of graphs, tables, pictures and flow charts into your learning management system, use them in student handouts, or copy them into your lecture presentations.

FOR THE STUDENT

MINDTAP

MindTap is the next-level online learning tool that helps you get better grades!

MindTap gives you the resources you need to study – all in one place and available when you need them. In the *MindTap Reader*, you can make notes, highlight text and even find a definition directly from the page.

If your instructor has chosen *MindTap* for your subject this semester, log in to *MindTap* to:
• Get better grades
• Save time and get organised
• Connect with your instructor and peers
• Study when and where you want, online and mobile
• Complete assessment tasks as set by your instructor

When your instructor creates a course using *MindTap*, they will let you know your course link so you can access the content. Please purchase *MindTap* only when directed by your instructor. Course length is set by your instructor.

PRINTED STUDENT TEXTBOOK

The Workbook for *Professional Beauty Therapy* is designed to work in conjunction with the student textbook, confirming the essential knowledge presented throughout the text. Purchase this text at au.cengage.com/student.

PREFACE

The *Professional Beauty Therapy Workbook*, fourth edition, is the companion to *Professional Beauty Therapy*, fourth edition, and was developed to complete the delivery of the current Australian Beauty Therapy training package. It reflects the beauty and spa industries and can be used to assess or to provide evidence of participation for each beauty qualification, with activites that are mapped to the performance evidence and knowledge evidence for every unit.

Pedagogy is made easy for instructors with more advanced information, learning activities and assessment tools accessed via the instructor and training resources. *Professional Beauty Therapy*, along with its associated e-resources, can be used alone or be coupled with your training organisation's resources.

For students, the book has a realistic approach to beauty therapy and leaves a strong impression of what to expect when working in the industry. Beauty students learn their craft from quality texts, though learning can be structured with the rigidity of the rules and regulations of the training package.

This text is an innovative way to professionally deliver the competencies of the training package and with a creative touch.

The practice of beauty can be found here in one conveniently cross-referenced text.

How to use this workbook

The *Professional Beauty Therapy Workbook* has been written to be used in conjunction with the student textbook, *Professional Beauty Therapy*. Each chapter in the workbook corresponds to a chapter in the student book, and both are written to cover the units of compentencies in the SHB Beauty Training Package in their entirety. Developed to confirm the practical knowledge in the student textbook, the workbook for *Professional Beauty Therapy* enriches student learning across the Beauty Services qualifications. Through a series of activities, the students' understanding of the theory and concepts will be expanded and enhanced. The workbook covers the SHB30119 Certificate III in Beauty, the SHB40119 Certificate IV in Beauty and the SHB50119 Diploma of Beauty Therapy in their entirety.

The workbook has been structured to mirror the *Professional Beauty Therapy* textbook, by keeping learning outcomes and section headings consistent across both texts, thereby providing students with a streamlined learning experience. The workbook has been further enhanced with Performance Evidence and Knowledge Evidence icons to indicate where content in the workbook links to assessable material in your course, highlighting the underpinning knowledge needed for practical content.

ABOUT THE AUTHORS

Andrea Day has over 21 years of beauty industry experience in both Australia and the UK. Her training and assessing experience span 17 years. Throughout her career, she has kept her hand in the industry, working in beauty salons and spas in Melbourne and Central London. She is currently Senior Educator of Hair, Make-up and Beauty at Victoria University Polytechnic, Melbourne, after her 16th year as head teacher, assessor and coordinator. As a qualified vocational and educational training (VET) practitioner, she is focused on quality, inclusion and adaptability as she actively connects with industry through RTO and beauty trainer networks, colleagues, industry peers and students for feedback and inspiration.

Melissa Phillips has over 12 years of experience in the fields of education and science. She has a Master of Teaching, which provides her with a strong pedagogical framework for structured content delivery and assessment. Melissa has a Bachelor of Science (Honours) and worked as a geologist; she then transitioned to education, where she enjoys teaching secondary students and is currently working as a coordinator and science/maths teacher. Melissa also has a Bachelor of Multimedia, which developed her ability to create engaging activities for assessment and participation in the blended sphere.

Melissa would like to thank her husband Josh and her daughter Maggie for supporting her and allowing her the space to write.

Lorraine Nordmann has over 20 years of experience in the beauty therapy industry – an industry she has seen grow significantly in status, technology and specialties during her roles as lecturer, assessor and external verifier for City & Guilds. She aims to ensure that this high profile of the beauty therapy industry is maintained. Lorraine is also the author of the Habia official guides, *Beauty Basics and Beauty Therapy – The Foundations*, covering the latest industry standards at NVQ/SVQ Levels 1 and 2.

ACKNOWLEDGEMENTS

Cengage and the authors wish to acknowledge the following reviewers, whose feedback has helped to shape this new edition:

- Renee Burns – North Metropolitan TAFE
- Vikki Arnold – TasTAFE
- Kirsten Fagan – Holmesglen Institute
- Caterina Harb – TAFE NSW
- Sharon Campbell – South Metropolitan TAFE
- Bianca Wicks – TAFE NSW
- Ellen Lewis – TAFE NSW
- Michelle Cario – Tasmanian Academy of Hair and Beauty
- Aikaterina Vlahos – Canberra Institute of Technology
- Katrina McIntyre – Australian College of Beauty Therapy
- Sarah Chircop – Sunraysia Institute of TAFE
- Karalyn Smith – TAFE NSW

The authors would also like to thank Mallysa Neeland for her work on Chapter 16. Additionally, Andrea would like to thank her colleagues at VU Polytechnic and her VET industry peers at other RTOs for their counsel throughout the writing process.

PART 1

THE BUSINESS OF BEAUTY

SECTION 1
THE BEAUTY PROFESSIONAL   2

1   Comply with organisational requirements within a personal services environment   3

2   Provide salon services to clients   19

3   Advise on beauty products and services   32

4   Sell to the retail customer   43

5   Research and apply beauty industry information   50

6   Conduct salon financial transactions   62

SECTION 2
HEALTH AND SAFETY   71

7   Apply safe hygiene, health and work practices   72

8   Maintain infection control standards   102

SECTION 1

THE BEAUTY PROFESSIONAL

1 Comply with organisational requirements within a personal services environment
2 Provide salon services to clients
3 Advise on beauty products and services
4 Sell to the retail customer
5 Research and apply beauty industry information
6 Conduct salon financial transactions

CHAPTER 1: COMPLY WITH ORGANISATIONAL REQUIREMENTS WITHIN A PERSONAL SERVICES ENVIRONMENT

LEARNING OBJECTIVES

After completing this chapter, you should be able to:

LO1.1 understand your employment rights and responsibilities, as well as specific employment agreements you may have with your employer and how to comply with them

LO1.2 follow salon policies and procedures by working within the organisational requirements of your salon

LO1.3 support the salon team, and understand the importance of courteous, helpful and non-discriminatory behaviour

LO1.4 demonstrate an ability to maintain personal hygiene and personal presentation

LO1.5 show employability skills through effective work habits, such as prioritising, following instructions promptly and seeking advice and direction when required.

INTRODUCTION

The beauty industry is built on a vast range of products and services that help the clients to look their best. Skincare, cosmetic and spa consumers are increasingly savvy, driving an expectation of industry innovations and more advanced information and skills. The most valuable aspect of a beauty salon or spa is the staff; that is, the professional skills, knowledge and attitude that a salon or spa aims to project to clients. In this chapter, you will learn about general codes of conduct, employability skills, communication with colleagues and clients, ethical and legal considerations and other attributes that bring these professional qualities to your work.

The learning activities in this chapter allow you to demonstrate your knowledge and understanding of the unit of competency *SHBXIND003 Comply with organisational requirements within a personal services environment*, using the performance and knowledge evidences.

LO1.1 YOUR EMPLOYMENT RIGHTS AND RESPONSIBILITIES

Knowing your rights and entitlements as a worker means that you will get what you deserve and that your employer can't treat you unfairly. Likewise, you need to understand your responsibilities so that you don't find yourself in the wrong, or making false accusations.

Sources of information

Performance Evidence

Knowledge Evidence

EMPLOYER ASSOCIATIONS

Employer associations are organisations that help members of a specific industry to liaise and negotiate through any disputes or issues. Two of the major employer associations in the beauty industry are the HBIA, ABIC and APAN. Do some research online to answer the following questions.

1   What do the acronyms HBIA, ABIC and APAN stand for?

_____

2 What groups do each of these associations represent?

_____

3 Research online to see if you can find another major industry association in the hair and beauty industry. What is its name, and who does it represent?

_____

4 If you were a nail technician working in a salon and had a grievance with your employer, which association could you turn to for help? How might it be able to help you in such a situation?

_____

_____

FAIR WORK OMBUDSMAN

Look at the website of the Fair Work Ombudsman (**https://www.fairwork.gov.au**) to answer the following questions.

5 What is the name of the Award that covers the beauty industry?

_____

6 Who is not covered by this Award?

_____

_____

_____

7 In the pay guide, what would a beauty therapist level 3, who is employed casually, get paid for a shift they complete on a Saturday between 9 a.m. and 12 p.m. (before tax)?

_____

8 Use the leave calculator and select the Award relevant to the beauty industry. As a beauty therapist level 3, how much annual leave would you have if you:

- had started working exactly one year earlier
- had taken no leave
- worked full-time, Monday to Friday, 7 hours a day.

_____

FAIR WORK COMMISSION

Look at the website of the Fair Work Commission (**https://www.fwc.gov.au**) to answer the following questions.

9 Bring up the details of the Award relevant to the beauty industry. If you were a beautician who holds a Certificate III in Beauty Services (or equivalent), what classification would you be? (Hint: look at Schedule B of the Award).

_____

10 What is the minimum weekly wage for someone employed full-time at this classification?

_____

11 What is the average number of hours of work that is considered ordinary in a week?

_____

Basic aspects of employment laws

Performance Evidence

Knowledge Evidence

FAIR WORK ACT 2009: MINIMUM WORKPLACE ENTITLEMENTS PROVIDED BY THE NATIONAL EMPLOYMENT STANDARDS (NES)

12 Having a Fair Work Information Statement and being given long service leave are two of the 13 minimum entitlements of the National Employment Standards (NES). List the other 11 minimum entitlements.

13 Which of these entitlements would you get if you were a casual employee?

STATE AND TERRITORY GOVERNMENT BOARDS AND COMMISSIONS FOR ANTI-DISCRIMINATION AND EQUAL EMPLOYMENT OPPORTUNITY (EEO)

Laws differ in some states and territories because there are federal laws and state and territory laws that do not always overlap, so it is important to check the laws that apply to you. The states and territories have individual government boards or commissions that enforce anti-discrimination and equal opportunity legislation and offer support services. Here is a list of the state and territory government boards:

- Australian Capital Territory: ACT Human Rights Commission
- New South Wales: Anti-Discrimination Board of NSW
- Northern Territory: Northern Territory Anti-Discrimination Commission
- Queensland: Anti-Discrimination Commission Queensland
- South Australia: Equal Opportunity Commission
- Tasmania: Equal Opportunity Tasmania
- Victoria: Victorian Equal Opportunity & Human Rights Commission
- Western Australia: Equal Opportunity Commission
- National: Australian Human Rights Commission
- New Zealand: Human Rights Commission (New Zealand).

14 Search your state or territory government board or commission for various types of support services in relation to anti-discrimination and equal employment opportunity that exist in your state or territory and list them in the following table.

One source of information may be: **https://www.humanrights.gov.au/about/links-human-rights-organisations-and-resources**

TYPE OF ANTI-DISCRIMINATION OR EEO	ORGANISATION	ONE KEY FUNCTION
Age		
Carer responsibilities		
Criminal record		

TYPE OF ANTI-DISCRIMINATION OR EEO	ORGANISATION	ONE KEY FUNCTION
Disability		
Domestic and family violence		
Race		
Gender		
Bullying		
Pregnancy and working parents		

Industrial awards for hairdressing and beauty employees

There are aspects to the Hair and Beauty Industry Award that are relevant to your job role; for example, if you work only on nails and are qualified at Certificate II level, you can expect to be paid less than a person who is qualified at Diploma Level and is able to perform specialised facials and permanent hair removal.

Knowledge Evidence

You are a qualified beauty therapist who has just finished a Diploma of Beauty Therapy. Look up your Industry Award (**https://www.fwc.gov.au**) and answer the following questions.

15 What is the hourly rate for a weekday?

_____

16 What is the hourly rate for a Saturday?

_____

17 What is the hourly rate for a Sunday?

_____

18 What is the hourly rate for a public holiday?

_____

19 What is the hourly rate for overtime?

_____

20 What is the hourly rate for working on a rostered day off (RDO)?

_____

21 After one year of experience, does your wage go up?

_____

LO1.2 WORKING WITHIN ORGANISATIONAL REQUIREMENTS

Performance
Evidence

Each salon will have its own requirements that each employee must ensure they follow. It is the employee's responsibility to make sure they understand and comply with the salon's requirements. There are several resources available for employees outlining their rights and responsibilities that can be researched online to help answer the below questions.

Sourcing information

STAFF HANDBOOKS

Staff handbooks give the salon staff guidance and the skills to follow the salon policies and procedures. They should include information on:

- health and safety
- appointment scheduling
- communication skills
- how to deal with problem clients
- telephone communication
- refund procedures
- training
- leave/rosters
- personal presentation
- job descriptions
- trade unions.

1   Select one aspect of the staff handbook and describe what you would expect to find.

_____

_____

_____

_____

_____

_____

TRADE UNIONS

Knowledge
Evidence

A trade union is a group of employees and union officials who come together to promote the members of the union and their common interests.

2   What is the name of the trade union that members of the beauty industry would belong to?

_____

3   What role do unions play in the development of industry Awards?

_____

_____

The trade union can offer advice specific to the Hair and Beauty Industry Award. You need to pay a regular fee to join a union, and they should provide information relating to:

- contracts of employment
- rosters and leave.

4 Find out two pieces of information your trade union can offer relating to:

- types of breaks
- penalty rates.

LO1.3 SUPPORTING YOUR WORK TEAM

Performance
Evidence

Employer workplace rights and responsibilities

Performance
Evidence

RELEVANT STATE OR TERRITORY ANTI-DISCRIMINATION OR EEO LAW

Anti-discrimination laws are protected by overarching legislation. Research the anti-discrimination laws that are relevant to your state or territory at the Attorney-General's Department website: **https://www.ag.gov.au/RightsAnd Protections/HumanRights/Pages/Australias-Anti-Discrimination-Law.aspx**

Answer the following questions.

1 If you were to make a complaint about your employer, who you thought was acting unlawfully and in a discriminatory manner, where would you lodge the complaint in your state or territory?

2 What is the time limit for a complaint in your state or territory?

Refer to **http://www.humanrights.gov.au/employers/good-practice-good-business-factsheets/good-practice-guidelines-internal-complaint** to learn more about the characteristics of model staff complaint procedures and answer the following questions.

3 How could you make the complaints procedure more accessible to a staff member who speaks English as a second language?

4 What is the most appropriate way to discuss a staff complaint?

ROLE OF RELEVANT STATE OR TERRITORY BOARD OR COMMISSION IN MANAGING COMPLAINTS

There is an anti-discrimination commission for every state and territory of Australia. The websites are listed here:

- Tasmania: **http://equalopportunity.tas.gov.au**
- Australian Capital Territory: **http://hrc.act.gov.au**
- New South Wales: **http://www.antidiscrimination.justice.nsw.gov.au**
- Victoria: **https://www.humanrightscommission.vic.gov.au**
- Queensland: **https://www.qhrc.qld.gov.au**
- South Australia: **http://www.eoc.sa.gov.au**
- Western Australia: **http://www.eoc.wa.gov.au**
- Northern Territory: **http://www.adc.nt.gov.au**

5 Identify and list five services the commission in your state or territory provides.

_____

_____

_____

_____

_____

_____

_____

_____

_____

_____

_____

HARASSMENT PROVISIONS: TYPES OF DISCRIMINATION AND HARASSMENT, RIGHTS AND RESPONSIBILITIES OF EMPLOYEES AND EMPLOYERS, AND CONSEQUENCES OF NON-COMPLIANCE WITH THE LAW

6 Which law specifically protects the client, if they are an immigrant to Australia, from language and cultural discrimination?

_____

7 Refer to the section 'Non-discriminatory behaviours' in your textbook. Consider non-discriminatory actions that demonstrate a cultural understanding. Which of the following are examples of discrimination? (Select all that apply.)

A Making a joke about a person's (language) accent.

B Asking a person who speaks English as a second language to repeat a word so that you can understand it.

C Insisting all people perform all work duties irrespective of cultural or religious faith; for example, all women should provide treatments to men.

D Deciding to employ a Greek-speaking person in a salon over an equally qualified or capable therapist because the area has a high Greek population.

8 Complete the following table by stating if you feel the behaviour is discriminatory or not. If yes, list which type of discrimination it is.

	DISCRIMINATORY BEHAVIOUR?		LIST THE TYPE OF DISCRIMINATION; E.G. RACIAL, SEXUAL, ETC.
	YES	NO	
A beauty therapist informs a potential client over the phone that they cannot schedule an appointment as the lift is not working and there is no other wheelchair access to the salon.			
A beauty therapist is Indigenous Australian and asks permission to take leave to attend a funeral. He is given leave to stay away for three days to fulfil traditional obligations to his family.			
A beautician working at the make-up counter of a department store refuses to give the complimentary makeover to a girl who is Indigenous Australian, stating that she doesn't stock fashion colours suitable for dark skin.			
A salon has a gift voucher loyalty scheme that allows women to return for follow-up treatments. When a male client enquired, they were told they would be given the same voucher.			
An apprentice from Argentina complains to her manager that workmates make jokes about migrants from South America, and she finds them offensive. She asks the manager to speak to her colleagues.			
A manager refuses to take on clients who speak with a foreign accent because she says she can't understand them, and gets younger staff to take those appointments instead.			
A hair stylist is Muslim and is told she must work evening shifts through the period of Ramadan (a period of fasting in daylight hours that lasts for one month).			
A salon manager makes a point of refusing to permit clients to have accounts with the salon; all treatments must be paid in full at the time. However, he allows those who talk to him in the language of his country of birth to have an account.			

REPROVISIONS: RIGHTS OF EMPLOYEES AND RESPONSIBILITIES OF EMPLOYERS TO MAKE MERIT-BASED EMPLOYMENT DECISIONS

Your employer will at times be required to make decisions based on your work performance and qualifications. There are, however, discrimination laws that protect you from unfair decisions.

For each of the following scenarios, suggest a fair decision:

9 You have worked at the salon for one year and have requested to have every second Saturday off. The recently employed beauty therapist, equally qualified, has requested to take every Saturday off, but she asked first. What would be a fair decision?

_____

10 It is common to have a commission-based pay structure to encourage performance targets; for example, percentage on retail sales and percentage on treatment sales is calculated and added to income. You have been making more retail sales this year than other staff, but are underperforming in treatments. How should you expect to be rewarded?

_____

Organisational policies and procedures

Performance Evidence   Knowledge Evidence

ACCEPTING, DECLINING AND AMENDING ROSTERED HOURS

The following scenario relates to rosters in a salon. There are many reasons why staff may prefer to work different shifts, and working out rosters to accommodate every employee can be a real headache. The bigger the workplace the more complex it gets.

11 In this scenario, you are to work out a roster for a two-week cycle for the salon using the information provided:

- The salon is open from 9 a.m. until 6:30 p.m. from Monday to Friday, and on Saturday when it's open until 5 p.m.

- Between 12 p.m. and 2:30 p.m. is the busiest time, with many clients coming in during their lunchtime.

- Thursdays and Saturdays are the busiest days, so three members of staff are required to work on those days.

- Grace is the salon manager. She needs to be in the salon on Tuesdays, Wednesdays and Thursdays to get the most administrative work done. She prefers to be there for closing to ensure the books are balanced.

- Two staff are needed to work 9 a.m. to 10:30 a.m. and two to work 10.30 a.m. to 6.30 p.m. Grace asked her staff about their preferences and received the following responses:

  - Matt: 'I'd rather start early and finish early so I can get plenty of rest before Saturday, and have Saturdays off for football.'

  - Liz: 'I have to help in the family restaurant at night. I can't work past 5 p.m.'

  - Frieda: 'When I'm on late shift I have to pay someone to mind my kids after school, so I lose money on late shifts.'

Week 1

	MATT	LIZ	GRACE	FRIEDA
Monday				
Tuesday				
Wednesday				
Thursday				
Friday				
Saturday				

Week 2

	MATT	LIZ	GRACE	FRIEDA
Monday				
Tuesday				
Wednesday				
Thursday				
Friday				
Saturday				

12 While employers should try to meet the staff's personal preferences, especially when they relate to family commitments, employers must also meet certain requirements to get the job done. Who did you put on the Saturday shift? Why?

_____

_____

_____

13 If you wanted to change your personal roster in the salon, what steps would you take?

PERSONAL AND CARER'S LEAVE

Research Australian law in relation to personal and sick leave at the Fair Work Ombudsman website: **https://www.fairwork.gov.au/leave/sick-and-carers-leave**

14 What is sick leave?

----

15 Who is entitled to carer's leave?

----

16 What other kinds of personal leave are there?

COUNSELLING AND DISCIPLINE

Your employer is responsible for managing your underperformance by following fair procedures. Underperformance should be followed up by the employer in the following steps:

1 counselling: a meeting to discuss the underperformance
2 warning: a verbal warning.

Research performance management and warnings at the Fair Work Ombudsman website: **https://www.fairwork.gov.au/employee-entitlements/managing-performance-and-warnings**

17 What is underperformance?

----

18 What should an employer do first when they notice an employee is underperforming?

19 What should be documented?

20 What is serious misconduct?

GRIEVANCES

21 What is a grievance?

22 A grievance can stem from either the employee or the employer. List two scenarios of each.

EQUAL EMPLOYMENT OPPORTUNITY

23 Under the *Equal Opportunity Act 2010*, what do businesses need to ensure they do when hiring staff? What is an example of an adverse action in this?

DISCRIMINATION AND HARASSMENT

24 Give some examples of how age discrimination can manifest itself in the salon.

TERMS AND CONDITIONS OF EMPLOYMENT

25 What is the timeframe in which an employer should provide a contract of employment to a new employee?

26 Give definitions of the following items that might be found in a contract:

a performance-based pay scale

_____

_____

_____

b entitlements

_____

_____

_____

c salon rosters

_____

_____

_____

Typical terms and conditions of employment

Knowledge Evidence

Terms and conditions of employment are negotiated with your employer when you sign the contract of employment. They include things such as:

- employment as a permanent, casual, full-time or part-time worker
- hours of work
- wages, salary levels or remuneration packages
- any other terms and conditions offered to you at the start of employment.

It does not include basic Award conditions that apply to every beauty therapist employed in Australia or New Zealand.

27 Which of the following options are terms and conditions of employment?

- Flexible work conditions
- Sick leave
- Access to training
- Dress code
- Rostered hours
- Overtime Award rates

General role boundaries for industry staff

Knowledge Evidence

CHAIN OF COMMAND

28 Review Figure 1.3 Job description in Chapter 1 of the textbook that describes the position description of a beauty therapist. Also review the job responsibilities outlined in Figure 1.1. With the assistance of your teacher, draw the hierarchy of the chain of command for the following team members:

- Nail technician
- Salon owner
- Beautician
- Beauty therapist
- Assistant manager
- Salon manager – business performance

Figure 1.1   Job responsibilities in the salon

JOB TITLE	RESPONSIBILITY
Nail technician	Performing nail treatments, and adhering to legal requirements and salon policies and procedures
Salon owner	Business registration, compliance, staffing
Salon manager	Business performance, rosters
Beautician	Performing beauty treatments to Certificate III–IV level, and adhering to legal requirements and salon policies and procedures
Beauty therapist	Performing beauty treatments to Diploma-level, and adhering to legal requirements and salon policies and procedures
Assistant manager	Assists the salon manager and can act as salon manager in their absence; often a beauty therapist as well

LO1.4 MAINTAINING PERSONAL PRESENTATION

Performance Evidence

Applying workplace dress, hygiene and personal presentation requirements

1   Look at Figure 1.2. List three things that are shown in the images that represent a good uniform for a beauty therapist.

_____

_____

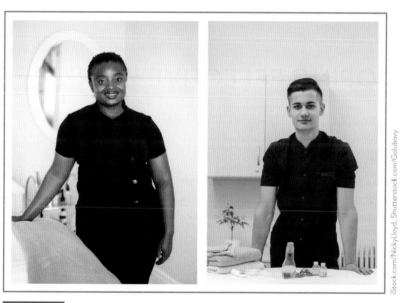

iStock.com/NickyLloyd; Shutterstock.com/Golubovy

Figure 1.2   Female and male beauty uniforms

2 How should a professional beauty therapist's nails look?

_____

3 A client arrives at the salon for a massage and is introduced to the massage therapist who will carry out the treatment. The therapist has long, dirty fingernails, her uniform is crumpled, and she smells like she hasn't showered. The client complains and demands a different therapist.

a Why are the aspects of the massage therapist that upset the client against the personal presentation and hygiene rules of a salon?

_____

_____

_____

b Is the client being unreasonable? Why?

_____

_____

_____

c What steps might the salon supervisor take to remedy the situation so the therapist doesn't put any future clients off their treatment?

_____

_____

_____

_____

LO1.5 DEVELOPING EFFECTIVE WORK HABITS

Performance Evidence

1 In the following scenarios, give an example of how the therapist could have handled the situation in a way that shows effective work habits:

a Janelle has been working at a new salon for a week when her colleague asks her to finalise the sales for the end of day. The system at the new salon is different from any that Janelle has ever used, but she thinks she can handle it. The next day, Janelle's manager is opening up the point of sale system and realises that the system was not finalised properly the day before.

_____

_____

b Tim was cleaning his work space after a massage by taking the towels to the designated area in the salon. His manager stopped him to tell him that he needed to finish the cleaning later and go straight to the counter to welcome clients. Tim figured that since he had almost cleaned his area, he would do that first before going to the front counter, only to hear the front service bell ringing to alert the salon there was a customer unattended at the front counter.

_____

   c At the end of the work day, Cara decided to check her schedule for the next day, rather than clean up her work space from her last client. When her manager was ready to lock up the salon not long after, Cara was still trying to get her work space cleaned up appropriately for the end of day.

Working effectively as a team member

2 Why are courteous and helpful behaviours important in the team environment of a salon?

3 Give three examples of courteous and helpful behaviours.

4 For the following non-discriminatory attitudes, give an example of how you could implement each:

   a trying not to sensationalise a person's attribute in a favourable way

   b when participating in discussions, making sure that all views and opinions are equally shared and heard

   c showing respect for people with medical conditions

5 You are working in a salon with four other beauty therapists and a receptionist. The receptionist has called in sick, and one of the beauty therapists is running late due to a client's extra needs. How would you help to keep the salon running smoothly during the day? What would you need to be mindful of?

Planning and organising work activities

6 Employability is efficiently managing day-to-day tasks: working methodically to manage tasks and following instructions in a timely and proficient manner. What are the steps to working efficiently?

7 How would you prioritise the tasks in the following scenarios?

   a Your next client is waiting for you in the massage room. The room you just used for a full leg wax for the previous client has not yet been cleaned and put back as it should.

_____

_____

   b You notice there are products on display that need to be topped up and the display needs fixing just as the salon assistant tells you your next client has arrived.

_____

_____

   c You are setting up for a treatment and you realise the product you are about to use has almost run out and has been discontinued; furthermore, the machine is faulty. Which should you attend to first?

_____

_____

CHAPTER 2: PROVIDE SALON SERVICES TO CLIENTS

LEARNING OBJECTIVES

After completing this chapter, you should be able to:

LO2.1 welcome clients, respond to their needs, take and record relevant information to the beauty service and direct them to the appropriate area for treatment

LO2.2 develop rapport, anticipate contingencies, promote products and services, and process sales, returns and refunds

LO2.3 effectively schedule and manage appointments

LO2.4 respond to complaints in an appropriate manner to ensure client satisfaction

LO2.5 identify and respond to clients with special needs.

INTRODUCTION

The beauty therapist is skilled at creating a client experience that is individual and dedicated to skincare, wellness or enhancing the client's appearance. Receive, consult and interact with clients in a professional and courteous manner to become a valuable asset to the beauty business.

The learning activities in this chapter allow you to demonstrate your knowledge and understanding of the unit of competency *SHBXCCS002 Provide salon services to clients*, using the performance and knowledge evidences.

LO2.1 RECEIVING CLIENTS

Performance Evidence

The level of customer service a beauty salon can deliver maximises client satisfaction and ensures better skincare results.

Appropriate verbal and non-verbal communication

Knowledge Evidence

1   Practise your communication techniques with other students. As a group, spend a minute reading through the table below, then go around the group and find someone who fulfils each criterion, making sure not to double up on any one person (e.g. if you have found someone was born overseas, you must find a different person who can play a musical instrument). You can use your own name in only one row.

Take note of how you speak to each other and use open questions to find out information from your classmates. Be respectful and use positive body language.

SOMEONE WHO...	NAME
... was born overseas	
... can play a musical instrument	
... loves Brussels sprouts	
... owns a cat	
... doesn't have a passport	
... can change a car tyre	
... knows how to knit	

VOICE TONALITY AND VOLUME

2 How does smiling when on the telephone enhance your telephone technique?

_____

3 What is a technique you can use with the pitch of your voice on the telephone?

_____

BODY LANGUAGE

| Figure 2.1 | Verbal and non-verbal cues

4 The above images show examples of positive and negative body language. What exactly is the positive and negative body language being displayed here?

_____

_____

_____

5 Think of some positive and negative cues that you could use when dealing with clients and give three examples of each.

_____

_____

_____

TELEPHONE

6 Complete the following table by stating how you would talk or respond over the telephone to customers in each situation. The first one has been done for you.

CUSTOMER SITUATION	SALON WORKER'S RESPONSE
e.g. Initial answering of the telephone in the salon.	e.g. 'Good morning, Body Works Beauty Salon, Georgina speaking. How may I help you?'
Customer says, 'I have an appointment tomorrow that I need to cancel.'	
Customer says, 'I'm trying to find somewhere that will tint my eyebrows.'	
A general lull in conversation.	

Customer says, 'I need to speak to your manager, I'm very unhappy about the service I received yesterday.'	
Customer says, 'Do you do nail art? I've had terrible experiences at other salons and am looking for a better service.'	
Farewelling the customer at the end of the call.	

Correct techniques using phone and email

Knowledge
Evidence

FUNCTIONS AND USE OF TELEPHONE

7   The following guidelines should be followed when taking telephone calls in the salon. Write next to each guideline what it means, and why it is important.

   a Answer quickly:

   _____

   _____

   This is important because …

   _____

   b Introduce yourself:

   _____

   _____

   This is important because …

   _____

   c Be prepared:

   _____

   This is important because …

   _____

   _____

   d Be welcoming and attentive:

   _____

   This is important because …

   _____

8   There are many methods of communication through which clients can access a business. What are the benefits of a phone call over contacting a salon through its social media pages?

   _____

   _____

   _____

EMAIL/LETTERS

9 Email and letters are a good way to connect with clients for marketing purposes. List at least five features you think an email or letter should include to make it stand out among other emails. (It might help to look for similar types of emails in your inbox, or ask your tutor.)

10 When would a letter be more appropriate than an email?

Face-to-face communication techniques

When communicating, it is important to engage with the client to build rapport. The best way to engage with people is to:

- Smile. Smiling shows interest and engagement. Add nods and smiles throughout the conversation.
- Make eye contact. It shows you are fully focused on the client, and not distracted.
- Stop talking and listen. It allows the client to say all that they have in mind and for more information to be shared.
- Ask questions. It confirms understanding or indicates that more information would be helpful.

11 Give examples of times you have engaged with people in these four ways, and describe how it made a positive difference to the interaction.

a Smile:

b Make eye contact:

c Stop talking and listen

d Ask questions:

Personal grooming and presentation

Knowledge Evidence

12 List six things that demonstrate bad personal hygiene. (Refer to the section 'Personal hygiene' in Chapter 1 of your textbook.)

_____

_____

_____

_____

Receiving clients and making appointments

Knowledge Evidence

RECEIVING CLIENTS

13 After greeting a client when they first enter the reception area, what questions would you then ask them?

_____

_____

14 Part of welcoming the client is speaking to them and behaving in a respectful manner. If their therapist is running behind in appointments and there is going to be a wait, you would need to start the conversation in an apologetic and friendly manner, such as, 'Hi Helen, how are you today? Your therapist is currently running a little behind, I'm sorry ...'

Can you think of something you could offer them?

_____

_____

LOCATION OF WORKPLACE AREAS AND SECTIONS

When receiving clients, in the interests of health and safety, you need to direct them around the salon with the assumption that they do not know their way around. If a person makes a wrong turn or trips, the salon will most likely be liable for the consequences. There are a number of ways you can direct a client around the salon:

1 Position clear and appropriate signage.
2 Give clear, verbal directions.
3 Physically usher the client through (for clients who are unfamiliar to the environment or with mobility needs).

15 Of the above three ways, which would be most appropriate for a regular client who needs to go to the toilet during a facial treatment?

_____

FUNCTIONS AND USE OF APPOINTMENT SYSTEM

16 What interval of time should the appointment system be divided into? Why does it matter?

_____

_____

17 A client has finished their treatment and returns to the reception area to pay for their treatment and a product they are purchasing. What questions would you ask them in relation to booking their next appointment?

_____

_____

_____

Farewelling techniques

18 When farewelling a client, it is important to be polite. What are three common phrases you can use to farewell the client?

Knowledge Evidence

_____

_____

19 Ethical behaviour and practice needs to be evident in all your professional activities; for example, when you usher the client to their treatment room or ask them to pay. Read the section 'Ethical practice' in Chapter 1 of your textbook for ideas and list six ways you can behave in an ethical manner when dealing with a client.

_____

_____

_____

_____

_____

_____

_____

LO2.2 DEVELOP RAPPORT, ANTICIPATE CONTINGENCIES, PROMOTE PRODUCTS AND SERVICES, AND PROCESS SALES, RETURNS AND REFUNDS

Knowledge Evidence

Anticipate contingencies

1 It is important to have a contingency in case of an unexpected event in the salon. How might using a contingency plan ensure the day runs smoothly?

_____

_____

Payment arrangements

2 It is good to be aware of all the ways payments can be processed. Refer to Chapter 6 Conduct salon financial transactions to learn more about paying by cash and about non-cash transactions.

List what you need to know when processing the following payment arrangements:

a (Gift) Voucher:

_____

_____

_____

b Credit card:

_____

_____

Product returns

3 What are the differences between a return, a refund, an exchange and a credit?

Sales and refunds

4 What does Australian Consumer Law state about the payment method used in a refund?

---

5 Complete the table below by writing in the resolution for each of the complaint scenarios given in the first column. Choose from the following resolution options:

- sample of products
- complimentary service
- referral to supervisor
- exchange
- fuller explanation
- credit

SCENARIO OF COMPLAINT	COMPLAINT RESOLUTION
Client has bought a product, only to take it home and find out the product had been tampered with inside the packaging and is now discoloured.	Refund
Client has realised the cleanser is not what they had been prescribed by the beauty therapist – they have not opened the product.	
Client has changed their mind about the product purchased, but store policy is that there is no refund on purchases for this reason alone.	
Client is unhappy with the quality of the service.	
Client does not understand why she cannot have a heat treatment after being in the sun.	
Client is not satisfied with the complaint resolution offered.	
Client got a breakout after using a salon product. She likes the product line but needs to know which ingredient is causing the reaction.	

Products, services and special packages

Knowledge
Evidence

6 When promoting products and services at your salon during client bookings, what are three of the
five things your client should be made aware of?

_____

_____

_____

7 What could you offer a regular client who likes the facial, lash tint and massage package on offer for new clients,
but it doesn't meet their requirements?

_____

Accessing and maintaining client records

Knowledge
Evidence

RECORD KEEPING

8 Client cards are used for every treatment performed in the salon; however, the type of client card used will depend
on the treatment. What information would remain the same regardless of treatment type?

_____

_____

9 The storage of client cards needs to be secure. What are the two types of client cards, and what are the main ways
they are stored in the salon?

_____

_____

_____

PRIVACY

10 When client records are kept well they help to ensure the smooth running of the salon. There are ethical concerns
over the type of information gathered on a client card. Explain what this means.

_____

_____

_____

11 A client refuses to tell you any information intended for their client card until you answer their questions on the
ways the salon uses that information. How would you proceed?

_____

_____

_____

_____

LO2.3 SCHEDULING APPOINTMENTS FOR CLIENTS

Scheduling client appointments

Knowledge
Evidence

1   The particulars of scheduling appointment systems in salons vary from business to business. However, what columns or information must the scheduling appointment system have?

_____

_____

2   List four reasons why a client's appointment might need to be rescheduled.

_____

_____

_____

_____

TELEPHONE

3   When taking a message over the phone, what details should you ensure you have?

_____

_____

_____

EMAIL

When emailing a client, it is important to remain professional and efficient. Things to keep in mind are:

- Be concise and 'to the point'.
- Draw the line between informal professional conduct (acceptable) and friendly informal text, such as emoticons or acronyms (not acceptable).
- Have a meaningful subject line.
- The person CC'd into the email does not have to do anything; they are just informed.
- Identify yourself, and your salon role.
- Read over the email to check for spelling or grammatical errors.
- Assume others may end up reading it. Emails are often used in legal disputes.
- Respond promptly to emails.
- Assume clients won't read attachments; paste images to the email if you want them seen.

4   Write a mock email to your tutor, in response to the client below:

Hi there,

I would like to make an appointment for a full leg and bikini wax. I can do Tuesday, Wednesday or Friday lunchtime. How long will it take? I would prefer sugar wax if possible.

Many thanks,

Meghan

_____

_____

_____

_____

_____
_____
_____
_____

MESSAGES TAKEN IN PERSON

When taking a message from a client in person, it can be easy to forget simple details. When taking the message, be sure to note:

- the time
- the person's full name
- the person's contact details; i.e. mobile number and/or email
- the person's query, e.g. product enquiry
- the intended recipient's name (e.g. the salon manager 'Attn: Sally')

Pass on the message as soon as you can.

5   A client has approached reception with a query about a new treatment that you are unaware of. They insist they have received the treatment overseas with the same products you stock. How do you pass the query to the salon manager?

_____
_____
_____
_____
_____

LO2.4 RESPONDING TO CLIENT COMPLAINTS

Resolving complaints with remedial actions

Knowledge
Evidence

CONSUMER PROTECTION

1   When providing advice to clients about products, what specific regulations and legislation should the salon's policies and procedures include consideration of?

_____
_____

HANDLING AND RESOLVING COMPLAINTS

2   What are four types of complaints that could be presented at a salon?

_____
_____
_____
_____

3   Why is it so important for salons to resolve complaints?

_____
_____

4　Some complaint resolutions can only be actioned by the salon manager (according to the specific salon's policy and procedures). List four possible examples of these resolutions.

_____

_____

_____

_____

5　What does it mean to turn a client's complaint into a high-quality customer service experience? What are ways the salon can ensure this has happened?

_____

_____

_____

_____

Work health and safety

6　Refer to Figure 8.1 in Chapter 8 of your textbook and find the legislation that covers the three topics outlined in the table below. Under the table, make short notes on how you are expected to maintain work health and safety during treatments.

Knowledge Evidence

TOPIC	FEDERAL LEGISLATION	STATE OR TERRITORY LEGISLATION
Personal protective equipment		
Manual handling		
Young people's safety		

　a　Personal protective equipment

_____

_____

_____

   b Manual handling

_____

_____

_____

   c Young people's safety

_____

_____

_____

Ethics of professional behaviour

Knowledge
Evidence

7 Every organisation should have a set of ethics that guide its practice. Research your training facility to see what they outline in their standards. Note: this might be referred to as an ethical guide, a code of conduct, or a charter. Answer the following questions:

   a How does the ethical guide of your college protect the wellbeing of all those who attend the college? Give at least two examples.

_____

_____

   b Can you see how these ethical guidelines might transfer to a beauty therapy workplace? Give at least two examples of when this would be relevant.

_____

_____

LO2.5 RESPONDING TO CLIENTS WITH SPECIAL NEEDS

Knowledge
Evidence

Read through the information found at **https://humanrights.gov.au/our-work/lgbti/about-sexual-orientation-gender-identity-and-intersex-status-discrimination** in regards to sexual orientation, gender identity and intersex status. It is unlawful to discriminate against a person based on these things.

1 The pronouns that a person is referred to by can have a big impact on whether they feel safe to be themselves in a space. If a client whom you have always referred to as she/her and has used the name Tracy tells you they would like to now use the name David and use the pronouns of they/them, what is the best way to respond?

_____

_____

_____

_____

_____

2 How might you discreetly assist a client who has need of mobility assistance?

_____

_____

3   Without the use of AUSLAN, there are ways you can communicate respectfully to clients from the Deaf community. What are they?

_____

_____

_____

4   Some clients might request a female therapist for cultural reasons. What should the response to these requests be?

_____

_____

_____

Dealing with difficult or abusive clients

5   Activity: Think about difficult clients, or situations in the salon that might make clients emotional, abusive or hard to deal with. Come up with a scenario for each as a group. In pairs, role-play each scenario, swapping between being the difficult client to being the salon worker. Spend 3–4 minutes, and then swap roles. Pay special attention to how you talk to difficult clients, and always try to turn the situation around to be a positive experience.

Effective questioning and active listening techniques to establish client needs

6   How would you greet or offer to help clients with the following special needs?

  a Intellectual disability:

_____

_____

  b Heavily pregnant:

_____

_____

  c Visual impairment:

_____

_____

_____

CHAPTER 3: ADVISE ON BEAUTY PRODUCTS AND SERVICES

LEARNING OBJECTIVES

After completing this chapter, you should be able to:

LO3.1 develop product knowledge by sourcing information on products and services, understand the competitor and share the information with clients and colleagues

LO3.2 recommend salon services by assessing client requirements through consultation

LO3.3 prepare the client for a product demonstration, including cleansing the skin

LO3.4 demonstrate beauty products and services by applying products, respond to client queries, and evaluate and follow up with the client.

INTRODUCTION

The beauty therapy experience begins with a consultation and personalised advice from a beauty expert. While salon-quality beauty products may be purchased and treatment bookings can be made online, expert recommendations based on the client's physical condition can guarantee that the client's needs are satisfied. The client's characteristics, needs and requirements are also used to develop an in-salon treatment plan.

The learning activities in this chapter allow you to demonstrate your knowledge and understanding of the unit of competency *SHBBCCS005 Advise on beauty products and services*, using the performance and knowledge evidences.

LO3.1 DEVELOPING PRODUCT KNOWLEDGE

Knowledge
Evidence

Sourcing product information

1 Each of the items listed below is a good way to discover information about a new product or range. Give an example of the different information you obtain from each item

   a Product leaflet, brochure or booklet

   b Internet site for product or range

   c Product label

   d Price list

   e Product and service manual

   f Discussion with customers, staff members and product suppliers

Features and benefits

Knowledge
Evidence

Performance
Evidence

2   After a treatment, you take the client to the reception area to discuss products. What could you be focusing on if you were recommending a moisturiser and describing its features, effects and benefits?

_____

_____

_____

_____

_____

3   Think of two products for each of the following, and list the effects and benefits as you would for a client.

   a   Nail care

      product 1: _____

      effects: _____

      benefits: _____

      product 2: _____

      effects: _____

      benefits: _____

   b   Make-up

      product 1: _____

      effects: _____

      benefits: _____

      product 2: _____

      effects: _____

      benefits: _____

   c   Skin care

      product 1: _____

      effects: _____

      benefits: _____

      product 2: _____

      effects: _____

      benefits: _____

   d   Cosmetics

      product 1: _____

      effects: _____

      benefits: _____

product 2: ....

effects: ....

benefits: ....

e Hair products

product 1: ....

effects: ....

benefits: ....

product 2: ....

effects: ....

benefits: ....

4 As it is not always necessary to memorise the cost of every item in a range, if you were to try to remember a few items, how would you choose which ones to memorise?

....

5 Most salons will have an agreement with one brand to stock their range only. If you work somewhere that has more than one brand, you will need to know the options from each brand that are comparable. Go online and find two brands for the following products. For each one, list its features, benefits and cost for a product of a similar size.

a Moisturiser

....

....

b Toner

....

....

c Exfoliant

....

....

Clients who will benefit from treatments

Performance
Evidence

When a client comes in for a salon treatment, it is good to initially gauge the type of product or treatment that you think will be of most interest or benefit to them. Different client types are described at the beginning of every treatment chapter; for example, refer to the section 'Considerations' in Chapter 9 of your textbook.

6 In the table below, circle which client type you think would benefit the most from the product or treatment listed.

CLIENT TYPE (CIRCLE YOUR CHOICE)	PRODUCT OR TREATMENT
New/regular clients	Facial skincare trial pack (small size cleanser, toner, moisturiser and exfoliant)
Mature/young clients	Deep cleanse facial treatment
Clients with time/busy clients	Bath luxury range
Male/female clients	Direct high frequency for after waxing (as a routine)

There are legal requirements relevant to giving advice about and selling beauty products and services. The consumer protection legislation is governed by the ACCC (Australian Competition & Consumer Commission).

Your local, state or territory consumer protection agency (sometimes called 'consumer affairs') can provide you with information about consumer rights and options. In the case of a complaint or disagreement, they can be contacted by consumers to help negotiate a resolution between the salon and the consumer.

7   What is your local consumer protection agency that advises for the ACCC?

The salon will have policies and procedures regarding selling products and treatments that both benefit the consumer and also protect the health and rights of consumers, such as:

• The customer should be advised of complementary products or services according to their identified need.

• Staff personal sales outcomes are reviewed to maximise future sales.

8   For each of the procedures above, what is a consequence that might occur if the procedure is not observed?

ORDERING PROCEDURES

Knowledge
Evidence

9   Visit a salon and ask the staff the following questions about their ordering protocols. Write down the answers.

   a   Who does the ordering in the salon? Does it change or is one person generally responsible?

   b   How are the orders processed? Is it through a company representative who comes to the salon? Is it done over the phone, or electronically?

   c   How much extra stock do they order of different items? Do they differentiate between popular and less popular items when ordering?

   d   Where do they store the products that don't fit out on the shelf? What conditions are they kept in (e.g. tidy, dark, open, airy, disorganised, etc.)?

PRODUCTS IN WORKPLACE RANGE

10   All salons will have product ranges that they use and products that they sell (usually these are from the same product range). What variations of products, policies or treatments might you encounter if you started working at a different salon?

SHELF LIFE AND USE-BY DATE/STOCK AVAILABILITY

11   What is a normal shelf life for a skincare product?

12   If an item's use-by date is indecipherable, how would you go about figuring out what it might be?

13 What does FIFO stand for, and why do you need to follow this method?

AVAILABILITY

Knowledge
Evidence

14 If the product that would suit the client best wasn't in stock, what would you do?

STORAGE REQUIREMENTS

15 What storage advice would you give for the following products? Moisturiser, exfoliator, cuticle oil, hand cream.

WARRANTIES

Knowledge
Evidence

16 What is a warranty according to the ACCC? What is the ACCC and why is it important?

17 Activity: Research laws that protect the consumer with regards to:

- trade practices

- product labelling

- recommended retail pricing

- safety of cosmetics

- privacy

- distance selling (i.e. online).

LO3.2 RECOMMENDING SALON SERVICES

Performance
Evidence

Ability to identify client needs and constraints

1 The following table lists different client needs or constraints. Mix and match the attributes by choosing one from each column. Once you have a description for each of your six 'clients', describe the services they would benefit from, along with the products you would advise they purchase. Explain your decisions.

Mature skin	Works outdoors
Teenager	Facial scarring
Pregnant	Combination skin
Busy mum	Works in an office, dehydrated skin
Client who is short on funds	Prone to acne
Male client	Heavy smoker

Selection of suitable products or treatments

Knowledge
Evidence

2 Refer to the relevant treatment chapters to match the effects, benefits and contraindications to the following treatments.

TREATMENT	EFFECT	BENEFIT	CONTRAINDICATION
Manicure and pedicure			
Facial services			
Body treatments			
Hair reduction services			

3 When recommending a service that involves skin penetration – for example, body piercing – what legal requirements do you need to consider?

4 Waxing can be performed on the skin of people under the age of 16, with caution. The skin's follicles are smaller and the skin is thinner. It is considered the same thickness at the age of 5 as it is at the age of 60. Referring to 'Factors that affect the treatment plan' in Chapter 11, is there any legal documentation to complete before proceeding?

Presentation of features and benefits

Knowledge
Evidence

5 After a consultation with the client to determine their needs, you will be directing them back to reception to make their initial booking. How would you advise the client at this point in relation to the bookings needed for the treatment plan?

6 If the client has budgetary or time constraints, would you tell them you couldn't complete the treatment for them?

_____

_____

LO3.3 PREPARING A CLIENT FOR PRODUCT DEMONSTRATION

Cleansing the skin

Knowledge Evidence

1 If you don't have running water nearby when performing a product demonstration for a client, how many bowls should you typically have on hand, and why?

_____

2 What are the seven steps to cleansing the face in a seated position?

_____

_____

_____

_____

_____

_____

_____

_____

_____

Laura May Grogan

LO3.4 DEMONSTRATING BEAUTY CARE PRODUCTS

Performance
Evidence

Product demonstrations are usually performed outside of your treatment area. It therefore takes additional thought and consideration to ensure you are applying the correct work health and safety practices and maintaining infection control.

1   When setting up a product demonstration, what considerations should you make for the following:

   a client and therapist positioning

   _____

   _____

   _____

   _____

   _____

   _____

   b layout of work space

   _____

   _____

2   If the client asks about a product, how would you respond if:

   a you know the answer?

   _____

   b you are unsure of the answer?

   _____

   _____

   _____

   _____

Legal requirements

Knowledge
Evidence

3   Research the federal, state or territory and local health and hygiene regulations relevant to product demonstrations, and list at least two things you would need to consider in relation to the following:

   a water supply

   _____

   _____

b product dispensing

_____

_____

4   What is the best way to ensure you don't discriminate against any client when advising on products
    in the salon?

_____

5   If a client with a particular skin tone was told they couldn't buy any products to match their skin because the salon
    doesn't stock it, what sort of law is that breaking?

_____

Specialised product knowledge

Knowledge
Evidence

6   To ensure the success of your salon, selling products is vital. In the following table, suggest things that
    could go wrong if a client is not given important information about the product.

INFORMATION YOU SHOULD KNOW ABOUT PRODUCTS	WHY IS IT IMPORTANT?	EXAMPLE OF WHAT COULD GO WRONG
Contraindications	So that a person with a medical condition does not have an adverse reaction to the product.	
Causes of skin sensitivities or allergies	Product ingredients can cause adverse reactions and allergic reactions, including anaphylaxis.	
Basic ingredients or materials in product	To know the product ingredients and materials that the client is indicated or contraindicated to.	
Complementary products and services	To meet the client's needs to the best of your ability, and maximise sales.	
Features and use of products	For product safety, it is a legal requirement that you ensure the client uses the product in the way the product manufacturer has intended.	

Suitability of products and services

Knowledge
Evidence

CLIENT ABILITY TO FOLLOW HOMECARE ADVICE

7   If a client says they are unlikely to follow the correct homecare advice, what would you do?

_____

_____

_____

CONTRAINDICATIONS

8   If a client has a contraindication to a particular treatment, what advice would you give them?

_____

_____

9   In the following table, list a treatment type, and give a contraindication that might present for that treatment. Review each treatment chapter to find a contraindication to match. Give 10 examples.

TREATMENT TYPE	CONTRAINDICATION
e.g. Massage	Sunburn

ALLERGIES AND SENSITIVITIES TO INGREDIENTS IN SKINCARE

10   What is an SDS? How does it relate to a client who might display signs of skin sensitivities or allergies?

_____

_____

_____

11   The following ingredients in skincare products are the most common cause of sensitivity. Fill in the possible likely reactions for each one.

   a   BHAs

_____

   b   Retinoids

_____

   c   Pineapple and papaya enzymes

_____

   d   Crustaceans and other sea extracts

_____

   e   Egg, nuts and seeds

_____

12 If a client told you they had some mild allergies, what would you perform? What steps would you take to perform it?

_____

_____

_____

_____

_____

_____

ETHICAL CONSIDERATIONS

13 Provide three examples of ethical requirements that clients might have for their treatments.

_____

_____

_____

CHAPTER 4: SELL TO THE RETAIL CUSTOMER

LEARNING OBJECTIVES

After completing this chapter, you should be able to:

LO4.1 establish customer needs, by developing rapport and using communication techniques

LO4.2 provide advice on products and services by tailoring options, explaining features and benefits, advising of promotions, addressing objections, using competitor comparisons, collaborating with the customer, up-selling and cross-selling

LO4.3 facilitate the sale of products and services with the techniques to close the sale, directing client to POS, farewelling client and providing after-sales services.

INTRODUCTION

Selling is a skilled way of communicating with the client in order to achieve the client's objectives. It requires you to do more than reel off product company scripts. Selling involves using communication skills and other strategies to determine the client's skin concerns so you can match a product or service with their needs. You then need to follow through and gain the client's confirmation.

The learning activities in this chapter allow you to demonstrate your knowledge and understanding of the unit of competency *SIRXSLS001 Sell to the retail customer*, using the performance and knowledge evidences.

LO4.1 ESTABLISHING CUSTOMER NEEDS

Performance Evidence

1 What are three ways you could build rapport with a client in the salon?

_____

_____

2 Based on this, what do you think the meaning of 'rapport' in the salon is? Why is it important?

_____

_____

_____

3 Using the information found at **https://www.business.qld.gov.au/running-business/consumer-laws/customer-service/improving/principles**, which looks at the principles of good customer service, what are the five things you should do to ensure you provide the best customer service?

Knowledge Evidence

_____

_____

_____

_____

_____

Communication

BODY LANGUAGE

Knowledge
Evidence

4 What three things describe someone who is actively listening?

_____

_____

_____

OPEN AND CLOSED QUESTIONS

Knowledge
Evidence

5 Fill in the following words to describe the differences between open and closed questions.

   a Asking _____ questions will result in a one-word response, and is used for gaining _____ information from the client.

   b Asking _____ questions prevents one-word answers, and encourages the client to share _____ information.

6 Give examples in the table below to show when you would use closed and open questions.

OPEN QUESTIONS	CLOSED QUESTIONS
E.g. How has your skin felt after using your current products?	E.g. Do you currently use a toner?

VERBAL AND NON-VERBAL CUES

Verbal cues are a way of communication between two people where words, tones and expressions are used to convey understanding, listening, curiosity, or a host of other messages. Non-verbal cues are how we communicate with each other without making sounds; when we communicate with our bodies.

Knowledge
Evidence

7 Give examples for each of the non-verbal cues listed below:

   a reluctant

_____

   b eager

_____

   c impatient

_____

EFFECTIVE COMMUNICATION

8   Read through the information found at **https://www.and.org.au/articles.php/12/9-tips-for-assisting-** Knowledge
    customers-with-disability and respond to the following situations in a salon retail context. Evidence

   a   A woman comes into the salon with an assistance animal. You can see her looking at the products on display,
       so you know she has some vision. The woman speaks quietly to her assistance dog, so you can tell that her
       speech is not slurred.

   _____

   _____

   _____

   _____

   b   Two men walk into the salon, and they are clearly using sign language to communicate. They smile your way
       and walk directly to the counter. One of the men says 'Hello', which you can understand though the word
       seems slightly unusual.

   _____

   _____

   _____

   _____

   c   You see a woman in a wheelchair push the salon door open and then look around the salon, seeming slightly
       unsure of what to do next.

   _____

   _____

   _____

9   Visit **https://blog.amt.org.au/index.php/2018/01/31/cultural-safety-for-massage-therapists**. This blog post
    references the four principles of cultural safety in a health setting. What are they?

   _____

   _____

   _____

   _____

10 The beauty salon can provide a wide array of services that involve touching another person. Why is it so important
    to be aware and sensitive to other cultures and their needs in this setting?

   _____

   _____

   _____

11 How could the concept of building a good rapport with a client cross over with being sensitive to cultural
    diversity?

   _____

   _____

LO4.2 PROVIDING ADVICE ON PRODUCTS AND SERVICES

Performance
Evidence

Consumer law

Knowledge
Evidence

1   What does ACCC stand for?

_____

2   What can happen if you sell a product or service to a person and withhold health or safety information?

_____

_____

3   A regular client comes in for a facial, and as they are leaving the treatment room they tell you they would like to purchase their regular moisturiser, which they have bought from the salon in the past. You know the manufacturer is having troubles with shipments, leaving the salon short on that product. What are two options you could take to maintain a sale in this example?

_____

_____

_____

_____

_____

4   Your salon receives an email stating that a client had come in the day before for an eyebrow tinting treatment, and they are now experiencing an adverse reaction. They are threatening to go to the ACCC about it. What might be the best course of action?

_____

_____

_____

_____

_____

_____

5   The way that customer and client information is collected, used, stored and disclosed in a salon is governed by the *Privacy Act 1988*. List two times when personal information is utilised in a salon retail setting, in which you would need to maintain privacy for the client.

Knowledge
Evidence

_____

_____

6   An example of the standards that a state or territory will have in place can be found at the NSW Fair Trading website: **https://www.fairtrading.nsw.gov.au/buying-products-and-services/buying-services/beauty-services**. Use this information to answer the following questions.

   a   When should product ingredients be made available to the customer?

   b   How should ingredients appear on the container?

   c   Do testers or samples need to have ingredients lists on them or supplied?

   d   What does it mean if a customer asks for proof of purchase? Do you have to give it to them?

Customer and industry expectations

Knowledge
Evidence

7   List four expectations a client would have of the products, staff and salon in the beauty industry.

8   What is a feature of a product? What is a benefit of a product? Why is it important to know these in the retail environment?

9   At your college or local beauty salon, take a good inventory of the products that are used or sold. How could you categorise them? For example, could they be classified as luxury or fundamental? Or are they corrective versus preventative? You can use a different classification, but you need to be able to defend your stance to your group.

10 What are four things you should remember when dealing with customer questions or objections about a product or service?

Knowledge
Evidence

11 When comparing a product your salon stocks to a competitor's product, why is it a good practice to be able to point out the good things about the competitor's product?

_____

_____

12 Complete the table below to describe how each skill shows you are collaborating with a client, rather than dictating what they should receive or buy.

ATTRIBUTE	EXPLANATION
Listening and responding	
Sharing information	
Being prepared for rejection	

13 What is the difference between up-selling and cross-selling?

_____

_____

14 A customer says they want to change their morning skincare routine and want to know what sort of cleanser you stock. Give an example of how you could up-sell this sale, and how you could cross-sell this sale.

_____

_____

_____

15 Your salon is having a promotion on a particular type of moisturiser. Which of the two statements below would you choose to alert the customer to this promotion? Explain why.

A 'You should buy the bigger moisturiser. It is cheaper at the moment, so it's a no-brainer.'

B 'I know you usually buy the smaller bottle of this moisturiser, but we currently have a promotion on for the larger bottle. Would you like that one instead?'

_____

_____

LO4.3 FACILITATING THE SALE OF PRODUCTS AND SERVICES

Performance Evidence

Knowledge Evidence

Principles and techniques of sales

1 Approaching a customer can be one of the more daunting aspects of starting a role in a retail environment. Over the next few days, notice how retail staff approach customers. Fill out the table with the type of store they are in (this will matter, because a salon is a different selling environment to a supermarket), and record their opening lines.

STORE TYPE	OPENING LINE

2 A customer has picked a product off the shelf and is holding it in one hand while she uses her other hand to start looking closely at labels of other products. How do you know this is a buying signal? What could you do?

3 A customer tells you he isn't sure exactly what he is after, but knows he needs a cleanser that will help maintain moisture in his skin throughout the day. Your salon sells two different moisturisers, and both have moisture retention for dry skin as a feature, but you know that one of those products will make a better profit for the salon. How might you answer the customer?

4 What is meant by complementary products?

5 Some retail settings have customer loyalty cards or programs. Why is this a good practice to promote sales? How might a salon set up such a loyalty program?

6 Give an example of the following closing techniques.

METHOD	EXAMPLE
Closed question	
Direct close	
Assumptive	
Choices	
Payment alternatives	
Create urgency	

7 What is a good example of how to direct a client towards the point of sale (POS) area when a decision has been made?

CHAPTER 5: RESEARCH AND APPLY BEAUTY INDUSTRY INFORMATION

LEARNING OBJECTIVES

After completing this chapter, you should be able to:

LO5.1 source beauty information using research skills and apply it to enhance work performance

LO5.2 research legal and ethical issues by sourcing and interpreting the information and applying it to the salon workplace

LO5.3 update and share your knowledge of the beauty industry, products and services, including current issues of concern.

INTRODUCTION

Beauty therapists and the entire salon team should make time to research information in their industry in order to remain competitive. Knowing your industry encourages career growth; and for the salon, business growth.

The learning activities in this chapter allow you to demonstrate your knowledge and understanding of the unit of competency *SHBBRES001 Research and apply beauty industry information*, using the performance and knowledge evidences.

LO5.1 SOURCING AND USING INFORMATION ON THE BEAUTY INDUSTRY

Performance Evidence

Knowledge Evidence

Industry associations

1   Industry associations can be a source of current industry information. Which industry associations are relevant to your job role? How would you go about accessing their current information for the beauty industry?

_____

_____

_____

_____

_____

2   Find the trade union, employer group and professional association best suited to you, and find out what information they list specific to the beauty industry.

_____

_____

_____

_____

_____
_____
_____
_____
_____
_____

3 Suggest a trade union, employer group or professional association that can best advise you on your working conditions.

_____

Trade magazines

Knowledge
Evidence

Some trade magazines are targeted towards the beauty industry as a whole, and some can be aimed more squarely at spa, nails or other niche areas of the industry. They can be informative, feature press releases for product and treatment innovations, or recognise talent and achievements in skills or business appreciation.

4 Go online and find the most informative trade magazines relevant to you. List them below.

_____
_____
_____
_____

5 Use 'Step-by-step: Develop research skills' in Chapter 5 of your textbook to determine how reliable the magazines you identified are as a source of information to keep you up to date on the most recent trends. Summarise your thoughts below.

_____
_____

6 List at least four ways you could evaluate the accuracy of the information provided in trade magazines.

_____
_____
_____
_____

7 List four trade magazines that relate to the beauty industry. For each one, list something that you like about it, and something that you don't.

_____
_____
_____
_____

Improve work performance

8   Read the section 'Information for work performance' in Chapter 5 of your textbook. Complete the table below with your own performance analysis as shown in Figure 5.2 in that section of the textbook.

PERFORMANCE ANALYSIS	
STRENGTHS	WEAKNESSES
WHAT YOU CAN WORK ON	SOURCES OF INFORMATION

9   List ways you can apply your knowledge to the beauty industry. Use the prompts in the table below to help you identify your learning needs, then list where you can find information to help you with this and what you hope to learn.

KEY AREA IN BEAUTY	SPECIFIC AREA YOU WOULD LIKE TO STRENGTHEN (CIRCLE ONE TO SELECT)	SOURCES OF INFORMATION	WHAT I HOPE TO LEARN
Client service skills	• staying calm under pressure • being caring and respectful of various cultures and backgrounds • being responsible and reliable in performing good-quality services as well as standing by the quality of the products • being empathetic to client needs • time efficiency		
Product retailing skills	• commercial awareness • knowing what makes the business profitable • communication skills: reading the client's verbal and non-verbal cues, such as body language • ability to explain benefits and effects of the product or treatment • numeracy – able to make quick calculations • problem solving – able to be flexible and adaptable		
Treatment techniques	• the type of salon you wish to work in • training requirements • industry trends • profitability		

LO5.2 RESEARCHING LEGAL AND ETHICAL ISSUES

Performance
Evidence

Knowledge
Evidence

Environmental issues and requirements

1   Research environmental and sustainable solutions for businesses from two different sources and list two ways in which a salon could minimise its environmental impact for each of the following:

   a   energy and water

   _____

   _____

   b   recycling

   _____

   _____

   c   waste minimisation

   _____

   _____

2   For the answers you came up with above, choose one from each category and design a sign that could be put up to remind your colleagues to do the right thing.

   a   Energy and water

   b   Recycling

   c   Waste minimisation

Industrial relations issues

Knowledge
Evidence

3   Research antidiscrimination laws in your state or territory. Write in the following table a scenario for each different type of discrimination that might occur in a salon.

DISCRIMINATION TYPE	SCENARIO
Disability discrimination	
Racial discrimination	
Sexual harassment	
Sex discrimination: pregnancy	
Sex discrimination: male therapist	
Age discrimination	

4   What does 'equal employment opportunity' mean?

   _____

   _____

5 Where in the salon would you find the equal employment opportunity information?

_____

_____

Industry expectations of employees

Knowledge Evidence

6 There are certain expectations for staff working in the beauty industry. For the following categories, give an example of what the expectations are for employees.

a Good work ethic

_____

b Environmental awareness

_____

c Customer service

_____

d Hygiene

_____

e Personal grooming

_____

7 For the following two examples, explain how discrimination can affect you, your colleagues and the salon.

a Not offering assistance to a client with mobility issues who needs the toilet.

_____

_____

b Talking about a colleague and referring to her as having 'bad skin for her age'.

_____

_____

Industry working conditions

Knowledge Evidence

PRIVACY

8 Review the Privacy Principles from the _Privacy Act 1988_. How does Commonwealth legislation protect the following in relation to privacy?

a Clients

_____

b Employees

_____

c The salon

_____

9 If a client comes in and says they are worried about their 18-year-old daughter's activities, and wants to know if she's had any treatments and how she's paid for them, what would you say?

_____

_____

WORKPLACE RELATIONS

Workplace relations is about the professional relationship you have with your employer. It is common in the beauty industry to become friendly with your employer, so it can be easy to fail to understand the contractual arrangements at the beginning of employment. Doing so, however, can avoid disputes later.

10 Review Chapter 1 of your textbook to learn about industry Awards. If you are paid the Award wage, what does that mean?

_____

_____

_____

_____

11 Review Chapter 1 of your textbook to learn about employment agreements, known as contracts of employment. When do you sign a contract of employment?

_____

12 What types of things can be negotiated before you sign the contract?

_____

_____

_____

_____

Relationship with other related industries

Knowledge
Evidence

13 You will need to build networks and alliances with people and businesses in other related industries, as there are many other industries that rely heavily on, and have a close relationship with, the beauty industry. Do some of your own research to complete the table below, showing industries that have a relationship with the beauty industry, and how they might interact.

RELATED INDUSTRY	HOW IT INTERACTS WITH THE BEAUTY INDUSTRY
e.g. Fashion industry	Beauty industry provides services and treatments to the models, including make-up, skincare, hair removal and nails.

RELATED INDUSTRY	HOW IT INTERACTS WITH THE BEAUTY INDUSTRY

New products, technology, techniques and services

ADVERTISING

According to Australian Consumer Law, it is necessary to advertise products and treatments so that the client receives something that does what is says it will do. Otherwise, by law you are required to refund or exchange the goods or service. Consumers are protected by the Australian Competition & Consumer Commission (ACCC).

14 Refer to the section 'Products' in Chapter 5 of your textbook. When advertising a product, what seven things does the salon need to observe?

_____

_____

_____

_____

_____

_____

_____

_____

PRICING

The salon owner or salon manager typically decides the price of a product, though you might be asked to assist. Learn about how pricing strategies are developed on the website given below and use the information there to answer the following question. To find the pricing strategies information, visit **https://www.business.gov.au** and follow the 'Business Information', 'Products & services', then 'Pricing' links.

15 How would a salon go about figuring out a pricing structure for a new product, technology, technique or service? Describe the following pricing strategies:

a cost-based

_____

_____

b competition-based

_____

c value-based

_____

d product-based

PRODUCT RECOMMENDATIONS

16 It is important when recommending products to know about the new products that are available. What sort of information does the ACCC expect you to know? You can find information here: **https://www.accc.gov.au/accc-book/printer-friendly/29527**

TRAINING

17 Research at least one recent treatment innovation 'trend' that has forced many beauty therapists to up-skill to be able to stay abreast of current technology.

18 What might happen if the salon adopted one of these technologies without the proper up-skilling of employees?

Work ethic required to work in the industry

Knowledge
Evidence

19 What are three ways you can improve the ethical standards in the salon?

Consumer protection and trade practices/duty of care

Knowledge
Evidence

20 What does 'duty of care' refer to?

Public indemnity insurance providers cover a range of insurances to suit the beauty industry, such as:
- *business insurance*: for business assets, such as equipment, stock and buildings
- *public liability*: for incidents including slips, trips and spills
- *professional indemnity*: for poor-quality service and mishaps during treatments.

21 Choose one insurance provider and find out what types of work health and safety incidents it might cover you for.

22 Contact your local council and find out what licences/registrations a salon needs in order to operate. List them below.

_____

_____

_____

_____

_____

_____

_____

_____

_____

_____

23 What is the ethical way to proceed when you have overbooked and cannot treat a client? Ask a salon what their procedures are and list them below.

_____

_____

Hygiene

Knowledge
Evidence

24 Why is hygiene, and health and safety, so important ethically and legally in the beauty industry?

_____

_____

_____

_____

_____

25 Read the article at **https://www.forbes.com/sites/meimeifox/2021/01/26/how-two-top-beauty-salon-franchises-are-surviving-the-pandemic/?sh=3b6ff7b715d5**. List at least five tips you can take from the salon owners that have had to shut the doors due to the COVID-19 pandemic.

_____

_____

_____

_____

_____

Work health and safety

26 What is the *Work Health and Safety Act 2011*? Who does it protect in the salon?

_____

_____

27 Use the Internet to find the most recent laws governing workers and consumers in Australian businesses. See what information you can find on the topics listed below. For each, list the law that governs the topic, the governmental body that administers the law and the website you found the information on. On a separate piece of paper, write a scenario for each topic that shows how someone might encounter the issue, and what the person might do about it.

   a Equal opportunity employment

   b Consumer protection

   c Workplace relations

   d Duty of care

   e Hygiene

   f Work health and safety

_____

_____

_____

_____

_____

_____

LO5.3 UPDATING YOUR KNOWLEDGE OF THE BEAUTY INDUSTRY AND PRODUCTS

Career opportunities within the industry

1 Why is it important to be able to obtain new product knowledge, and knowledge of industry changes and trends, and workplace requirements?

_____

2 There are three key areas you can work on to achieve career goals. List them and briefly explain each one.

_____

_____

_____

_____

_____

3   The following questions are designed to help you clarify the direction of your career in the beauty industry. Discuss your answers with your instructor and colleagues.

   a   I undertook this beauty course because …

   b   From this course, I hope to learn mostly about …

   c   What I like best about the beauty industry is …

   d   What I like least about the beauty industry is …

   e   If I complete this course, I would like to see myself in the following roles …

      Keep this questionnaire for the duration of the course and review your vision for the future at its completion to see if your answers have changed. It is not unusual to change direction within the industry throughout your career.

4   Suggest two websites or recruitment agencies that can assist you in the search for work in the beauty industry, and explain the benefits of each method.

5   Choose a career pathway in the beauty industry, and follow it from the entry-level position to the role you would eventually like to be working in. At each employment stage, list a few dot points of what you think the role of that person in the salon would be.

Sourcing and sharing industry information

6   It is important to retain your confidence as a 'beauty expert' by staying up to date with the industry. List at least three places where you could regularly check for current beauty industry information.

_____

_____

_____

_____

_____

_____

_____

7   Of the current issues of concern listed below, choose one and research it to understand how the issue is of impact in your state/territory, if there's any kind of organisation or governing body that is attempting to address the concern, and give an example of something a salon could do to minimise the impact of the concern.

   a   New products and techniques

   b   Sustainability issues

   c   Emerging markets

   d   Industry expansion and retraction

   e   Skills shortage

   f   Infection control

8   Why is it important to share new industry information within the salon?

_____

_____

_____

CHAPTER 6: CONDUCT SALON FINANCIAL TRANSACTIONS

After completing this chapter, you should be able to:

LO6.1 operate point of sale (POS) equipment
LO6.2 complete POS transactions
LO6.3 complete refunds
LO6.4 remove takings from the register or terminal
LO6.5 reconcile takings.

INTRODUCTION

Learning to conduct financial transactions helps you to spend less time handling money and more time focused on clients. Professional and efficient processing of payments is appreciated by the clients, and by salon staff as they are able to start and finish the work day with ease.

The learning activities in this chapter allow you to demonstrate your knowledge and understanding of the unit of competency *SHBXCCS001 Conduct salon financial transactions*, using the performance and knowledge evidences.

LO6.1 OPERATING POS EQUIPMENT

Performance
Evidence

Knowledge
Evidence

Cash sales

Visit the web address **https://www.ato.gov.au/business/gst**. Using the information you find, answer the following questions.

1   What does GST stand for? Who governs it?

_____

2   What per cent of goods, services and other items sold or consumed in Australia is the GST?

_____

3   For the following items, state whether they would have GST applied to them, or not.

ITEM	WILL GST BE APPLIED?
Moisturiser	
Bananas	
Full leg wax	
Coffee beans	
Cuticle oil	
Baby nappies	
Milk	

CASH REGISTER

4 What are the components of a regular salon POS terminal?

5 Explain the different methods of storing cash in a manual till, an automatic till or a computerised cash desk.

Demonstrating secure payment handling procedures

Performance
Evidence

6 What is tender?

7 What might you have to do to transfer the tender?

8 When calculating cash and non-cash documents, what are the takings you need to add, and what are the items you need to subtract (before subtracting the float)?

9 Research salon software systems that handle the payment procedure for you. For example:

- **https://www.gettimely.com/timelypay**
- **https://www.fresha.com/for-business**
- **https://www.phorest.com/au/features/salon-pos-software**

List the key features of these salon systems.

Counting cash

Knowledge
Evidence

10 Describe the purpose of a cash float, and how you would use it to balance the register's takings at the end of the day.

LO6.2 COMPLETING POS TRANSACTIONS

Performance
Evidence

Non-cash sales

Electronic payment methods are linked to online accounts. This includes EFTPOS cards, 'tap-and-go' and 'buy now, pay later'.

CREDIT CARD/EFTPOS

1 Research the procedures of 'buy now, pay later' options. Weblinks have been provided for you to help answer the questions:

   a How do you process the payment with Afterpay® (**https://www.finder.com.au/afterpay**)?

---

     i How long does the client have to pay the money back?

---

     ii Do they need to pay the salon?

---

     iii What happens if they do not pay?

---

   b How do you process the payment with ZipPay® (**https://www.finder.com.au/zippay**)?

---

     i How long does the client have to pay the money back?

---

     ii Do they need to pay the salon?

---

     iii How does the money come out of their account?

'Tap and go' payments have a number of benefits to the salon:

- less risk than when handling cash
- efficiency/faster transaction time
- hygiene
- increased flexibility with where the POS can be; client can use card or device.

2 Research 'tap and go' brands online and list two popular brands.

Determining change required and denominations of change/tendering change

Performance
Evidence

3 List the step-by-step sequence for the following transactions:

a cash payments (6 steps)

b cheque payment (8 steps)

c electronic payment (8 steps)

4   If a transaction comes to $32.70, how would you let the client know this? What are two ways that your salon might require you to count the change back to the client if they pay with a $50 note?

_____

_____

_____

_____

5   Always maintain privacy and security when tendering change. There can be times when your client wishes to pay for the transaction while having a treatment (to save time), so you need to ensure the cash is secure and client personal information is private. Describe how you could conduct the transaction for a client while they are having a pedicure. (You might ask your tutor for advice.)

_____

_____

_____

Ensuring security of cash and non-cash transactions

Knowledge
Evidence

6   Look up the Payment Card Industry's (PCI's) Security Standard Council's 12 standards that are developed to maintain data security when making non-cash payments: **https://www.pcisecuritystandards.org/pci_security**. Complete the table below to show your understanding. Draw a line to match each goal with the standards that have been developed.

GOAL	PCI DATA SECURITY STANDARDS REQUIREMENTS THAT APPLY
Implement Strong Access Control Measures	1. Install and maintain a firewall configuration to protect cardholder data 2. Do not use vendor-supplied defaults for system passwords and other security parameters
Regularly Monitor and Test Networks	3. Protect stored cardholder data 4. Encrypt transmission of cardholder data across open, public networks
Protect Cardholder Data	5. Use and regularly update anti-virus software or programs 6. Develop and maintain secure systems and applications
Build and Maintain a Secure Network	7. Restrict access to cardholder data by business need-to-know 8. Assign a unique ID to each person with computer access 9. Restrict physical access to cardholder data
Maintain an Information Security Policy	10. Track and monitor all access to network resources and cardholder data 11. Regularly test security systems and processes
Maintain a Vulnerability Management Program	12. Maintain a policy that addresses information security for employees and contractors

7   Which PCI Data Security Standards requirements apply in the following situations?

  a   Every therapist log-in to the salon website should be ID and password protected.

_____

  b   The salon needs to install software to protect against theft of information, hackers and computer viruses.

_____

_____

8   What are three steps that should be taken to ensure the security of cash at the POS?

_____

_____

9   What security checks should you perform when receiving payment by credit card?

_____

_____

_____

10 Look up the legislation that is relevant to you and list the requirements for work health and safety at the POS in a salon.

_____

_____

11 In your salon, what types of things do you need to consider at the POS to prevent health and safety incidents?

_____

_____

_____

LO6.3 COMPLETING REFUNDS

Performance
Evidence

Refunds/exchanges

Knowledge
Evidence

1   What is the difference between a refund and an exchange?

_____

_____

2   What is an option, other than a refund or exchange, that a business could provide to a customer who is unhappy with something they have purchased? In general, would this third option be useful for a salon?

_____

_____

3   Write down a scenario where a client comes in to the salon and complains about a faulty item. Try to think about the following in your scenario:

    a   What steps would you take to keep the situation calm?

    b   How would you talk to the client?

    c   In your scenario, what are the best options to remedy the situation and keep the client happy?

    d   Why is it important to keep the client happy?

_____

_____

_____

4   Read information about refunds and Afterpay® (**https://help.afterpay.com/hc/en-au/articles/360018322171-I-am-a-Merchant-How-does-Afterpay-In-Store-Work-**) and answer the following question.

Can a client have a refund if they paid by Afterpay®? If so, what is the process?

LEGISLATION

5   Go online to find the consumer law information about refunds and exchanges at the following website: **https://www.accc.gov.au/consumers/consumer-rights-guarantees/repair-replace-refund**

What does it say about approaching the product manufacturer for a refund?

6   Research the different legislation regarding consumer protection that exists in your state or territory. Try comparing this legislation to another state's laws to see how they might differ. (It is important to know the laws of the place that you work in. If you move state and something goes wrong, ignorance of different laws will not be a valid excuse.)

LO6.4 REMOVING TAKINGS FROM REGISTER OR TERMINAL

Performance Evidence

Knowledge Evidence

Maintaining cash float

1   How do you separate the float at the end of the day?

Knowledge
Evidence

Recording takings/securing cash and non-cash transactions

2   Complete the remainder of the table below, coming up with scenarios for each row.

TRANSACTION ERROR REPORT				
DATE	CLIENT NAME	TRANSACTION ERROR	CORRECTIVE ACTION	RESOLVED? (THERAPIST)
2/6/2019	Mrs Jones	Overcharged EFTPOS	Reverse payment	Yes (Joanna)
7/7/2019	Mr Tan	Incorrect price ticket	Contact client for refund	No

When conducting transactions, confidentiality and security can be assured by correctly managing personal financial information, including transaction information from financial institutions (banks).

The Office of the Australian Information Commissioner (OAIC) has published 'Ten tips to protect your customers' personal information'. Find them at **https://www.oaic.gov.au/agencies-and-organisations/business-resources/ privacy-business-resource-9**

Using the 10 tips, answer the following questions (covering only the tips relevant to your work and to financial transactions).

3   Name one personal information handling process or procedure in your salon that you have had to be trained to do.

4   Who is responsible for the privacy of client transaction history information, according to the *Privacy Act 1988*?

5   Which of the following information is relevant and not relevant to:

   a   a financial transaction for a client wishing to pay for a manicure with a credit card? *Credit card number; Client full name; Date of birth; Address; Medical history; Card expiry*

RELEVANT	NOT RELEVANT

b a financial transaction for a client wishing to pay for a wax with cash? *Financial institution (bank) name; Address; Full name; Mobile phone number; Date of birth; Emergency contact*

RELEVANT	NOT RELEVANT

6   Can you expect to be able to store the credit card information for use every time the client visits?

_____

_____

_____

_____

7   Can you foresee a situation where you would need to destroy personal information securely? How would you do it?

_____

_____

_____

_____

_____

LO6.5 RECONCILING TAKINGS

Knowledge
Evidence

1   What is the correct process for calculating the register balance at the end of the work day?

_____

_____

2   What are three reasons there might be a discrepancy in the register balance?

_____

_____

_____

SECTION 2

HEALTH AND SAFETY

7 Apply safe hygiene, health and work practices

8 Maintain infection control standards

CHAPTER 7: APPLY SAFE HYGIENE, HEALTH AND WORK PRACTICES

LEARNING OBJECTIVES

After completing this chapter, you should be able to:

LO7.1 protect yourself against infection risks with personal protective equipment, handwashing, basic first-aid dressings and sharps handling and disposal

LO7.2 apply salon safety procedures, including risk assessments, with regards to hazards, manual handling and unsafe work practices that can cause accidents and incidents

LO7.3 use electricity safely, preventing incidents, using and storing equipment, and reporting faults and unsafe work practices

LO7.4 minimise infection risks, including through personal hygienic practices and minimising cross-infection, and know skin penetration guidelines

LO7.5 follow infection control procedures, preventing spread of blood-borne disease and contamination of products, materials, work surfaces, equipment and linen

LO7.6 follow procedures for emergency situations, including fire procedures and knowing the responsible personnel for first aid and evacuation procedures

LO7.7 follow salon cleaning protocols, mixing and storing chemicals, cleaning equipment and disposing of waste hygienically and sustainably.

INTRODUCTION

Health and safety in the salon is primarily important to comply with work health and safety (WHS) legislation. In the event of an accident or incident, it is imperative that the salon is protected from legal action by the fact that it is a safe, healthy and risk-free environment. Employees and clients should expect to enter a salon that is clean and safe.

In this chapter, you will learn about how to establish and maintain a safe environment, consult with staff, assess and control risks, and work with policies and procedures to prevent hazards and record hazards that may occur.

The learning activities in this chapter allow you to demonstrate your knowledge and understanding of the unit of competency *SHBXWHS001 Apply safe hygiene, health and work practices*, using the performance and knowledge evidences.

LO7.1 PROTECTING YOURSELF AGAINST INFECTION RISKS

Performance Evidence

Contingency and infection control procedures

Knowledge Evidence

1 What is PPE? Give five examples of PPE in the workplace.

2   Choose two of the PPE items that you listed above and give the do's and don'ts of usage in the salon.

_____

_____

_____

_____

_____

_____

_____

_____

_____

_____

_____

_____

_____

_____

_____

_____

_____

_____

_____

_____

_____

3   What is the code and name for the Australian and New Zealand standard that governs the way used needles are dealt with in the salon?

_____

_____

4   In case of needlestick injury, what steps should be taken?

_____

_____

_____

5   Why must towels and linen be changed after every client?

_____

6   How often should fresh laundry be used?

_____

7   What five things should you consider when choosing single-use coverings to use in the salon?

_____

8 List the three reasons it is vital to practice good handwashing techniques.

_____

_____

_____

9 If there is no running water available, what handwashing options are available?

_____

10 What is meant by the term 'adequate first-aid procedures' when dealing with a therapist who has an open wound?

_____

_____

_____

LO7.2 APPLYING SALON SAFETY PROCEDURES

Performance Evidence

Knowledge Evidence

Hazards, near misses, incidents and accidents

Both accidents and hazards are to be recorded in the Accident Report book. Other terminology used in this context are 'near misses' and 'incidents'.

- An _accident_ is an incident that may or may not result in injury.
- An _incident_ is a general name for a near miss, hazard or accident.
- A _near miss_ is a dangerous incident that could have led to a serious injury.
- A _hazard_ is a risk; anything with the potential to cause damage to a person, property, the environment or equipment.

1 In the table below, list an example hazard that could be found in your workplace. Then state how you would remove the hazard, to whom you would report the hazard and how. You are required to formally report a hazard by noting it in an Accident Report book. Include the following aspects of the work environment:

- broken or faulty tools and equipment
- fire
- slips, trips and falls
- chemical spills
- needlestick injury
- spills and leakage of materials
- cleanliness of the general salon area.

ASPECT OF WORK ENVIRONMENT	HAZARD	REMEDIAL ACTION/ HOW WILL YOU REMOVE IT?	TO WHOM AND HOW WILL YOU REPORT IT?
Broken or faulty tools and equipment			
Fire			
Slips, trips and falls			
Chemical spills			

ASPECT OF WORK ENVIRONMENT	HAZARD	REMEDIAL ACTION/ HOW WILL YOU REMOVE IT?	TO WHOM AND HOW WILL YOU REPORT IT?
Needlestick injury			
Spills and leakage of materials			
Cleanliness of the general salon area			

2   When consulting about WHS processes and risk assessment, you need to be actively involved in a management strategy. List the three main processes you need to be involved with in your workplace.

_____

_____

_____

3   For the three processes you listed in the above answer, give three examples of how you have (or can in the future) get involved in your workplace.

_____

_____

_____

You must learn how to work safely with tools, equipment and hazardous substances during treatments to ensure everyone's health and safety.

4   Review the examples listed in Figure 7.15 under 'Standard and additional precautions for treatments' in Chapter 7 of your textbook. In the table below, suggest two other examples of standard and additional WHS precautions for beauty services that may include PPE and improvements to your techniques.

BEAUTY PRACTICE	ELEVATED RISK	STANDARD AND ADDITIONAL PRECAUTION

5   In the table below, write an example of each of the different types of unsafe work practices.

UNSAFE WORK PRACTICE	EXAMPLE
Failure to report obstruction to corridor, stairwell or fire exit	
Failure to report a spill or break	
Security breach	
Health (infection control) breach	
Bullying and harassment	

Bullying behaviour is defined as:

- unfair and excessive criticism
- publicly insulting victims
- constantly changing or setting unrealistic work targets
- undervaluing employees' efforts at work.

6   What three things do you need to do in the case of bullying and harassment?

_____

_____

_____

Risk assessment and control

Many products and equipment used during treatments and for cleaning are potentially harmful. It is essential that they are used and stored and monitored correctly and in accordance with the manufacturer's instructions.

Performance Evidence

Knowledge Evidence

7   There is a hierarchy of risk control. List the five steps.

_____

_____

_____

_____

_____

8 Complete the following table to show your understanding of the WHS regulations in relation to the risks given in the first two columns. _Note:_ you can easily locate hazard information in the safety data sheets (SDSs) for acetone and sodium hypochlorite (bleach disinfectant).

RISK	RISK TYPE	WHAT MAKES IT A HAZARD?	CORRECT USE AND HANDLING	CORRECT STORAGE	CORRECT DISPOSAL/ CLEANING
Acetone	Fire Chemical spill				
Sodium hypochlorite (bleach disinfectant)	Chemical spill/ spill or leakage of materials				
Hypodermic needle	Needlestick injury				
Blunt cuticle nippers	Broken or faulty tools and equipment				

RISK	RISK TYPE	WHAT MAKES IT A HAZARD?	CORRECT USE AND HANDLING	CORRECT STORAGE	CORRECT DISPOSAL/ CLEANING
Faulty machine in the corridor	Slips, trips and falls				

9 Determine risk control measures in the salon for waxing treatments. Complete the risk assessment form below to list potential hazards and dangerous goods according to priority and with a safety precaution procedure. Use the risk assessment form shown in Figure 7.13 in Chapter 7 of your textbook as a guide, and for further assistance refer to **https://www.safeworkaustralia.gov.au/covid-19-information-workplaces/industry-information/ office/risk-assessment**. The four hazards identified in the wax room that need prioritising are:

- wax heater faulty – temperature always too high, wax can burn clients
- wax solvent spray often found on the shelf and leaking near candles
- magnifying lamp globe has blown – straining to see the hairs under the room's lighting
- paint peeling from the walls; walls cannot be cleaned properly.

HAZARD	SAFETY PRECAUTION PROCEDURE	PERSON RESPONSIBLE	DATE TO BE IMPLEMENTED	REVIEW DATE
HIGH RISK/PRIORITY				
MEDIUM RISK/PRIORITY				
LOW RISK/PRIORITY				

10 Most companies will have various templates for reporting specific things to a manager. Give two examples of when you might have to use a template in a health and safety context.

_____

_____

_____

_____

_____

SAFETY DATA SHEET (SDS)

11 When handling chemicals you must ensure safety precautions are taken. How do you find out about the chemicals you use and how to use them safely?

_____

_____

12 Contact your supplier to obtain an SDS for one of the hazardous chemicals you are using. Complete the blank Risk Management Sheets below using the information from the SDS. You need to:

Step 1: Identify the type of hazard and describe the risk for the chemical.

TYPE OF HAZARD (CIRCLE APPROPRIATE)	DESCRIPTION OF RISK
BIOLOGICAL	
microbial/disease blood/body fluid food handling	
CHEMICAL	
hazardous chemical non-hazardous chemical	
CRITICAL INCIDENT	
emergency evacuation lockdown other disruption	
ENERGY SYSTEMS	
electricity gas pressurised gas systems	
ENVIRONMENT	
UV exposure water weather temperature noise animal/insect	

TYPE OF HAZARD (CIRCLE APPROPRIATE)	DESCRIPTION OF RISK
PLANT AND EQUIPMENT	
building equipment treatment rooms furniture tools	
MANUAL TASKS	
repetitive or heavy tasks restricted space restricted light	
PEOPLE	
clients staff children animals visitors physical psychological/stress	
OTHER HAZARDS (LIST)	

Step 2: Assess the risk level by determining:

- the seriousness
- the likelihood of the hazard occurring.

SERIOUSNESS	DESCRIPTION
INSIGNIFICANT	No treatment required
MINOR	Minor injury requiring first aid treatment
MODERATE	Minor cuts, bruises, bumps
MAJOR	Injury requiring medical treatment or involving lost time
CRITICAL	Serious injury requiring specialist medical treatment or hospitalisation

LIKELIHOOD	SERIOUSNESS				
	INSIGNIFICANT	MINOR	MODERATE	MAJOR	CRITICAL
ALMOST CERTAIN	Medium	Medium	High	Extreme	Extreme
LIKELY	Low	Medium	High	High	Extreme
POSSIBLE	Low	Medium	High	High	High
UNLIKELY	Low	Low	Medium	Medium	High
RARE	Low	Low	Low	Low	Medium

RISK LEVEL	DESCRIPTION	ACTION
EXTREME	If an incident were to occur, it would be likely that a permanent, debilitating injury or death would result.	Immediate and/or significant changes need to be implemented to prevent injury.
HIGH	If an incident were to occur, there is some chance that an injury would result.	Changes need to be implemented to prevent injury.
MEDIUM	If an incident were to occur, it would be likely that an injury requiring medical treatment would result.	The problem needs to be fixed before the activity is repeated.
LOW	If an incident were to occur, there would be little likelihood that an injury would result.	Continue with the activity with the existing risk management controls in place.

Step 3: Select one or more control methods and describe your action; these are in the hierarchy of risk control.

RISK CONTROL METHOD	HOW YOU CAN CONTROL THE RISK
Elimination: remove the hazard completely from the workplace or activity	
Substitution: replace a hazard with a less dangerous one (e.g. a less hazardous chemical)	
Redesign: make a machine or work process safer (e.g. raise a bench to reduce bending)	
Isolation: separate people from the hazard (e.g. safety barrier)	
Administration: put rules, signage or training in place to make a workplace safer (e.g. induction training, highlighting trip hazards)	
Personal Protective Equipment (PPE): protective clothing and equipment (e.g. gloves, hats)	

Step 4: The risk control is reviewed and monitored regularly. Record the evaluation of the review for reference in a Risk Evaluation Form.

This risk evaluation and review is conducted in accordance with the risk assessment attached, implementing the control measures outlined in Step 3. Changes will be made to the activity, if required, to manage any emerging risks to ensure health and safety.					
Person responsible (contact name):				Position:	
Date:				Phone:	
Have the current control measures fixed the problem associated with the identified hazard?	Yes	No	Maybe	Notes:	
Have there been any changes to the current control measures?	Yes	No		Details:	
Are further control measures required in the future?	Yes	No	Maybe	Notes:	
Further notes:					
Next review date:				Signed:	

13 Where are the SDSs kept in your salon?

Safe work practices

It is important to know your WHS responsibilities. The key people in the workplace include you and the employer, clients and visitors to the salon.

The employer may be a PCBU, who is a person conducting a business or undertaking.

Safe Work Australia developed federal 'model' WHS laws for implementation across all states and territories nationwide. The model WHS Act places the primary duty of care on the PCBU. The term PCBU is an umbrella concept used to capture all types of working arrangements or structures. A PCBU can be a company, an unincorporated body or association or sole trader/self-employed person. Individuals who are in a partnership that is conducting a business will individually and collectively be a PCBU.

Information for your state or territory authority can be found at the following organisations:

- NSW: SafeWork NSW
- ACT: WorkSafe ACT
- WA: WorkSafe WA
- Qld: Workplace Health and Safety Queensland
- SA: SafeWork SA
- Tas: WorkSafe Tasmania
- Vic: WorkSafe Victoria
- NT: NT WorkSafe.

Comcare is a government body with a responsibility to provide expert advice and services to promote safety for the purposes of rehabilitation and compensation. It is an excellent resource for general health and safety information to prevent accidents and incidents in the workplace.

Use links to information at your state or territory authority websites and Comcare to answer the following questions.

14 What are the PCBU's three responsibilities for using PPE in the workplace?

15 Is it the worker's responsibility to maintain work health and safety in the salon?

16 If you are doing work experience, are you also responsible for maintaining work health and safety in the salon?

17 If workers are made aware of the work health and safety policies and procedures, whether they are reasonable or not, do they need to follow the instructions? For example, are new employees expected to vacuum and mop floors despite having the same workload as everyone else? See **http://www.comcare.gov.au/preventing/governance/procedures**

18 What ramifications could you or your salon face if you cause harm to your client or put them at risk?

When conducting a risk assessment regarding manual handling, you need to ensure people have been following WHS and OHS regulations.

19 List the six things that a risk assessment of correct manual handling techniques needs to consider.

20 Explain the best way to lift heavy boxes.

_____

_____

21 Using the Safe Work Australia website, find the definition of:

a bullying

_____

_____

b sexual harassment

_____

_____

_____

_____

LO7.3 USING ELECTRICITY SAFELY

The contingency plan provides response procedures intended to protect human health and the environment in the unlikely event that an emergency occurs; for example, a list of the emergency procedures in the event of a bushfire. The plan is put in place so that staff know what to do.

Knowledge Evidence   Performance Evidence

Emergency procedures

When devising an action plan for any emergency event, consider:

- what you plan to do beforehand to be ready
- what you plan to do during the emergency to keep people safe and minimise the effect on the salon
- what you plan to do after the emergency to help recover and return to work as soon as possible.

1 In the table below, complete the second column with your notes about a contingency plan for each stage of an emergency; for example, in the event of a flood in the salon. It might help to think of a specific situation that is more likely to apply to your workplace: some areas are more prone to floods; some are more prone to bushfires.

CONTINGENCY PLAN	EMERGENCY ACTION PLAN
Prepare for the emergency event (list at least five things)	

CONTINGENCY PLAN	EMERGENCY ACTION PLAN
Prepare the salon (list at least two things)	
DURING THE EMERGENCY EVENT	
Keep people safe (list at least three things)	
Recover from emergency event (list at least five things)	

Electrocution

Electrocution is electric shock that can cause death. Electric shock can also have serious health effects:

- *Burns:* lower voltage shocks can cause skin burns; high voltage shocks (i.e. >500 V) can cause internal burns.
- *Damage to nerves:* primarily in the heart, the current surges through to the nerves of the heart muscle causing fibrillation (spasm); if the current moves towards the head, the person becomes unconscious.
- *Muscle spasm:* when severe, these can cause fractured bones and dislocations.
- *Respiratory system:* paralysis; causes the heart to fibrillate or to stop beating.
- *Lymph and blood:* oedema and blood coagulates causing muscles to swell up at the point electricity enters the body.
- *Kidney failure:* an imbalance of fluid and electrolytes causes blood pressure to drop, which affects the flow of fluids through the kidneys.

Knowledge
Evidence

The minimum current that a person can feel is 1 milliampere (1 mA). The current is felt at this level during galvanic treatments in specialist facials and in body treatments. It takes 100 mA to cause electrocution or fibrillation (heart attack).

2 What is the difference between electric shock and electrocution?

_____

3 How can fibrillation (heart attack) occur as a result of electric shock?

_____

4 How much current is considered enough to cause electrocution?

_____

ELECTRICAL CIRCUITS

The main electrical supply to the salon, often called the 'mains', has a panel that separates the salon from the large wires that supply the premises from the power company. These wires contain virtually unlimited electrical power, so the panel in the mains must have circuit breakers or fuses to limit the power and control or 'funnel' that power to the various circuits that branch from it.

CIRCUIT BREAKERS

Circuit breakers and fuses limit the power to what the wiring system can handle. Too much current through the circuit is known as an 'overload'. When too much power has been put through a circuit; for example, when you turn too much stuff on, the circuit breaker or fuse snaps open and cuts off the circuit, stopping the flow of electricity.

It is important to let an electrician assess the circuit when a circuit breaker has tripped to check for any potentially dangerous problems.

SHORT CIRCUIT

If electrons take a path short of the complete circuit, it is called a short circuit. In the case of a short circuit, the resistance is low because the flow of electrons has travelled along an unintended path (e.g. to your device or worse, a person). View the video at the following link to understand more about short circuits: **https://www.youtube.com/watch?v=TUpPMZ2WH08**

Circuit breakers are installed in all safe devices to prevent damage from short circuit. Overheating after a short circuit is a common cause of fires. Short circuits can occur due to:

- electrical fittings not having been installed properly
- faulty power sockets
- faulty electrical equipment
- faulty insulation
- unearthed equipment
- unsafe electrical practices.

5 A fault in the electrical equipment can create a short circuit. In your own words, describe how this might happen.

_____

_____

6 If a current bypasses the load, what does that mean has happened with the electrons?

_____

_____

7 What two events can cause overloading? Why is overloading a circuit a problem?

_____

_____

RCCB, RCD AND EARTH LEAKAGE CIRCUIT BREAKERS

If the current leaks from your electrical equipment, there is most likely a problem with the circuit that could cause electric shock. A residual current device (RCD) earth leakage circuit breaker detects the leak (especially on the metal casing of a machine or lamp) and interrupts the circuit if the voltage is dangerously high. A residual current circuit breaker (RCCB) is both a circuit breaker and an RCD. Earth leakage circuit breakers (ELCBs) are no longer used – the RCDs have replaced them.

SURGE PROTECTORS

A surge protector limits the voltage supplied to the electrical equipment by either blocking or short circuiting the current to earth. Earthing is known as 'grounding'. An electrical device with an earth has a wire that travels to the ground, which has a net zero charge or voltage, so it absorbs an unlimited amount of current. The switch found on multi-adaptors is an example of a surge protector.

8   What is the difference between a circuit breaker, a fuse and a surge protector?

_____

_____

_____

HOW TO PREVENT AN OVERLOAD

1   Always replace faulty power points immediately and use a qualified electrician.
2   Don't plug double-adaptors or extension cords into multi-adaptors or other double-adaptors.
3   Know which machines and electrical equipment use the most power. The main culprits are:
    – steamers
    – autoclaves
    – infra-red lamps
    – incandescent and halogen light lamps
    – G5 and vibromasseurs
    – Wood's lamps
    – wax pots.
4   Understand the circuit branches of the electrical system from the mains supply. That way, you can know how much you can plug into the various power points.
5   Maintain your electrical equipment:
    – have it tested regularly
    – keep it clean
    – keep it away from water and humidity.

9   Give the reasons for the following storage protocols for electrical equipment:

   a Flammable liquids: store away from flammable substances.

_____

   b Stability: keep on a level surface and sit upright.

_____

10 List three examples of things you should always do when using electrical equipment safely.

_____

_____

_____

11 Look around your training facility and make notes in the following table regarding using and storing three different pieces of electrical equipment.

EQUIPMENT	USAGE	STORAGE	ELECTRICAL TESTING
e.g. Hot wax pots	Used to keep wax at a constant temperature	When not in use, wax pots are covered and kept on a trolley out of the main walkway in a treatment space	The wax pots are tested and tagged every year by a licensed electrician

12 If you were to discover that the electrical equipment you were to use that day had exposed wires in the cord, what would you do?

_____

MINIMISING STATIC ELECTRICITY IN THE SALON

Watch the video at **https://www.youtube.com/watch?v=yc2-363MIQs** and answer the following questions.

13 Explain how static electric charge builds up in people and objects to eventually cause an electric shock.

_____

14 List four causes of static electricity in the salon.

_____

_____

_____

_____

15 List five ways you can minimise excess static electricity in the salon.

_____

_____

_____

_____

_____

_____

LO7.4 MINIMISING INFECTION RISKS IN THE SALON ENVIRONMENT

Performance
Evidence

1   List two possible ways that you can maintain high levels of personal hygiene in the salon for each category:

    a body

    _____

    b feet

    _____

    c teeth

    _____

    d hair and nails

    _____

    _____

2   For each of the following in the table below, explain how to manage the risk to yourself and clients of exposure to infection.

Health and safety	
Personal hygiene	
Cuts on the hands	
Cross-infection	
Use hygienic tools	
Disposable products	
Working surfaces	
Gowns and towels	
Laundry	
Waste	
Eating and drinking	

Smoking	
Drugs and alcohol	

Local legislation

3 Look up the 'Skin penetration guidelines' in your state or territory to answer the following questions:

    a The skin penetration guidelines exist to regulate body decorating and grooming practices that are not carried out by health professionals. What are four examples of these?

_____

_____

_____

_____

    b Is it the responsibility of the client to tell their therapist before a skin penetration procedure if they have a blood-borne virus?

_____

_____

    c What does the term 'aseptic procedure' refer to? Why is it important with any skin penetration procedure?

_____

_____

_____

LO7.5 FOLLOWING INFECTION CONTROL PROCEDURES

1 When cleaning a space, it is always important to use the sequence provided. Why are floors always the last to be cleaned in a workspace?

_____

2 In your training college, there should be a roster to show the cleaning that has been done. What sort of information can be found on such a roster? *Note:* if you can't see one, you will often see such rosters in public toilet spaces.

_____

Infections in the salon

There are different types of microbes that can cause infection. They can be categorised as either bacteria, fungi, viruses and parasites. It is important to learn about the harmful microbes to help prevent the spread of their illnesses. Not all microbes cause harm to the body, and in fact some are beneficial.

When you suspect an infectious disease, you need to assess whether it puts any people at risk of cross-infection. If so, refer the client to a GP for diagnosis and possible treatment. For more about the diseases, refer to Appendix A Contraindications and restrictions.

Knowledge Evidence

The immune system helps to protect the body against harmful microbes when the body is infected. For example, when a person has the flu, the glands swell due to the increased production of white blood cells and antibodies, which attack the virus.

BACTERIA

Bacteria are single-celled organisms that survive in a variety of environments and conditions. The more dangerous, virulent strains of bacteria can destroy the body's cells, which causes disease.

Some bacteria are harmless, and some may be beneficial to the body. Beneficial bacteria in the gut, lactobacillus acidophilus, for example, function to break down food, provide nutrients and eliminate harmful microbes.

FUNGI

Fungi are plant-like organisms subclassified by their structure as moulds (multicellular) and yeasts (unicellular). Fungi produce spores, which are tiny single cells with a hard, outer shell. Fungi prefer a warm, moist environment, and their spores can lay dormant for long periods while the conditions are unfavourable.

Some fungi may be useful, such as mushrooms, compost fungi and yeasts used to make bread. However, all fungi secrete digestive enzymes and absorb nutrients from their host, and this also damages the tissues of the body, resulting in disease. Fungi are similar in structure to human cells, so antifungal treatments are often only 'fungistatic', meaning it prevents growth, rather than 'fungicidal' due to the potential toxic side effects. Treatments are often prescribed for a period of time to allow for uncasing of 'dormant' spores.

VIRUSES

A virus is a tiny 'non-living' organism. It has no ability to undergo metabolism, grow or reproduce on its own. It invades a host cell where it releases its genetic material, which causes the cell to produce many viral clones, which causes the cell to eventually rupture, thus releasing its clones that subsequently infect surrounding body cells. The body's immune system is the most reliable way for the body to fight a virus.

PARASITES

A parasite is an organism that lives in or on a host to obtain nutrients. They are almost always harmful to humans because they consume nutrients, damage body cells and leave toxic waste. Some parasites are visible to the naked eye, such as lice or worms, and some microscopic parasites, such as protozoa, travel the bloodstream to the organs of the body.

Adapted from Dylan Webb's work in Victoria University's *Infection in the Salon: Student Guide 2015.*

Diseases of the skin

Knowledge Evidence

BACTERIAL DISEASES OF THE SKIN

DISEASE	CAUSATIVE AGENT	CHARACTERISTICS
Impetigo	*Staphylococcus aureus; Streptococcus pyogenes* (occasionally)	Superficial skin infection characterised by isolated pustules (round elevations on the skin containing pus) that become encrusted.
Erysipelas	*Streptococcus pyogenes*	Reddish patches on skin caused by toxins; often with high fever. Usually preceded elsewhere in the body by a streptococcal infection (e.g. 'strep.' sore throat).
Pseudomonas dermatitis	*Pseudomonas aeruginosa*	Superficial rash (self-limiting) often associated with pools and saunas.
Infected burns	*Pseudomonas aeruginosa*	Infection with blue-green pus commonly of second- and third-degree burns.
Acne vulgaris	*Propionibacterium acnes*	Inflammatory lesions originating with accumulations of sebum that rupture a hair follicle.
Foot odour	*Corynebacterium spp.*	Pitting on sole of the foot, 'smelly sock' odour; underarm region, strong body odour.
Folliculitis (pimples) stye	*Various bacteria*	Infection of the hair follicle (infection of eyelash follicle) – stye.
Furuncle (boil)	*Staphylococcus aureus*	Localised region of pus (abscess) surrounded by inflamed tissue.

FUNGAL DISEASES OF THE SKIN

DISEASE	CAUSATIVE AGENT	CHARACTERISTICS
Tinea	*Microsporum spp., Trichophyton spp., Epidermophyton spp.*	Group of dermatomycoses (fungi of skin and hair). Spreading red rash, loss of hair.
Candidiasis (thrush)	*Candida albicans and other Candida spp.*	Infection of mucous membranes and moist skin areas. Causes persistent redness and/or ulceration and itchiness.
Pityriasis	*Pityrosporum ovale, P. orbiculare*	Mottling of skin.

VIRAL DISEASES OF THE SKIN

DISEASE	CAUSATIVE AGENT	CHARACTERISTICS
Warts (verruca vulgaris)	*Papillomavirus*	Horny projection of the skin caused by proliferation of cells.
Chickenpox (varicella)	*Herpesvirus (varicella-zoster)*	Usually contracted in childhood. Pussy vesicles confined, in most cases, to face, throat and lower back.
Shingles (zoster)	*Herpesvirus (varicella-zoster)*	Vesicles similar to chickenpox typically on one side of waist, face and scalp, or upper chest. Sufferers are usually elderly, sick or immunocompromised (virus remains dormant in body after chicken pox episode). Severe pain is common.
Herpes simplex	*Herpesvirus (herpes simplex type 1)*	Most commonly appears as cold sores – vesicles around the mouth. Can also affect other areas of skin and mucous membranes.
Measles (rubeola)	*Paramyxovirus (measles virus)*	Rash of small, raised spots first appearing on face and spreading to trunk and extremities.
Rubella (German measles)	*Togavirus (rubella virus)*	Mild disease with measles-like rash, but less extensive and disappears in three days or less. Severe birth defects common if contracted by pregnant women in the first trimester of pregnancy.

Parasites (macroparasites) of significance in the salon

Knowledge
Evidence

DISEASE	CAUSATIVE AGENT	CHARACTERISTICS
ECTOPARASITES		
Head lice	Head louse (*Pediculus humanus capitis*)	Initially, small red spots at site of bite; followed by itchiness and diffuse redness indicating established infestation. Eggs (nits) adhere to base of hairs.
Pubic lice	Pubic or crab louse (*Phthirus pubis*)	A species of lice generally limited to the pubic regions, but can also infest the armpit and torso.
Scabies	Itch mite (*Sarcoptes scabei*)	Typically occurs on hands, wrists, elbows, feet, genitals, buttocks, breasts and armpits. Itchiness develops approximately one month after initial infestation. Scratching may result in a secondary staphylococcal infection followed by eczema.

ENDOPARASITES		
Enterobiasis (threadworm or pinworm)	Threadworm (*Enterobius vermicularis*)	Restlessness and irritability due to itchiness, particularly at night. This is due to the migration of female worms to the anal area at night to lay their eggs. The sticky eggs lodge under fingernails when the area is scratched. Transmission is oral.

Adapted from Dylan Webb's work in Victoria University's *Infection in the Salon: Student Guide 2015*.

3 Form groups of two or three and research types of diseases caused by the different microbes listed below. Use material in this book or found online. For each type of microbe, name and describe seven possible diseases:

a bacteria

b fungi

c viruses

d farasites

The description must include:

• common name and scientific name of the microbe

• signs and symptoms of infection

• causes/how the disease can be transmitted

• treatment.

_____

_____

_____

_____

_____

_____

_____

_____

_____

_____

_____

_____

_____

_____

_____

_____

_____

_____

_____

_____

_____

4 In the table below, list the symptoms and required action you would take in the salon for the following infectious contraindications.

INFECTIOUS CONTRAINDICATION	SYMPTOMS	REQUIRED ACTION
Erysipelas	'St Anthony's fire'. A superficial form of cellulitis, a bacterial infection of upper layers of the dermis.	Recommend the client see a GP for diagnosis and treatment as soon as possible.
Herpes		
Impetigo		
Paronychia		
Pediculosis capitis		
Scabies		

Transmission

Knowledge
Evidence

Passing a pathogenic microbe from one person to another can occur either:

- directly:
  - through person to person physical contact and droplet spray
  - via contact with animals and/or their shed and discharged particles.
- indirectly:
  - via inanimate objects that have been contaminated.

5 In the beauty industry, what instruments could be indirectly contaminated?

_____

_____

6 Decide which of the following transmissions are direct or indirect:

   a A work surface was not cleaned down.

   _____

   b Someone coughs in the space of another person.

   _____

   c The therapist has an uncovered open wound on their finger, and is performing a piercing.

   _____

7 Draw a line to match each situation with the way cross-infection might have occurred.

SITUATION	CROSS-INFECTION
Therapist comes to work when they have a runny nose and treats clients	Client to operator
Client does not disclose that they have tinea on their feet	Operator to client
Therapist touches the phone while wearing gloves in the middle of a treatment and doesn't wipe it down	Operator to operator

HUMAN CARRIERS

Some people are carriers of dangerous pathogens while at the same time exhibiting no outward sign (or symptom) of the potential disease they harbor. They can and do shed the pathogen in the course of their everyday activity. They are often unaware that they are a source of infection and contamination. These people are referred to as 'carriers' and 'silent shedders'.

ANIMAL CARRIERS

No animals of any description are permitted in the salon. No fish, no birds, no cats, no dogs, no snakes, no insects. Animals can be carriers of pathogens and are said to act as vectors or 'carriers'. For this reason, no animals of any description are permitted to be directly inside the salon. There are three exceptions:

- guide dogs for the blind (seeing eye dogs) and hearing dogs for the deaf. A person who is neither blind nor deaf may be training a dog
- police and military dogs on a legally authorised inspection (not just visiting friends).

   Please note: persons or animals become _infected_ but inanimate objects become _contaminated_.

ENTRY AND EXIT POINTS

An entry point of a microbe into the body can be through the:

- respiratory tract: mouth, nose
- anus
- lungs

- urinogenitals
- broken skin: wounds, eczema and dermatitis
- eyes
- ears.

Exit points will involve all the above and more. This includes all body wastes and fluid discharges, such as:

- saliva
- nasal and mouth fluids
- vomit and bile
- blood, lymph and tissue plasma
- faeces and rectal mucus
- urine
- semen and vaginal discharge
- tears
- ear waxes
- shed skin scales
- hair
- open damaged skin and weeping wounds.

All of the above may be produced in your salon at some time in the future and you will come into contact with them. Consider that some of your future clients could have medical conditions they may not be willing to share with you.

Diseases transmitted via the blood

Knowledge
Evidence

DISEASE	CAUSATIVE AGENT	CHARACTERISTICS
Hepatitis	Hepatitis B virus (HBV) is the cause of serum hepatitis; also C & D (the latter only if infected with HBV)	Malaise, jaundice, chronic liver disease, commonly severe liver damage, high mortality rate with HDV.
Acquired immuno-deficiency syndrome (AIDS)	Human immunodeficiency virus (HIV)	HIV destroys the T4 lymphocytes of the immune system leaving the body vulnerable to a large number of opportunistic infections as well as the cancer Kaposi's sarcoma. The majority of these diseases are not found in people with an intact immune system.

8   Draw a line to match the infection with the mode of transmission.

INFECTION
Tinea pedis
Hepatitis B
Conjunctivitis
Influenza

SALON SITUATION (MODE OF TRANSMISSION)
Contaminated cuticle nippers
Not wearing a face mask
Extractions in a facial with a needlestick injury
Bathing both eyes in lash tint with one wipe

9   In case you are exposed to blood or body fluid from another person, your workplace will have a plan of action. What plan of action should be taken if you have an open wound where contamination has occurred?

The body's defences

Knowledge
Evidence

Normal healthy humans have various levels of defence against invasion by pathogenic (both transient and opportunistic) organisms. These defence systems operate at three 'levels':

Natural resistance (non-specific)

1   First line of defence (superficial)

2   Second line of defence (internal)

Acquired resistance (specific)

3   Third line of defence (cell-mediated immunity and antibody-mediated immunity)

Acquired resistance is either a response initiated by either:

* the immune system
* an immunisation.

Acquired immunity (specific immunity) develops through a person's lifetime and responds to specific triggers. This is our third and last line of defence. If this line is overwhelmed, you stand a very high chance of dying unless some medical intervention occurs and is successful. Activation of immune system defence takes time as the body learns how to deal with the infectious agent or foreign body and then sets up the processes needed to advance the fight against the offending agent. Once 'equipped', the body can then respond rapidly on second and subsequent entries and invasions of the agent.

Adapted from Dylan Webb's work in Victoria University's *Infection in the Salon: Student Guide 2015.*

IMMUNISATION

There are precious few drugs that are virucidal. Our immune system is the best chance of survival, as well as staying fit and eating healthily. One form of protection is immunisation. The National Health and Medical Research Council recommend vaccination specific for the occupation. For beauty workers the categories that apply include:

* people who work in specific remote areas
* people exposed to human tissue, blood, body fluids or sewage.

Depending on the location of your workplace, you should consult a GP for the recommended vaccinations. Communicable disease found in 'remote areas' varies from state to state. Hepatitis B is the only vaccine available for category 2 occupation and is highly recommended for all beauty therapists.

10 Research what other vaccines are recommended for beauty therapists and list them below.

_____

_____

As viruses can easily be transmitted in droplet spray and body fluids, it is important that salon owners and operators ensure protective measures are implemented. A person who is a carrier of Hepatitis B appears no different from a non carrier and they may not even be aware that they are carriers.

Knowledge
Evidence

Contamination

11 In the table below, list three reasons why micro-organisms can grow in products, three ways to identify a contaminated product, and three ways to prevent contaminating a product.

TOPIC	ANSWER
Reasons that micro-organisms can grow in products	
Ways to identify a contaminated product	

TOPIC	ANSWER
Ways to prevent contaminating a product	

12 How often should you change towels and linen between clients? Why?

_____

_____

LO7.6 FOLLOWING PROCEDURES FOR EMERGENCY SITUATIONS

Performance
Evidence

The way you are expected to respond in an emergency to an evacuation varies according to the nature of the emergency and the size and nature of your business. You may be required to:

- know how to use a fire extinguisher
- break the glass covering a fire alarm to alert the fire department of the emergency.
  Consider:
- people with mobility issues and other special needs
- that fire wardens are responsible for mobilising everyone in an emergency evacuation. You must follow the warden's advice and know what the warden hierarchy means. The warden in a small salon is typically the highest in the chain of command; for example, the salon manager, and if not, the assistant salon manager. Check your WHS policies and procedures manual.

Managing emergencies

Knowledge
Evidence

Watch the YouTube video at the following link to understand the first response to an emergency evacuation:
https://www.youtube.com/watch?v=UkO6iqJasDs

1 After responding to an emergency evacuation drill in your organisation, complete the following self-reflection questions:

a What did the warden ask you to do? For example, did you have to leave straight away, or did you have a first warning?

_____

_____

_____

b Did you have an evacuation meeting point? Where was it?

_____

_____

2 At your college or training premises, identify where the emergency evacuation procedure is. Write down the following important points:

a Who are the responsible personnel during an emergency?

_____

b Where is the emergency evacuation point?

_____

c Where is the closest fire hydrant?

_____

d Who do you contact in case of emergency?

_____

LO7.7 CLEANING THE SALON

Selecting, preparing, using and storing cleaning and disinfection equipment

Knowledge
Evidence

1 List the steps to disinfect biohazardous waste; for example, a blood spill or blood-soaked tissue, gloves and other single-use items used during extractions.

_____

_____

_____

_____

_____

_____

_____

_____

_____

_____

_____

_____

2 What is the difference between physical, thermal and chemical disinfection? List the benefits of each.

Physical:

_____

_____

Thermal:

_____

_____

Chemical:

_____

_____

3　From your knowledge of a sharps container, complete the following sentences:

　　a　The sharps container must be _____.

　　b　A _____ is never re-capped.

　　c　The sharps container should never _____.

　　d　The container should be placed _____ above the floor.

4　What are the five questions you need to ask yourself when deciding on the correct waste disposal method?

　　_____

　　_____

　　_____

　　_____

　　_____

5　List at least two places you can put general waste.

　　_____

6　Write one sentence giving further information on each of the following cleaning recommendations:

　　a　Use the correct cleaning agents.

　　_____

　　_____

　　b　Clean with the required frequency.

　　_____

　　_____

　　c　Use the correct equipment.

　　_____

　　_____

　　d　Take care of cleaning materials.

　　_____

　　_____

　　e　Observe a high standard of personal hygiene.

　　_____

　　_____

　　f　Establish a cleaning schedule.

　　_____

　　_____

7 How do you know how to dilute, use and store a potentially dangerous chemical product in the salon?

_____

_____

_____

8 Your training facility or home will have all sorts of cleaning products. Find three different types and fill in the table with the product type, and its directions for correct storage.

CLEANING PRODUCT	DIRECTIONS FOR STORAGE

Sequencing cleaning and disinfection procedures

Refer to Appendix C Treatment area and equipment in your textbook and confirm your salon procedures with regards to cleaning protocols.

Knowledge Evidence

9 What is your salon's cleaning procedure for the manicure bowl?

_____

_____

_____

_____

CHAPTER 8: MAINTAIN INFECTION CONTROL STANDARDS

After completing this chapter, you should be able to:

LO8.1 comply with health and safety laws, regulations and guidelines, know your legal responsibilities and follow standard salon policies and procedures for infection control

LO8.2 monitor hygiene of premises, including the removal and disposal of contaminated material, preparation of linen and disposable coverings, and measuring, using and storing cleaning products

LO8.3 maintain infection control when performing skin penetration procedures, including sharps handling, PPE, single-/multi-use equipment and skin preparation

LO8.4 use sterilising equipment, monitor and validate the process

LO8.5 maintain and show awareness of the clinic design for the purposes of infection control.

INTRODUCTION

The purpose of infection control is to remove or destroy pathogenic microbial contaminants from the salon to prevent the transmission of infectious disease. It should be assumed that every client who walks into the salon is potentially infectious, so it is important to maintain infection control standards at all times to minimise health and safety risks.

The learning activities in this chapter allow you to demonstrate your knowledge and understanding of the unit of competency *SHBBINF001 Maintain infection control standards*, using the performance and knowledge evidences.

LO8.1 COMPLYING WITH INFECTION CONTROL LEGAL OBLIGATIONS

Performance Evidence

Infection control risk assessment procedure

Knowledge Evidence

1 You have learnt how to participate in WHS processes and risk assessments in Chapter 7. Choose one area of the salon (or simulated salon environment) and perform an infection control risk assessment on the area. Use Figure 7.13 under 'Risk assessment' in your textbook. Include in your answer impacts of infection control on any of the following:

a layout of the work area

b equipment

c receptacles, containers, etc.

d benches and trolleys

e ventilation

_____

_____

_____

Risk management processes

Knowledge
Evidence

As detailed in Chapter 7 of your textbook under 'Risk management' in the 'Applying salon safety procedures' section, the risk management procedure for all workplaces is:

1   Find the hazard.

2   Assess the risk.

3   Fix the problem.

4   Regularly review the risk.

2   Case study: An infection control hazard has entered the salon – a client with influenza who is coughing and sneezing. It has been a long winter and you happen to know numerous people who have been unwell for more than five days with colds or the flu. Step 1 in the risk management procedure is done – the hazard has been identified. Describe what you must do for the following three steps to complete the risk management process.

   a   Step 2: Assess risk

   i   Assess the potential risks associated with the hazard.

   _____

   ii   Identify who is at risk from the hazard.

   _____

   iii   Identify how any risk is to be minimised or eliminated.

   _____

   _____

   _____

   b   Step 3: Fix the problem

   _____

   _____

   _____

   _____

   _____

   c   Step 4: Regularly review the risk

   _____

   _____

Licensing and registration

Knowledge
Evidence

3   In relation to the salon's compliance with infection control and skin penetration procedures, what are the licensing and registration requirements for your salon business (or salon simulated environment)?

   _____

   _____

WHS inspections

Work health and safety inspections are conducted by your WHS authority inspector to help to identify, assess and eliminate or control hazards in the salon.

4   Who is your salon's WHS representative?

5   How do you perform regular WHS checks or 'audits' to ensure a WHS inspection will be passed?

Infection control guidelines

Australian Guidelines for the Prevention and Control of Infection in Healthcare (2010) are regularly updated by the Australian Government's National Health and Medical Research Council (NHMRC) to control risk of infection across the population.

Knowledge Evidence

6   Research the NHMRC's website to find the B1 Standard Precautions and its additional recommended precautions at **https://www.nhmrc.gov.au**. List the Standard Precautions as they apply to beauty therapy.

7   What is one infectious agent that handwashing is known to significantly prevent?

8   What is a personal protective equipment (PPE) item that can prevent you from transmitting a cold or flu to another person?

Industry codes of practice for infection control

Knowledge Evidence

Codes of practice are voluntary industry standards of conduct. They are set guidelines for infection control (and other professional activities) and for the way you deal with colleagues and your clients. You are required to promote the infection control code of conduct for your salon. Codes of practice can relate to a professional individual, a single salon business or represent the beauty industry as a whole.

9   Your salon or simulated salon environment will have a code of conduct. From that code, list three codes (do's or don'ts) relevant to infection control.

_____

_____

_____

_____

_____

_____

10   List a beauty industry body that issues codes of conduct for you, and may serve as a source of industry-specific business information in the future.

_____

LO8.2 MONITORING HYGIENE OF PREMISES

1   Complete the following table with the definitions of each term.

TERM	DEFINITION
Sterilisation	
Thermal disinfectant	
Chemical disinfectant	
Washing or cleaning	
Antiseptic	

2   Who is the EPA? What is their relationship to contaminated waste?

_____

_____

3   If the manufacturer instructed you to dilute their bleaching product to a 1% solution, how much water and how much bleach would that include?

_____

4   What is an appropriate way to store corrosive chemicals?

_____

5   How often should you be cleaning treatment table surfaces? What would you use to clean them?

_____

_____

6   What is included in the term 'linen' in a beauty salon? What is included in the term 'protective clothing' in a beauty salon? What is included in the term 'disposable coverings'?

_____

_____

LO8.3 MAINTAINING INFECTION CONTROL FOR SKIN PENETRATION TREATMENTS

Performance
Evidence

Naming organisms, naming diseases

A distinction must be drawn between organisms and diseases. Confusion between the two can lead to serious misunderstandings and potentially disastrous outcomes. Infectious diseases are caused by pathogenic micro-organisms and some macro-organisms such as worms, lice and mites.

NAMING A DISEASE AND DISEASE SYMPTOMS

Medical science describes diseases by the symptoms they produce in the host. Medical practitioners develop skills in recognising multiple signs or symptoms that a patient is showing. The greater skill is in linking those symptoms to the causative agent and then mapping a course of treatment.

SINGLE DISEASE: MULTIPLE CAUSES

These symptoms are the signs you display when ill. An infection of the lungs that brings on severe breathing difficulties is commonly called 'pneumonia'. Treatment requires correct diagnosis of the exact cause of your particular type of pneumonia. There are approximately 26 different pathogens than can cause infectious pneumonia, including a worm.

There are approximately 48 species of fungi that cause the diseases we call 'tinea', for some the cure of which may require expert identification of the exact causative species. Three main types *Trichophyton rubrum*, *Trichophyton mentagrophytes* and *Epidermophyton floccosum* can be treated relatively easily.

SINGLE ORGANISM: MULTIPLE DISEASES

It is entirely possible for a single species of pathogen to produce quite different sets of symptoms. Thus – different diseases. One organism may cause multiple diseases depending on the individual host response, the variety of that pathogen and the location and entry path of that pathogen.

The virus *herpes varicella zoster* (HVZ) causes chicken pox in children, zoster (a form of shingles) in young adults and shingles in older adults. 'Golden staph' may cause school sores, boils, carbuncles and furuncles, food poisoning, acne, 'scalded skin syndrome' and 'toxic shock syndrome' (TSS).

Syphilitic bacteria have been called 'the great pretender' because the variety of symptoms they can create mimic so many other diseases. Pinning them down is often very difficult.

SCIENTIFIC NAMES

Using the scientific names of organisms is preferred. These names are 'agreed names' that are settled on by an international jury and avoid the confusion that often arises with common or local names. Wherever you are in the world, if you use the scientific name then there can be no confusion as to the exact organism you are talking about. There is a standard format for scientific name writing: The *generic* name is written first and the *specific* name is written second, and there may be more than one member of species in a generic group. Both names are in a different script to the common text. To highlight the unique identification of this organism use either:

- italics: *Staphylococcus aureus*
- bold face: **Staphylococcus aureus**
- underlined: <u>Staphyloccus aureus</u>

Note: The generic name always begins with a capital letter (*Staphylococcus*). The species name always begins with a lowercase letter (*aureus*).

Once you have established the name of the organism, you can then refer to it in scientific shorthand. For example, *Staphylococcus aureus* can then become *S. aureus*. Italics/bold/underline still applies.

COMMON NAMES

Common names are still sometimes used for some old familiar 'friends'. For example, of the *Staphylococcal* group of bacteria, the most common and dangerous is *Staphylococcus aureus* which literally means 'golden round balls growing in little clumps'. All health workers call it 'Golden staph' and often abbreviate it to *S. aureus*.

Disease transmission and control

As a practising beauty therapist or beauty worker, you are going to be regularly confronted with the possibility that you may be responsible for the transmission of diseases or you may come into contact with them from your clients or fellow workers. Knowing about these diseases and how to prevent their transmission is an essential part of being a professional.

Knowledge Evidence

WHAT IS DISEASE?

A disease can be regarded as an abnormal physiological functioning of the cells, tissues or organs of the body. Two broad categories exist:

1 non-infectious diseases: no microbes involved
2 infectious diseases: microbes alone being the causative agent.

Infectious disease

Depending on how easily they are transmitted, these diseases are regarded as:

* non-contagious – difficult to nearly impossible to transfer human to human
* contagious – easily transferred human to human
* very highly contagious – extremely easy to transfer from human to human.

In all cases, the infectious disease that causes illness is caused by a microbe that must gain entry into the body and establish itself. It then goes on to breed and thus spreads itself further within that individual host before exiting to seek a new host. Should the combined invasion and breeding effects overwhelm our body's ability to fight back, then we say we show the signs and symptoms of that disease.

An infectious disease is caused by a pathogenic (disease causing) micro-organism (microbe). Not all microbes are pathogenic. In fact, the vast majority of microbes in our world are non-pathogenic. They are responsible for organic breakdown, decay, decomposition and nutrient recycling. Not all pathogens affect humans. Many are highly specific as to the organisms they attack. Many are pathogens of plants.

For a microbe to affect you:

* you must be exposed to the pathogen
* it must gain entry
* it must be of a species that can live in or on humans (nobody has ever been diagnosed with dog distemper). Some diseases can be caught from animals, but these cannot be passed on to other humans. Cryptosporidium from birds is an example
* it must be of a particular virulence or 'strength'
* you need to be exposed to at least an 'infective dose'. One bacterial cell is unlikely to bring on an illness – but 100 000 or more have a chance
* you must be susceptible to the microbe; that is, your defence mechanisms fail for one reason or another (you are immuno-compromised). 'Bird flu' only seriously affected those with already weakened immune systems.

Adapted from Dylan Webb's work in Victoria University's *Infection in the Salon: Student Guide 2015*.

1 What does contagious mean? Name one contagious disease you may encounter in a salon.

2 What is a pathogenic microbe?

3 What five things does a pathogenic microbe need in order to start an 'infection'?

Microbes using the body

Not all microbes are harmful. The most successful microbes cause us no harm at all and our bodies are covered with billions of harmless microbes. Microbes can be divided into three groups:

1 normal body flora
2 opportunistic flora
3 pathogenic flora.

NORMAL BODY FLORA

These microbes commonly live on and in us. We all have these and they are our own unique microbial community – the *microbiome*.

They are either harmless (*commensal*) or beneficial to the body (*symbiotic*), such as the gut-inhabiting bacteria. They often have specific food and chemical requirements and like particular 'habitats' of the body.

You can never get rid of your microbes. No amount of washing, cleaning or scrubbing can completely remove them. They grow back to recover your teeth, skin and hair within 20 to 40 minutes. Some form protective gelatinous coatings called *biofilms* that help them to wait out the bad times of hostile chemistry, dehydration and poor food supply.

OPPORTUNISTIC FLORA

Members of this category are usually normal flora or are harmless visitors, but they can gain access to certain parts of the body (usually by puncture or wound, i.e. they get into places they shouldn't really be) or if your body defences are 'run down' because you are:

- physically stressed or run down
- emotionally or psychologically stressed
- intaking poor nutrition
- on medication
- already infected with a primary disease
- suffering a non-infectious illness
- recovering from illness or injury.

Once they have gained access, opportunistic flora can then take advantage of you. It is not uncommon for a female taking antibiotics for a primary 'strep. throat' infection to come down with a secondary ('thrush') candidiasis infection. This is a fungal (not a bacterial) infection and is the result of the *Candida* yeast fungus taking advantage of the antibiotic suppression of the offending bacteria.

A number of opportunistic infections are endogenous infections. This means that you keep re-infecting yourself. Folliculitis after waxing and persistent acne lesions (both *Propionibacterium acnes*), boils and carbuncles (*Staphylococcus aureus*) are others.

These bacteria are some of the many that can form protective biofilms. When conditions turn favourable (for the microbes that is – not you) they emerge from this protective wrapper and attack. Disease ensues. Some can remain under their biofilms for decades (e.g. *Streptococcus pneumoniae*).

PATHOGENIC FLORA

Organisms in this group are normally transient visitors that during their stay cause trouble. Hopefully our body's defence systems overcome them and they are wiped out before any serious damage ensues. They have not learnt to use us successfully. They effectively 'bite the hand that feeds them'. They are too virulent (strong and aggressive) and we have not developed resistance to them.

Two thousand years ago, the rubella virus killed adult Europeans. Today it is damaging to the developing foetus inside the first trimester of pregnancy. Nevertheless, populations that had no history of rubella were decimated when European explorers arrived in South America, North America, the Pacific Islands and Australia. It is estimated that 60 per cent of the population of Aboriginal and Torres Strait Islanders were wiped out after 1788 as European viral diseases picked up from the Sydney Cove communities spread. Common killers were measles and influenza, which spread like wildfire from community to community before the British explorers even crossed the Blue Mountains.

Some pathogens may remain quietly resident in some people as if they are part of the normal body flora. These people are 'carriers' and spreaders of the pathogen though they show no outward sign of infection. They are 'silent shedders' of infection.

A person who is infected with certain types of pathogens that cause *notifiable disease* (or who is a carrier and thus shows no outward sign of infection) is required to be identified to the health authorities so that they can provide early warnings of outbreaks. Examples of notifiable diseases in Australia include HIV, chlamydia and Ross River virus.

Bacteria

Bacteria are a heterogeneous group of unicellular prokaryotic organisms. Prokaryotic refers to their primitive cellular organisation in that they lack a defined nucleus and nuclear membrane; have a single chromosome; and have no mitochondrial organelles. They are, however, numerous and prolific. They have been on the Earth for around 3.8 billion years and can be found in every type of environment. Human bodies are just one such environment.

By far the vast majority of bacterial species are non-pathogenic to humans and in fact our lives are enriched by their activity through decomposition and fermentation. The tiny few bacterial species that are pathogenic are of serious concern to us both in our daily lives and in the salon.

BACTERIAL STRUCTURE

Being microscopic, bacteria can only be seen with the aid of a microscope. Enormous clusters of individual cells form colonies. Colonies have distinct shape, colour and food requirements that aid in the identification of the particular species. Individual species vary in:

1   cell shape. These include:
    –   coccus (round)
    –   bacillus (rod)
    –   spirillus (spiral shape)
    –   vibrios (comma or blob shape)
2   motility, including:
    –   presence or absence of flagella
    –   number and position of flagella
3   adhesion and penetrative structures on the outer surface
4   cell wall: structure, composition and thickness
5   capsule presence or absence (surrounding the cell wall). A capsule can provide additional protection and adhesion
6   spore formation (only in *Bacillus spp.* & *Clostridium spp.*). *Clostridium botulinum* (used in Botox) was originally a spore forming, food poisoning bacteria
7   specific environmental requirements:
    –   acidity or alkalinity (pH) of fluid environment
    –   dissolved salt levels
    –   specific nutrient requirements
    –   oxygen levels (aerobic/facultative anaerobic/anaerobic)
    –   temperature/warmth
8   biofilm formation ability.

SIMPLIFIED CLASSIFICATION OF BACTERIA ACCORDING TO SHAPE

Figure 8.1 and Figure 8.2 group bacteria according to their shape (to the far left): bacillus, coccus, spirillus and with no cell wall. The species are listed on the far right, and are described below. Figure 8.1 shows gram positive bacteria. Gram positive bacteria are distinguished by an outer cell wall with gel material that is highlighted by a crystal violet dye. Gram positive bacteria are relatively easy to control and many are non-pathogenic.

Gram negative bacteria, shown in Figure 8.2, have an additional layer of protection to their cell wall, so their cell membrane does not take up the gram stain. Most are pathogenic and they are more resistant to drugs and antibiotics.

All bacteria require moisture; however, some can survive on the relatively 'dry' and salty surface of the skin (e.g. *Staphylococcus spp.*).

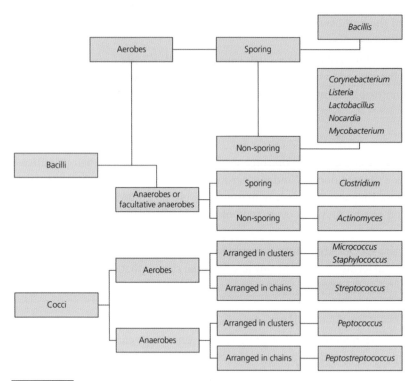

Figure 8.1 Gram positive bacteria

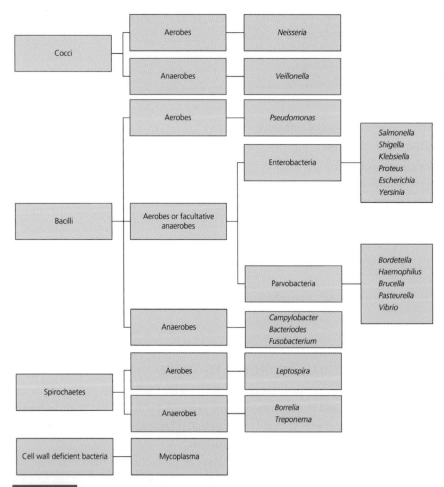

Figure 8.2 Gram negative bacteria

REPRODUCTION

Asexual reproduction. The bacterium (single cell) usually reproduces by a process called binary fission to split into two new cells that are identical to the parent cell.

Under ideal conditions this can happen every 20 minutes, causing an explosion in the number of growing and dividing cells that may soon overwhelm the host. This is an asexual (vegetative) reproductive technique.

Sexual reproduction. Sexual reproduction also occurs in bacteria by a variety of methods. The end result is a new cell(s) that are different from the parent cell and may be able to survive changed environmental conditions, thereby possibly producing a strain of bacteria with resistance to antibiotics.

BACTERIAL PRODUCTS

As living organisms, bacteria may produce various chemicals that leak out of their cell membranes, then through the cell wall and capsule (if they have one) and into the surrounding environment. Some of these products are:

- *enzymes* such as proteases, lipases and hyaluronidases, which dissolve and liquefy tissues of the body; that is, in necrosis
- *enterotoxins,* which interfere with cell function; for example, *Clostridium tetani* releases tetanus toxin causing muscle rigor
- gases, which cause bubbling and frothing gas gangrene.

Bacteria in the salon

Bacteria can be transmitted in the salon in numerous ways:

- on contaminated instruments and articles
- from infected persons (direct and indirect contact)
- animals (domestic and feral) and their residue
- droplet spray and airborne particles
- inadequate personal hygiene (endogenous re-contamination and oral-faecal route).

The living (vegetative) bacterial cells are reasonably easy to kill with heat treatment and the use of effective chemical disinfectants.

In order to do this the organisms must be exposed to the treatment and for the correct length of time. What is that correct time of exposure? Be sure to read the instructions and the safety data sheets (SDSs) for all cleaning products and equipment.

Scrubbing and cleaning must always precede exposure to heat or chemicals. *Pseudomonas auriginosa* can escape even powerful disinfectants, such as iodophors, thanks to their protective capsules. *Staphylococcus aureus* can handle quite salty conditions, such as sweaty hands inside protective gloves. The ability of bacteria like *S. aureus* to form biofilms makes them very difficult to remove without strict adherence to cleaning and disinfecting protocols and the correct use of PPE.

Surfaces of benches and cabinets are quickly re-contaminated by bacteria that cling to the oily skin scales that constantly flake off the body and blow around the rooms wherever people are. Horizontal surfaces become more quickly contaminated than vertical surfaces. Bacteria can ooze and be squeezed through apparently clean tissues, toilet paper and towels. We make the assumption that our clients have acceptable toilet habits and wash their hands properly. Clean laundry is essential for each client.

Bacterial spores are by far our biggest concern because these are not destroyed by the process of disinfecting or boiling. *Clostridium tetani*, which causes the disease tetanus, produces spores that can survive two hours in open boiling water (100 °C). To kill them requires the process of *sterilisation*. That is, the addition of adequate heat, pressure, time and/or pH of 1–2 or 12–14.

Bacteriostatic agents (most disinfectants) stop bacteria reproducing. Some chemicals are *bactericidal* (kills or poisons them.) Antibiotics are drugs that stop or kill bacteria. They are usually totally ineffective against viruses and fungal infections. As seen in Figures 8.1 and 8.2 earlier, gram positive bacteria are generally more susceptible than gram negative. A course of antibiotics *must* be followed to the end or else you are just encouraging the growth of drug resistant strains. Possibilities include:

- *S.aureus* causing *Pemphigus neonatorum* (also known as staphylococcal scalded skin syndrome)
- *Folliculitis* caused by *S.aureus*
- MRSA = Methicillin Resistant *Staphylococcus aureus*.

A special word on a non-life threatening but very concerning bacteria: *Propionibacterium acnes*.

This bacterium is strongly associated with the disease called acne vulgaris. It is a gram positive bacterium and is a common commensal bacteria. It is a non-contagious infectious disease. You cannot catch acne from another person. You can endogenously infect yourself.

It appears to be an *opportunistic microbe* with certain people and often appears in cyclical events. It is now known to form biofilms within the hair follicle and sebaceous glands. It releases a number of inflammatory chemicals – propionic acid and cytokines. Many people have the bacteria but never get the disease. It seems to be strongly tied to hormonal conditions and skin surface chemistry and the changes in that surface chemistry.

Some women who have never had any acne will go on to develop it in their menopausal years (often with great distress). Every single study done that has examined diet in relation to acne has come up negative. (Chocolate does not cause acne, some of you will be comforted to know.)

Adapted from Dylan Webb's work in Victoria University's *Infection in the Salon: Student Guide 2015*.

4   Are all bacteria pathogenic?

5   What are the five specific environmental requirements of bacteria?

6   Many bacteria seem to like a fluid environment, so how do you think they grow on the skin?

7   How can bacteria be spread in the salon? (List at least three ways.)

8   How can bacteria be killed in the salon?

9   How can bacteria be controlled medically in or on the body?

10 How high does the temperature need to be in order to kill bacterial spores?

Fungi

Of the 700 000 currently known species of fungi (singular: fungus), only a little over 100 are known to be pathogenic to humans. Fungal infections are rarely life-threatening (but can be) and as such tend to be regarded more as annoying and persistent. Their treatment is often difficult especially in the case of superficial (topical) skin infections because elimination requires regular and often long-term application. Bacteria and fungi often have opposing requirements, especially in terms of acid and alkaline habitats. Where one thrives, the other struggles.

The study of fungi is called *mycology*. A person who studies fungal biology is a mycologist. All fungi are eukaryote organisms, which means they have a nucleus and complex cell organelles. All require food in order to grow and multiply. They are not plants. They do not undergo photosynthesis. They must 'eat'. All produce digestive enzymes that leak out of the cells and dissolve the surrounding material, which is then absorbed and utilised. All fungi are aerobic or facultative anaerobes (i.e. they can get by when oxygen is in short supply but don't really thrive this way).

Many fungal organisms produce toxic chemicals. Fungi undergo sexual reproduction to form dormant resistant spores by the million. These spores hatch (or 'germinate') when they find themselves in a favourable environment. The usual distribution is by air and water.

TYPES OF FUNGI

Fungi come in two basic types:

1 unicellular: the yeasts (*Candida*)
2 multicellular: the moulds, the fleshy fungi (mushrooms, toadstools, powdery mildew, puffballs and tineas).

Yeasts

- *Sarcomyces* (Brewer's yeast)
- *Candida albicans*
- *Malessia furfur* (Budding yeast – mother and daughter cells):
  - these are single cells with a cell wall usually of a material called chitin (pronounced *kite–in*). They often grow in colonies
  - they reproduce asexually by budding from a 'mother' cell
  - they grow strings of unseparated buds from 'pseudohyphae', which often embed into superficial skin cells
  - colonies often have a 'wet' look because of a gelatinous sheath they produce. Chitinous cell wall contains a chemical (ergosterol) that inhibits antibiotics.

Some species of note

Candida spp. are common components of normal body flora resident in the intestine. Most candidiasis infections are *endogenous* (i.e. you re-infect yourself).

The majority of candidiasis is caused by one species (*Candida albicans*) and accounts for around 98 per cent of adult female infections. The other two per cent are caused by other species of the same genus (*Candida*) and each often requires a specific antifungal drug. Examples of species include *Candida albicans, C. glabrata, C. stellatoidea, C. parapsiosis, C. tropicalis, C. pseudotropicalis, C. krusei, C. guilliermondii* and *C. trachomatis*.

By far the most common culprit is *Candida albicans*. It can infect you through cracks and breaks in the skin between the fingers and attacks fingernails under the eponychium and the lateral nail fold.

Transfer is by what is commonly referred to as the oral-faecal route. This involves hand-anus-mouth and questionable personal hygiene.

Moulds

Mould fungi are filamentous and consist of long strands of cells joined end to end in a thread – called a *hypha* (plural: *hyphae*).

Many threads matted together make a mat network called a *mycellium* (plural = *mycellia*), which is a visible mass of hyphae. This mycellium will contain the spreading, growing actively digesting cells (vegetative cells). When mature, it will develop reproductive structures called *sporangia* (spore bodies).

There are many different styles and shapes of sporangia and several different methods of making spores. These are useful for correct identification. (Note: the above ground structures we call *mushrooms* are the complex reproductive organs of that fungus. The vast bulk of the organism remains hidden as thin mycellial networks in the soil or rotting material it is feeding on. The dust it produces is made of billions of fungal spores, each capable of germinating into a new fungal mycellial network.)

The above note is made because the moulds that affect the skin operate in a similar fashion except that no huge mushroom is produced, but nevertheless spores are produced from tiny sporangia. These embedded spores make tineal infections so difficult to eradicate.

Mould fungi that infect skin are called *dermatophytes*. The diseases they cause are called *dermatomycoses*.

There are many species of fungi that cause the common dermatomycoses referred to as the 'tineal' (common name) infections. Tineal infections on the body are incorrectly called 'ringworm' – it is *not* a worm. In the days before microbiology, it was thought that a little worm burrowed into a circle from its initial point of entry.

Three major *genera* of mould fungi cause the *Tineas*. Their general preferred locations are indicated in the table below:

GENERA	PREFERRED LOCATION
Microsporum spp.	Hair, skin
Trichophyton spp.	Hair, skin, nails
Epidermophyton spp.	Skin, nails

In all there are some 48 species in these three genera (groups) that cause the tineal diseases (dermatomycoses). They all produce keratases (keratin dissolving enzymes). They all produce spores. These spores may be left embedded in skin or clothing. Spore coatings make them quite resistant to chemical penetration.

SPORES OF FUNGI

Like the spores produced by the spore forming bacteria, fungal spores are designed to withstand adverse environmental conditions. Both have resistant outer coatings. Unlike bacterial spores, fungal spores are products of sexual reproduction and are produced in multiples from a mother cell. Large numbers are produced so they can be distributed widely. Fungal spores are nowhere near as resistant to heat treatment as bacterial spores.

Temperatures above 75 °C will usually kill them. A holding temperature above 65 °C for at least 10 minutes will kill nearly all fungal spores. For this reason, should you be doing your own salon laundry, you will be required to purchase or lease a washing machine capable of heating the water up to 65 °C.

Fungal spores are still quite resistant to chemical attack. For antifungal chemicals (drugs and therapeutic agents) to be effective, the spores must be constantly exposed and saturated with the chemical. Any lapse in treatment – even for a single day – often means you are back to the beginning and you start again as if the previous treatment never occurred. This human failing has often meant that some people have had an ongoing battle against tinea for well over 10 years. Constantly moist feet, for example, is an invitation for tineal infection.

Antifungal treatment

The first truly effective antifungal drug was NYSTATIN (New York State Institute of Health). Other commercial products have since been produced:

- Canesten®
- Lamisil® (terbinefine)
- Clotrimazol®
- Daktarin®

They come in over-the-counter (non-prescription) strength and prescription strength.

Systemic treatments (that are internal; i.e. swallowing an antifungal tablet) are not very successful because of the high degree of toxicity of the antifungal (especially to the liver) and tissue conversion of the antifungal drug before it can reach the dermis or epidermis. Some systemics have short-term side effects, such as loss of smell and taste, for up to six months. While highly effective in most cases, the loss of life's greatest pleasures may be a little too taxing for most. Topical (surface/superficial) treatment is the best.

Adapted from Dylan Webb's work in Victoria University's *Infection in the Salon: Student Guide 2015.*

11 What is the difference between a yeast and a mould?

12 How can you get a candida infection on the hands?

_____

13 What is ringworm?

_____

14 How high does the temperature need to be to kill fungal spores?

_____

15 What types of antifungal treatments might the doctor recommend for tinea of the feet?

_____

_____

_____

_____

Viruses

Viruses are unusual organisms. Some experts classify them as non-living infectious particles and others classify them as living organisms. This strange situation is the consequence of viruses' highly specialised lifestyles. Here they will be regarded as living organisms, which possibly represent the highest achievement in a parasitic lifestyle.

Viruses differ from most organisms chiefly in what they lack. They have:

- no cytoplasm
- no cell membrane
- no organelles.

They also do not do a lot of 'essential' life processes:

- no respiration
- no growth
- no movement capability
- no feeding
- no excretion.

What, might you ask, do they do? What do they have that is 'life'?

They all contain that essential item of every living being: RNA or DNA (never both). They 'reproduce' but they need another living organism to act as a host to do the mechanics of reproduction for them. They also synthesise specific chemicals and structures for transferring their RNA/DNA into the host cells.

THE STRUCTURE OF A VIRUS

Viruses typically consist of a loop of genes in RNA or DNA form. This is surrounded by proteins (capsomeres) to form a nodular coat. This capsid coat protects the nucleic DNA/RNA from tissue fluids and enzymes of the external environment.

The pattern of proteins is characteristic for each virus and variations on the pattern are produced by different 'strains' as the virus mutates or changes with replication (reproduction with a host). Some mutate very quickly. Some viruses have an envelope made of various combinations of lipids, proteins and carbohydrates or remains of the host cell's cell membrane that wraps the virus as it bursts out of the unfortunate host cell.

The presence of an envelope allows these coated viruses to withstand quite a lot of exposure and drying in the external environment. Spikes may also exist as characteristic protrusions. As particles outside the host cell (viroids) they may appear as geometrical, almost crystal-like shapes – helical and icosahedral shapes are common.

The size of a virus

Viruses are extremely tiny. A polio virus is only 30 nm in diameter. Viruses cannot be seen with a light microscope. (1nm = 0.000000001m = 1/100 000 000 of a metre = 10^9m) A very well made (and expensive) light microscope can see at 40 000 times magnification. An electron microscope with magnification of up to 250 000 times is needed to see viruses.

View the video at the following link to compare the size of a virus to *E. coli* bacteria (commonly found in faeces), fungi (baker's yeast) and human skin cell and human egg (ova): **http://learn.genetics.utah.edu/content/cells/scale**

HOSTS

Individual viruses are normally highly selective in the type of organism they can infect. Thus, we have:

- animal viruses; e.g. canine distemper, parvovirus, feline immunodeficiency virus
- plant viruses; e.g. tobacco mosaic virus, strawberry root virus
- human viruses; e.g. herpes viruses, influenza virus, human wart viruses
- bacterial viruses; e.g. diphtheroid bacteriophages, meningococcal bacteriophages.

Occasionally, a virus will cross-infect other species. Variants (or mutant strains) can survive in other organisms – usually not successfully as far as the host is concerned. This seems to be the case with the retrovirus, HIV, that is quite deadly to humans. The evidence indicates that it comes from the SIV (simian immunodeficiency virus) in monkeys, which doesn't appear to affect monkeys nearly as much.

More recently, COVID-19 has, due to a mutation in an animal virus, colonised human hosts. The mutation of the outer surface of the virus is what enables it to identify our cells.

Viruses favour particular body cells – the outer surface of the COVID-19 virus identifies cells in the blood vessels of various organs. It causes flu-like symptoms, and is particularly dangerous for people with respiratory health problems, weakened immune systems and other health issues.

Herpes simplex viruses I and II prefer skin cells – epidermal keratinocytes. The cell membrane of keratinocytes have specific recognition sites on the outside that specific viruses 'recognise' because the outer surfaces fit like a lock and key. Herpes simplex I infects keratinocytes in the oral and facial area; herpes simplex II infects the epithelial cells of the genital region.

THE LIFE CYCLE OF A VIRUS

The typical lifecycle (lytic cycle) of a virus is illustrated at the following link: **http://www.hhmi.org/biointeractive/ viral-lifecycle**. It involves:

1 attachment
2 penetration
3 un-coating
4 biosynthesis
5 maturation and assembly
6 release – to start again.

A single virus may induce a host cell to produce several hundred replicas of itself. The host cell begins acting abnormally and usually dies as it releases the viruses in a violent rupture.

The accumulated waste material of millions of rupturing cells and the increasing number of misbehaving host cells produces the typical symptoms of the disease and resultant illness of the organism.

VIRAL INFECTION

The effect of the virus on the host occurs in four phases:

1 Initial infection. Incubation phase – a 'waiting phase'. This phase may be 24 hours, or it may be decades long.
2 Cytopathic effect. Disease symptoms appear; the evidence of damage. Active cell destruction overwhelms the body.
3 Crisis phase. Inflammation/immune activation. Special immunological defences attack the virus.
4 Convalescence or death. An overwhelmed immune system = death.

Some viruses have a latent phase after initial entry and do not become active for months or even years. This is the reason for the long waiting period between becoming infected with HIV and the onset of AIDS. Creutzfeld-Jacob virus (CJV) has a latent phase of nearly 40 years before it emerges to destroy the brain. This virus has a 100 per cent mortality rate; no-one has ever recovered.

Herpes varicella zoster virus (HVZ), which causes chickenpox in children, goes into hiding in your nerve cells and waits many decades to then emerge from hiding in your old age to torment you with the electric burning pain of shingles (zoster).

Other viruses retreat to the nervous system in the face of a hostile host defence system and simply wait for an opportunity to come out of hiding. Examples are HSV I and II as well as the closely related HVZ, which causes oral/ genital herpes, and chickenpox/shingles respectively.

There also exist infectious agents more mysterious than a virus. These are:

- virinos – naked fragments of DNA 'ghosts'/naked fragments of RNA
- prions – infectious proteins.

TRANSMISSION

Typically, viruses are best transmitted directly either by:

- aerosol droplet spray
- blood and body fluids.

Enveloped viruses can survive exposure to air and can therefore be indirectly transmitted on contaminated articles (e.g. herpes simplex virus, human papilloma virus).

SOME COMMON VIRUSES

VIRUS TYPES	DISEASE
Papilloma virus	Warts
Herpes virus	Oral and genital herpes; chickenpox and shingles; mononucleosis/glandular fever
Poxvirus	*Molluscum contagiosum*
Picornavirus	Hepatitis A; polio; head colds
Togavirus	Rubella; hepatitis C; Dengue fever
Orthomyxovirus	Influenza A, B and C
Paramyxovirus	Mumps; measles (rubeola)
Retrovirus	AIDS (acquired immunodeficiency syndrome)
Hepadnavirus	Hepatitis B
Rotavirus	Gastroenteritis, winter gastro

You must ensure the proper use of gloves and the disposal of contaminated waste/sharps, as well as instituting correct techniques of disinfection and sterilisation. An emergency bodily fluid/spill clean-up kit should be in every salon. Likewise, effective hand washing and drying must be performed and masks used to prevent droplet spray.

Hand-to-hand contact

Any infected and contagious disease-carrying person is a risk and liability to the functioning of and success of a salon. For example, a sick and sneezing client or operator is a danger and should be sent home to seek treatment and return when they are no longer a risk.

Adapted from Dylan Webb's work in Victoria University's *Infection in the Salon: Student Guide 2015.*

16 What is distinctive about the anatomy and physiology of a virus?

17 Which is larger, bacteria or a virus?

18 Are viruses always pathogenic?

19 Why do you think it is that when we are sick with a cold (we are the 'host' to the influenza virus), we know our cat will definitely not catch it?

20 How long is the 'incubation phase' of a virus before disease symptoms appear?

_____

21 Why does oral/genital herpes come back?

_____

_____

22 How are viruses transmitted?

_____

_____

_____

_____

Animal parasites

A small but important collection of animal organisms live on and in the body. Some are non-pathogenic while others are pathogenic. They come from a wide range of animal groups:

- protozoans
- cestode worms (tapeworms)
- trematode worms (flatworms)
- nematode worms (roundworms)
- insects (wingless)
- arachnids (mites).

    If they live internally they are called _endoparasites._

    If they live externally (i.e. on the epidermis) they are _ectoparasites._

    They range in size from the single-celled microscopic protozoans to the multicellular macroscopic worms (some of these are frighteningly enormous, weighing several kilograms).

    Many have complex lifestyles and involve more than one host from different animal groups. They often use _vectors_ to aid their transport from one host to another. These vectors are usually insects that bite or suck body fluids (usually blood). Others rely on either you swallowing the eggs of the parasite or the parasite burrowing into your skin.

PROTOZOANS

Protozoans are a class of single-celled animals. _Commensal (not harmful)_	_Pathogenic (harmful)_
Entamoeba gingivitis commonly found in your mouth – a harmless browser.	_Entamoeba histolytica_ causes amoebic dysentery and is transferred through poor hygiene.
Giardia lablia causes intestinal infection in children with subsequent immunity.	
Trichomonas vaginalis causes vaginitis and sexually transmitted common low-grade chronic infection (fishy smell).	

    Uncooked contaminated animal meat fed to the family pet could result in your dog silently defecating the eggs of the Hydatid tapeworm. If humans swallow the eggs from an infected (pet) dog (e.g. by letting the dog lick your face), cysts will form in your body. These may become huge and cause eventual death. Fortunately, most cysts stay minute. This is one of the main reasons why no dogs or cats are permitted in the salon.

Roundworm

A very common little roundworm is the pinworm. _Enterobius vermicularis_ infect when the eggs are swallowed. It is known as an oral-faecal route infection. They live in the rectum and colon. Females crawl out the anus at night to lay eggs around the anus. You can then scratch your bottom and swallow more eggs. Eggs also spread to other members of the family. Clients may also leave eggs on massage beds, towels and so on. An itchy bottom may indicate infestation. Use a torch at night to check.

In the case of pinworm, proper hygiene and removal of towels and linen after each client is essential. One comfort is that adults seem to develop some immunity to pinworm but it is not guaranteed.

Treatment usually requires that every member of the family be treated as they will all have swallowed the tiny micro-fine dust-like eggs.

INSECTS

Lice

Human lice were once much more common than they are now. Two types occur as ectoparasites in Australia:

1 *Pediculus humanus capitus* (head lice)
   - These wingless insects crawl around head hair and lay their eggs glued to base of hair shafts (nits). Lice bite and suck blood. Irritation may lead to inflammation and a secondary infection of bacteria or fungi. Lice prefer short, fine clean hair and young girls tend to be more susceptible than boys.
   - Treatment is either by combing with a fine comb to remove all the nits or using a pyrethrum-based hair shampoo as directed or both. Some overnight hair lotions are available but do not seem popular because they are greasy.
2 *Phthris pubis* (pubic lice)
   - The pubic (crab) louse is bigger and stronger than the head louse and prefers the thicker and stronger hairs of the groin. Its hooked claws are suited to more solid hairs, and so they will infect pubic hairs, beards, eyebrows and eyelashes.
   - They move slowly and cannot jump. Transferring from one individual to another requires a certain amount of body contact time, though they are not always sexually transmitted. Treatment is with insecticide-based lotions.

Demodex folliculorum (the eyelash mite)

In the past, this mite was considered harmless in all situations. Now there is evidence that with certain people there is an erythematous skin reaction that can be quite nasty. Everyone seems to have them, but few people react to them. An infestation can cause rosacea.

Fleas

Fleas bite and drink blood. They come from dogs, cats and rodents.

Adapted from Dylan Webb's work in Victoria University's *Infection in the Salon: Student Guide 2015.*

23 In the table below, match the following parasites to the organism by writing the parasite into the table.

Parasites: Lice, Demodex, Enterobius vermicularis, Entamoeba histolytica

Protozoans	
Worms	
Insects	
Arachnids	

24 What is the oral-faecal route?

_____

25 How can eggs be transmitted via the oral-faecal route?

_____

Skin penetration treatments

26 List the six steps to prepare the skin for a skin penetration procedure.

_____

_____

_____

_____

_____

_____

_____

_____

27 If you have any broken skin on your hands or arms, what would you need to ensure you've done to protect yourself before performing a skin penetration procedure?

_____

_____

Procedures for sharps and needlestick injury

Knowledge Evidence

28 For each of the following, specify if there is a single-use/disposable option available, reusable/cleanable option available, or both.

  a Metal implements

_____

  b Electrolysis needle

_____

  c Wooden spatula

_____

  d Cotton tips

_____

29 If you need to put a needle down in the middle of a treatment, what is the correct procedure?

_____

_____

30 What is the correct way to hold a sharps needle?

_____

Disposing of waste and sharps

Knowledge Evidence

31 Where is the most appropriate place to dispose of the following items?

a Hypodermic needle after extractions

---

b Cotton tips after cleansing eye make-up remover

---

c Electrolysis needle

---

d Empty container from skincare product

---

32 Describe the steps for disposing of a hypodermic needle after a facial extraction procedure.

---

33 What is the needlestick or sharps injury procedure for notification and response according to your state or territory's guidelines or regulations?

---

34 What overarching federal legislation protects staff, clients and visitors from risk of needlestick injury?

LO8.4 STERILISING EQUIPMENT AND MAINTAINING STERILISER

Performance Evidence

Cleaning, disinfection and sterilising procedures

Knowledge Evidence

1 Refer to Appendix C Treatment area and equipment in your textbook and complete the following table to indicate the best way to clean salon objects and surfaces listed in the first column.

ITEM	CLEANING PROCEDURE	HOW OFTEN
Wax pots		
Brush machine attachments		
Glass bowls		
Tweezers		
Trolley		

Spills

Incidents involving blood and body fluid spills are a serious infection transmission hazard. Refer to the section on 'Sharps and blood or body fluid spills' in Chapter 8 of your textbook and complete the following questions.

Knowledge
Evidence

2   If there is a blood spill in the salon – for example, after a sharps injury – what is your salon's step-by-step procedure that supports infection control measures and prevents infection transmission?

_____

_____

_____

_____

_____

_____

_____

3 What is the name of the hospital-grade disinfectant you use to disinfect the spill?

_____

Manufacturer instructions

4 Refer to the section 'Sterilising equipment and maintaining steriliser' in Chapter 8 of your textbook. Describe below how you operate your salon's steriliser, according to the manufacturer's instructions. Make notes on the following aspects:

Knowledge Evidence

   a packaging of items

_____

   b loading

_____

   c monitoring the process

_____

   d validation of the process – identify problems in the machine:

FAULT IN MACHINE	POSSIBLE CAUSE/REMEDIAL ACTION

   e documentation requirements

_____

Australian Standards

Australian Standards AS/NZS 4815 *Reprocessing of reusable medical devices in health service organisations* and AS/NZS 4187 *Office-based health care facilities – Reprocessing of reusable medical and surgical instruments and equipment, and maintenance of the associated environment.*

5   What types of reusable beauty therapy tools might you reprocess so that they may be used safely without risk of transmission of infectious agents?

_____

_____

_____

6   Research online and list three topics covered in the stipulations of the Australian Standard 4815.

_____

_____

_____

7   How and why do we need to monitor the sterilising process?

_____

_____

_____

8   When validating sterilisation, what are you looking for when inspecting the items?

_____

9   What do you need to record when sterilisation is complete?

_____

_____

_____

LO8.5 MAINTAINING AWARENESS OF CLINIC DESIGN FOR CONTROL OF INFECTION RISKS

1   It is important that salons are designed to easily allow for appropriate infection control along with performing treatments. Complete the following table to show what feature considerations would need to be made for infection control.

SALON DESIGN ITEM	FEATURE CONSIDERATION
Lighting	
Cleaning area	
Floors, walls, ceilings, trolleys and cabinets	

2   What would be kept in the clean zone, and what in the dirty zone?

_____

_____

_____

_____

_____

SECTION 3
NAIL TREATMENTS 127

9 Manicure and pedicure services 128
10 Nail art and advanced nail art 157

SECTION 4
HAIR TREATMENT AND REMOVAL 168

11 Waxing services 169
12 Lash and brow services 184
13 Eyelash extensions 197

SECTION 5
MAKE-UP 209

14 Design and apply make-up 210
15 Remedial camouflage make-up 226
16 Photographic make-up 235

SECTION 6
BODY TREATMENTS AND SERVICES 243

17 Cosmetic tanning 244
18 Provide body massages 257
19 Aromatherapy massages 278
20 Aromatic plant oil blends 293

SECTION 7
FACIAL TREATMENTS 306

21 Provide facial treatments and skincare
recommendations 307
22 Specialised facial treatments 348

SECTION 3

NAIL TREATMENTS

9 Manicure and pedicure services
10 Nail art and advanced nail art

CHAPTER 9: MANICURE AND PEDICURE SERVICES

After completing this chapter, you should be able to:

LO9.1 establish client priorities and design a treatment plan for the nails by consultation and analysis of the skin and nail area

LO9.2 prepare for a nail service according to relevant health and safety legislative requirements and salon protocol

LO9.3 provide nail services according to the client's requirements and treatment plan

LO9.4 review the service, showing knowledge of adverse reactions, post-treatment recommendations and recording updates

LO9.5 clean the treatment area in preparation for the next service.

INTRODUCTION

Manicures and pedicures are popular with men and women to achieve a well-groomed appearance and to prevent nail and skin disorders. Nail treatments can be adapted to a variety of settings.

The learning activities in this chapter allow you to demonstrate your knowledge and understanding of the unit of competency *SHBBNLS007 Provide manicure and pedicure services*, using the performance and knowledge evidences.

LO9.1 ESTABLISHING CLIENT PRIORITIES

Performance
Evidence

1   Is a consultation necessary each time a client has a service in the salon? Why?

_____

_____

_____

2   The initial consultation for a manicure must be conducted on a visual, manual and verbal basis. What does that mean?

_____

_____

3   What are the three parts to be analysed during this consultation?

_____

Areas requiring special attention and special needs

Some clients have special needs, and some areas need additional special attention:

• *Athletes* – Athletes are not so gentle with their feet, so they will more likely present with calluses, blisters, haemorrhage in the heel ('black heel') and athlete's foot (tinea pedis). Do not remove too much callus from the heel of the foot – calluses naturally build up to protect the foot from physical trauma. Nail disorders that commonly affect athletes include onycholysis (this often leads to tinea unguium), split nails and bruised nails.

- *The elderly*: The client will have thinner skin, slower circulation and a more fragile body. Service the client with more attention to comfort and assistance. Give a gentle nail treatment with minimal buffing and cuticle work.
- *People with disabilities*: Be clear on the extent of the disability and accommodate the client accordingly. Always ensure you record the alterations to the service on the client record card. Treat the client with the same respect you would give any other client and keep in mind that beauty therapy can have a profound effect on a client's self-esteem.

4   What aftercare advice would you give a client who goes running regularly and who has shown symptoms of tinea unguium on the heels of the feet?

_____

_____

_____

Contraindications

Knowledge
Evidence

It is your role to enhance the appearance of the client's hands and feet. Referral to the appropriate professional is required for diagnosis and treatment advice for nail disorders and diseases.

COMMON CONTRAINDICATIONS AND RESTRICTIONS

Contraindications can be local or general. Some disorders require you to alter the treatment, and they are called precautions; for example, for weak nails and other non-contagious conditions. Infectious conditions, such as paronychia, require medical referral.

A list of common disorders that may be seen on the hands or feet is provided in Figure A.9 in Appendix A Contraindications and restrictions. If you suspect the client has a disorder in the table, follow the suggested action procedure.

CONTRAINDICATIONS OF PRODUCTS

Always follow the manufacturers' instructions when using nail and beauty products. Combining the products or using the products for the wrong service can lead to adverse reactions and damage to the client's skin and nails.

Contact dermatitis

One of the adverse reactions to misuse of products is contact dermatitis. Contact dermatitis presents with inflammation, blisters, an itching rash or a burn. It will remain as long as the irritant is in contact with the skin and it will take days after the initial contact for the symptom to disappear. There are three types of contact dermatitis.

Irritant dermatitis can be caused by a chemical that is usually highly alkaline or acidic. Common examples used in the salon include:

- solvents (e.g. acetone)
- strong cleaning agents and detergents (e.g. sodium lauryl sulphate)
- natural latex gloves
- acrylic nails (liquid monomer). To prevent this reaction to acrylic nails, try to apply the product in from the cuticle on the nail plate. Do not apply the product too wet. Use all work health and safety (WHS) measures to ensure the product and its vapours are kept at minimum in the treatment area.

Allergic dermatitis is due to a hypersensitivity to an allergen. It is a common condition and as such there are a number of known allergens to be aware of in the salon:

- adhesives, including band-aids, nail glue, glue for false lashes
- fragrances, as used in skincare and cosmetics
- nail polish
- natural latex gloves
- some plant extracts that may be in products.

The symptoms do not necessarily appear on the hands. In the case of nail polish allergy, the symptoms often show up on the face, which the hands are continually touching.

Photo-allergic dermatitis occurs when an otherwise harmless ingredient or product becomes an irritant while the skin is stimulated with light. The ultraviolet (UV) light used for nail enhancements emits UVA radiation, which can aggravate the skin.

An allergic reaction can be avoided by opting for products that are low irritant, such as:

- latex-free gloves
- organic nail polish
- hypoallergenic skincare products
- hypoallergenic band-aids (in the first aid kit).

Contact urticaria

Contact urticaria (hives) presents with immediate red swellings and possible blistering. It occurs when skin is in contact with an irritant, but note that the appearance is very different from irritant dermatitis – it is not a type of dermatitis nor eczema. Natural latex gloves are a common irritant.

Treatment for contact dermatitis and contact urticaria:

1 Remove the irritant or allergen by flushing the area with water.

2 Apply cold compress for 30 minutes, three times per day.

3 Refer the client to a GP for medical diagnosis. Allergies must be taken seriously because, although uncommon, anaphylaxis can be triggered by known allergens.

5 Review Figure A.9 in Appendix A Contraindications and restrictions as well as Figure 9.6 in your textbook. In the table below, identify the disorder from the picture shown, its relevance to manicure and pedicure, and the action you would take.

	DISORDER NAME	RELEVANCE TO MANICURE/PEDICURE	ACTION
Alamy Stock Photo/Science Photo Library			
Shutterstock.com/Levent Konuk			

	DISORDER NAME	RELEVANCE TO MANICURE/PEDICURE	ACTION
Alamy Stock Photo/Science Photo Library/ David Parker			
Shutterstock.com/Pan Xunbin			

6 During the consultation, a client tells you they have the following issues. Decide if each one will prevent the treatment or will not impact the treatment.

a They are allergic to penicillin.

_____

b They have a sprained ankle and want a pedicure treatment.

_____

c They have had a COVID-19 positive test within the last two weeks.

_____

d They have bruising around the knee and want a manicure treatment.

_____

e They have had an operation on their wrist within the last six months and want a manicure treatment.

_____

Scope of practice

It is important to know when a client should be referred to a medical practitioner and how different branches of medicine relate to nail services. As a beauty therapist or someone working in this industry, you should never diagnose a condition as it outside the scope of your practice.

Performance Evidence

REFERRING A CLIENT

7   When would you refer a client to a medical professional for a contraindication you have observed? Give three examples for a manicure treatment, and three for a pedicure treatment, and include where you would refer them.

_____

_____

_____

_____

_____

8   Write two ways you could alert a client to a contraindication, without diagnosing the condition.

_____

_____

_____

_____

_____

_____

_____

_____

_____

Nails analysis

Nail science is the knowledge of the structure and function of the skin and nails of the hands and feet. We learn this so that we can understand how beauty products and treatments work to improve the appearance of the skin and nails. The principles of nail science are used in nail treatments and beauty treatments throughout the textbook. It is a legal requirement to apply some of this knowledge to practice and it will be a part of your workplace policies and procedures to apply nail science principles to your nail treatments.

THE SKIN ON THE HANDS AND FEET

The skin is an organ of three layers, the epidermis, the dermis and the hypodermis. Review Workbook Chapter 25 to understand more about the anatomical layers of the skin.

The epidermis is the layer you can see – it comprises five distinct layers of squamous (squashed) epithelial cells. The uppermost layer, the stratum corneum, contains generally 10 to 35 layers of hard, compact, protective cells. Skin thickness at this layer of the epidermis can be related to a person's age and cell turnover at the stratum germinativum. The stratum corneum is at its thinnest (in order) on the eyelids, face, torso, limbs; it is thickest on the palm of the hands and the soles of the feet. Here, the stratum lucidum is also at its thickest, with approximately five layers to make the skin waterproof. The skin in these areas is also marked by the absence of the pilosebaceous unit, and densely-distributed sweat glands. This means no hairs, no sebum and more eccrine sweat than in other parts of the body.

Healthy skin on the hands and feet is even in colour and tone, with good thickness, elasticity and firmness. Functions of the skin are covered in more detail in Workbook Chapter 25. In brief, the functions of the skin are:

- protection against bacterial invasion and mechanical injury (callus protects bony areas)
- homeostasis – the regulation of the body in its environment though vascular, nervous response and sweating

- production of vitamin D
- insulation and morphology (fat cells)
- waste exchange
- prevention of water loss
- sensory detection of pain, pressure, touch, hot and cold.

THE STRUCTURE AND FUNCTION OF THE NAIL

Nails are epidermal appendages that grow from the ends of the fingers and toes. The part of the 'nail' we commonly refer to is technically termed the nail plate.

The nail plate

The nail plate is composed of layers of closely compacted epithelial 'nail' cells. The cells (keratinocytes) lose their inner organelles and become translucent when the protein keratin forms within the cell (keratinisation). Horizontally, there are three distinct layers. The top and bottom layers contain soft keratin, and absorb lipids and some moisture from the skin and nail bed respectively. The middle layer is made of hard keratin and can become dry, split and brittle when exposed.

Functions of the nail plate include:

- adding strength to the fingertip, giving the skin more grip when grabbing objects
- scratching/scraping
- peeling objects (such as an orange)
- protecting the underlying bone, and the nail bed, which contains many nerves and blood vessels.

Growth and development of the nail plate

Cells divide in the matrix and the nail grows forward over the nail bed, guided by the nail grooves, until it reaches the end of the finger or toe, where it becomes the free edge. As they first emerge from the matrix the translucent cells are plump and soft, but they get harder and flatter as they move toward the free edge. The top two layers (strata) of the epidermis form the nail plate; the remaining three form the nail bed.

Fingernails grow at approximately twice the speed of toenails, but toenails are thicker. It takes about six months for a fingernail to grow from cuticle to free edge, and about 12 months for a toenail. Fingernails are, on average, 100 cell layers thick and toenails have 150 cell layers.

The normal rate of nail growth is approximately 3 mm per month for fingers, which equates to approximately 0.1 mm per day. For toenails the growth rate is halved, so likewise, clients tend to require a manicure every 2–3 weeks and a pedicure every four weeks. As we age, moisture and lipids deplete from the nail bed and so the grooves become more prominent on the surface.

For a client in good health the nail plate naturally curves in two directions:

- transversely – from side to side across the nail
- longitudinally – from the base of the nail to the free edge.

Factors that affect natural nail growth

The rate of nail growth can vary according to a person's genes, age, general health, the time of year and lifestyle factors. There are many nail products and treatments on the market to benefit nail growth. Always research these products and share your knowledge and experience with your salon team.

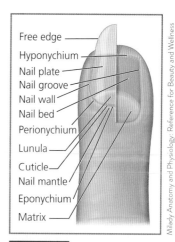

Free edge
Hyponychium
Nail plate
Nail groove
Nail wall
Nail bed
Perionychium
Lunula
Cuticle
Nail mantle
Eponychium
Matrix

Figure 9.1 The structure of the nail

Milady Anatomy and Physiology: Reference for Beauty and Wellness Professionals, 1e. Cengage Learning Inc. Reproduced by permission. www.cengage.com/permissions

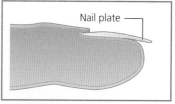

Nail plate

Figure 9.2 The nail plate

Milady Anatomy and Physiology: Reference for Beauty and Wellness Professionals, 1e. Cengage Learning Inc. Reproduced by permission. www.cengage.com/permissions

Figure 9.3 Factors that increase and decrease nail growth

FACTORS THAT INCREASE NAIL GROWTH	FACTORS THAT DECREASE NAIL GROWTH
Healthy diet/consuming supplements of zinc, calcium, vitamins D, A, C, and B – in particular B12, folic acid and biotin	Nutrient – poor diet/smoking/consuming alcohol or caffeine/some medications
Pregnancy	Breastfeeding
Exercise	Ageing
Massage	Poor circulation
Warm climate	Cool climate
Buffing and filing the nail plate	Systemic diseases and disorders/stress

The nail bed

The pink colour of the nail derives from the blood vessels that pass beneath the nail plate – in the nail bed. Also within the nail bed are many sensory nerve endings, more so in the fingertips than anywhere in the body, so without the nail plate to protect it, the nail bed is very sensitive. The nail bed is structured with longitudinal furrows corresponding to ridges of the nail plate to hold the nail in place and to guide its growth lengthways.

Functions of the nail bed include:
- to provide the nail matrix of the fingers and toes with nutrients
- via sensory nerves, to receive sensations such as pain, hot and cold
- to hold the nail plate in place.

The free edge

The free edge is the part of the nail that extends beyond the fingertip. This is the part that is filed and shaped. A healthy free edge appears white and smooth.

Functions of the free edge include:
- to protect the fingertip and the hyponychium
- as a part of the nail plate, to be used as a tool for grasping, peeling and scratching.

The matrix

The nail matrix, sometimes called the nail root, is the area of the nail with growing nail cells. It is an extension of the stratum germinativum layer of the epidermis, with cells undergoing mitosis (cell division) for growth and repair. The matrix lies under the eponychium and cuticle, at the proximal end of the nail (nearest to the body). Keratinisation begins at the matrix, a process by which keratin builds inside the cells as they move along the nail plate. The type of keratin that builds up after keratinisation is known as 'hard keratin', characteristic to the nail plate.

Functions of the matrix include:
- to produce new nail cells
- to determine the shape and thickness of nail:
  - a long matrix produces a thick nail
  - a flat matrix produces a flat nail
  - a curved matrix produces a curved nail.

The lunula

The crescent-shaped lunula is located at the base of the nail, lying over the matrix. It is white, relative to the rest of the nail, and there are two theories to account for this:
- newly formed nail plates may be more opaque than mature nail plates
- the lunula may indicate the extent of the underlying matrix – the matrix is thicker than the epidermis of the nail bed, and the capillaries beneath it would not show through as well.

Function of the lunula: none.

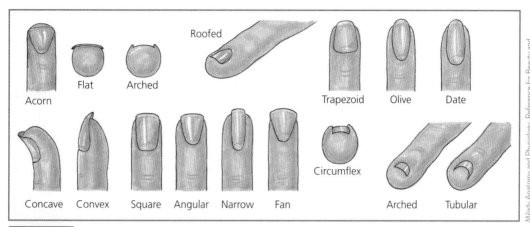

Figure 9.4 Nail shapes

The hyponychium

The hyponychium is part of the cuticle, specifically the portion of the epidermis under the free edge of the nail. The epithelial tissue on the nail bed surface grows outward from the nail bed underneath the nail plate. It is removed with a cuticle stick to prevent yellow staining underneath the free edge.

Functions of the hyponychium include:

- protecting the nail bed from microbial invasion
- holding the nail plate in place.

The nail grooves

The nail grooves are not a part of the nail but the nail bed epithelium. The grooves interlock with the ridges in the nail plate so that it can grow along a straight path from the matrix bed to its free edge.

Function of the nail grooves: to guide and keep the nail plate growing forward.

The eponychium

The eponychium is part of the cuticle and refers to the skin that attaches to the nail plate at the base (proximal end) of the nail. The eponychium is also in contact with the lunula of the nail plate. Trimming into the eponychium (cuticle) can cause overgrown cuticles, hangnail and subsequent infection, such as paronychia. See **http://www.fpnotebook. com/_media/dermNailBaseGrayBB943.gif**

Function of the eponychium: to protect the matrix from infection and dehydration.

The perionychium

The perionychium is the eponychium at the sides of the nail. Only loose edges of the cuticle should be cut after pushing.

Function of the perionychium: to protect the nail from infection and dehydration.

The cuticle

By definition, the cuticle comprises the eponychium, the perionychium and the hyponychium. It is the part of the skin that should be in contact with the nail plate, held in place by lipids. When in good condition, it is soft and loose; when overgrown or dry it is softened and cut to improve the appearance of the nails and to prevent tearing.

Function of the cuticle: to protect the matrix from infection and dehydration.

The nail walls

The nail walls are the folds of skin (epidermis, dermis and hypodermis) overlapping the sides of the nails.

Function of the nail walls: to cushion and protect the nail from damage.

Nail conditions: the effect of health and disease on the nails

If there is no nail disease, it is the work of the beauty professional to try to find the cause of a nail condition. Nails can become weak, brittle, discoloured or ridged for many reasons, and could relate to heredity or ageing. At the client consultation, try to understand the cause of the nail condition and find a nail care solution.

Knowledge Evidence

COMMON NAIL CONDITIONS

Weak nails

Softness, flexibility and thinness of nails can all be signs of nail weakness. To improve strength, you should treat the nails with a strengthener. Weak nails can be caused by genetics, illness or medication, and use of acrylic/gel nails and extensions. Excessive buffing can also weaken the nail as this removes the natural oils the skin produces to seal moisture in the nail.

Brittle nails

Brittle nails often peel and split due to a lack of moisture in the nail. Despite brittle nails often feeling strong, the two upper nail layers separate due to the dryness of the cement layer of the nail. Brittle nails may be caused by age, illness or medication or the incorrect nail care; for example, using nail products that dry the nail plate, neglecting the nails and excessive buffing. Nourishing and hydrating treatments such as hot oil and paraffin improve flexibility and protect the nail plate from nail polish ingredients by using a base coat.

Discolouration

Discolouration can be caused by many things, such as:

- bruising (black/blue), medical conditions, such as psoriasis (red) or fungal infection such as tinea unguium (yellow-grey), or mould (dark green); seek the advice of an appropriate medical professional
- lifestyle factors such as smoking (yellow)
- topically applied products, such as fake tan (brown), artificial nails or nail polish (nail colours in deeper shades of red contain more solvent ingredients, namely formaldehyde, which stains the nail plate yellow – advise clients to use a base coat).

Deep ridges

Healthy nails are naturally ridged, but relatively smooth in appearance. Deep ridges are most often due to heredity, ageing, illness or medication or they can be a result of injury. Ridges are generally vertical (longitudinal) but can be horizontal (transverse), as can be seen with Beau's lines (as detailed further in this chapter). Buff the nails if the nails are thick enough, but ridge filler is a product available to mask the appearance of deep ridges.

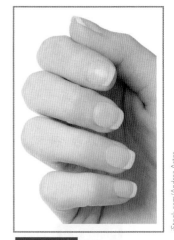

Figure 9.5 Normal appearance of natural nails

iStock.com/Andrea Astes

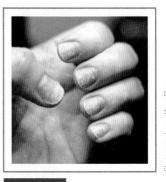

Figure 9.6 Nails damaged and brittle after the removal of nail enhancements

Shutterstock.com/Amy Planz

Applying the structure of the nail to practice

Knowledge Evidence

9   Describe the appearance of a normal, healthy nail.

_____

_____

10 What protein are the cells of the nail plate composed of?

_____

11 What part of the nail unit contains actively growing cells?

_____

12 Why are we not allowed to cut the skin around the base of the nail plate, even if the client requests it?

_____

13 Explain the difference between the nail plate and the nail bed.

14 Describe how the nail plate is formed.

15 What determines the thickness, width and curvature of the nail plate?

16 What does a longer matrix produce?

17 What would a highly curved matrix produce?

18 Normal healthy adults can expect their nails to grow at what rate per month?

19 Which nails grow the fastest – fingernails or toenails?

20 How long does it take for a nail to grow from the matrix to the free edge?

21 After losing a toenail, how long does it take for a new nail to take its place?

22 Why is the hyponychium important and what does it protect?

23 The nail plate consists of how many keratinous layers?

24 What are the 'tracks' the nail grows along?

25 How does water affect the natural nail?

26 What conditions favour the growth of fungal organisms?

_____

27 Name two common causes of onycholysis.

_____

_____

28 In what situation should a nail service not be performed?

_____

29 What is the most effective way to avoid transferring infections among your clients?

_____

30 If a client develops a nail infection, can nail technicians offer treatment advice for these conditions?

_____

31 Can nail technicians treat an ingrown toenail if there is no sign of pus or discharge?

_____

32 Which causes paronychia infections: bacteria or fungus?

_____

33 What is pterygium?

_____

34 Label the diagrams below using the following terms:

cuticle	eponychium	free edge
hyponychium	lunula	matrix
nail bed	nail groove	perionychium
nail plate (× 2)	nail wall	

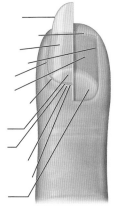
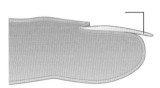

Milady Anatomy and Physiology: Reference for Beauty and Wellness Professionals, 1e. Cengage Learning Inc. Reproduced by permission. www.cengage.com/permissions

35 Draw a line to match the nail parts on the left with their correct description on the right.

NAIL PART	DESCRIPTION
1. Nail plate	A. folds of normal skin that surround the nail plate
2. Nail wall	B. the part of the cuticle under the nail plate, at the tip of the finger
3. Hyponychium	C. dead colourless skin attached to the nail plate
4. Free edge	D. nail part growing outward from the matrix, constructed in layers
5. Nail grooves	E. where nail growth begins; contains nerves and blood; if injured, irregular nail forms
6. Matrix	F. nail grows along these tracks
7. Eponychium	G. visible part of the matrix
8. Nail bed	H. extends beyond fingertip
9. Cuticle	I. portion of living skin on which the nail plate sits
10. Lunula	J. the part of the cuticle at the base of the nail plate in contact with the lunula

36 Insert the correct term in front of each definition in the below table.

TERM	DEFINITION
	Cuticle splits because of dryness
	Nail is ridged across (horizontally) – due to high fever, measles, zinc deficiency, pregnancy; lengthwise (vertically) – due to psoriasis, arthritis, poor circulation
	Flat or spoon nail, may be indicative of an iron deficiency
	White spots on the nail
	A viral infection that causes warts
	Dry, split and brittle nails – usually running vertically
	Nail plate loosens from the nail bed
	Nail biting
	An overgrowth of the nail in thickness but not painful (known as hypertrophy)
	Wasting away of the nail. Mainly caused by injury or disease
	Ingrown nails
	Peeling and pitted skin caused by a fungus

37 With your instructor, discuss and write down what the difference is between a nail condition and a contraindication or restriction.

_____

_____

38 When would you refer the client to a GP and when is it safe to treat the client?

_____

39 List three nail conditions that may be treated and suggest a nail treatment for each.

_____

_____

_____

Colours used

40 Match the following three special occasions and outfits with a flattering nail polish colour that you would recommend for the client's skin tone.

OCCASION	OUTFIT	SKIN TONE	POLISH COLOUR
Horse races	Black with white spots	Pale/cool	
Wedding	Pale blue	Tanned/cool	
High tea	Olive	Pale/ruddy/warm	

Client record management

41 What are the factors you should consider that will affect the type, duration or frequency of the treatment?

Knowledge
Evidence

_____

_____

_____

_____

42 Using the client card for nail treatments, design a treatment plan for a 52-year-old client who has diabetes, leuconychia and is requesting a natural manicure with short nails. She works in health care but would like to have polish on her nails.

_____

_____

_____

_____

_____

_____

_____

_____

_____

_____

LO9.2 PREPARING FOR THE NAIL SERVICE

In this section you will learn about nail products and post-treatment advice to confidently address the client's needs during nail treatments. Product advice is administered after correct assessment of the health of the skin and nails.

Performance
Evidence

Knowledge
Evidence

1 List at least three ways to maintain hygiene both before and throughout a manicure or pedicure treatment.

_____

_____

_____

_____

_____

2 After performing two manicure treatments in a row, you are finding that your wrists are feeling sore and fatigued. What could be the postural issue you have? How would you correct it?

_____

_____

_____

_____

Minimising damage to the skin and nails during treatment

Adverse reactions can occur as a result of a nail product or treatment procedure. Some may be unforeseen, but also there are nail tools and products that when carelessly used can damage the skin and nails. For example, select the correct grit nail file to suit the individual nail. Electric nail buffers can be very harmful if used incorrectly during artificial nail treatments.

Nail product safety

Some chemical ingredients are damaging to the skin and nails. Products with corrosive agents often have extreme pH levels. The pH of the skin can be between 4.5 and 6.2, and the nails are around pH 5. Some nail products can have a pH of around 12 and should not be used unnecessarily. Formulations that are around pH 5–5.5 are said to be safe and 'neutral' for skin and nail care.

Knowledge
Evidence

| **Figure 9.7** | Precautions when using some nail products |

NAIL PRODUCT	PRECAUTION
Nail polish remover	pH 11–12 – has a very drying effect on the skin. Use acetone-free where possible.
Nail bleach	Is usually a combination of hydrogen peroxide and/or citric acid and ammonia, which balances to a pH of about 7–8. It has been known to kill nail fungus. Use strictly for the manufacturer's set period of time.
Nail polish	The pH is not an issue as it contains no H+ or OH- ions, but red polishes contain formaldehyde and it does tend to stain the nail plate. Use only with a base coat underneath and check for prior allergies to nail polish.
Cuticle remover	Contains potassium hydroxide or sodium hydroxide, which can have a pH of up to 12. Use strictly according to the manufacturer's instructions and remove the product completely with water.

3   Use pH to explain why nail polish remover is not good for human skin and nails in large amounts.

_____

_____

Safe Work Australia develops the national policy on health and safety in the interests of workers' compensation. It provides advice about legal obligations in relation to product safety and other occupational hazards. The federal WHS regulations relevant to nail treatments are:

- Labelling of Workplace Substances [NOHSC: 2012 (1994)]
- Prevention of Occupational Overuse Syndrome [NOHSC: 2013 (1994)]
- Workplace Hazardous Substances [NOHSC: 2007 (1994)]
- Guidance Note for the Assessment of Health Risks Arising from the Use of Hazardous Substances in Workplaces [NOHSC: 3017 (1994)]
- National Code of Practice for the Preparation of Material Safety Data Sheets 2nd Edition [NOHSC: 2011 (2003)].

4   Ask your instructor where you would report an incident with products used during a manicure or pedicure treatment.

_____

_____

Products used in manicure and pedicure treatments

Knowledge Evidence

Understand the key active ingredients in products and the potential benefits to the client when recommending homecare. Advise the client of the recommended use of the product in the interests of safety and desired outcomes.

Figure 9.8   Key active ingredients in nail products

PRODUCT TYPE	INGREDIENT/ACTION	USE
Nail polish remover	• Acetone or ethyl acetate – solvent • Perfume • Colour • Oil – emollient to reduce drying effect of solvent	Removes nail polish Degreases the nail plate prior to applying polish Retail: Use non-acetone polish remover for nail enhancements
Hand lotion/oil	• Vegetable oils (e.g. almond oil) • Perfume • Emulsifying agents (e.g. beeswax or gum tragacanth) • Emollients (e.g. glycerine or lanolin) • Preservatives	Softens the skin and cuticles Provides 'slip' during hand massage Retail: Apply twice daily
Nail bleach	• Citric acid or hydrogen peroxide – bleaches the nail • Glycerine – emollient • Water	To whiten stained nails and the surrounding skin

Nail polish	• Solvent – creates a suitable consistency to apply, and dries at a controlled rate
	• Colour pigments – create nail polish colour
	• Resin – improves adhesion of polish to nail plate and flexibility
	• Nitrocellulose – film-forming plastic, holds colour
	• Plasticisers – provide flexibility after the polish has dried, reducing chipping and cracks
	• Pearlised particles – create a pearlised effect
	• Cytotoxic chemicals: It is not definitively known whether these chemicals are carcinogenic or likely to cause birth deformities when used in recommended concentrations (according to the SDS):
	  – toluene – solvent that dissolves ingredients in nail polish
	  – formaldehyde – film-forming plastic resin, improving adherence and flexibility
	  – dibutyl phthalate (DBP) – an oily plasticiser
		Colours the nail plate
		Provides some protection

PRODUCT TYPE	INGREDIENT/ACTION	USE
Base coat	• High in nitrocellulose • Resins • Solvents such as ethyl acetate and butyl acetate • Isopropyl alchohol – a non-toxic alternative to the preservative formaldehyde	Protects the nail plate from discolouration due to the formaldehyde in nail polish colours
Cuticle cream	• Emollients that sit on the skin surface (e.g. lanolin or glycerine) • Perfume • Colour	Softens the cuticles Retail: Apply morning and night to protect the cuticles and prevent hangnail
Cuticle oil	• Emollients that can be absorbed by the skin (e.g. oils such as jojoba and vitamin E)	Conditions the nail and surrounding skin Retail: Apply 2–3 times per day or as required
Nail strengthener	• Formaldehyde – film-forming plastic resin	Protects weak, brittle nails against breaking, peeling and splitting Retail: Apply as a base coat or clear polish
Cuticle remover	• Potassium hydroxide – a caustic alkali • Glycerine – a humectant added to reduce the drying effect on the nail plate	Softens the skin of the cuticles
Buffing paste	• Perfume • Colour • Abrasive particles (e.g. pumice, talc or silica) to remove surface cells	Shines the nail plate (used with a chamois buffer)
Nail polish drier	• Mineral oil – assists drying • Oleic acid or silicone – lubricant	Increases the speed at which the polish hardens
Nail polish thinner/ solvent	• Ethyl acetate – thins nail polish consistency • Toluene – solvent that dissolves	Thins nail polish that has thickened, restoring consistency
Ridge-filler	• A thicker formulation of a base coat, often containing 'filling' fibres, giving it a milky appearance	Provides an even surface to ridged nails, allowing a smooth polish application Retail: Use as a base coat
Exfoliant	• Granules or chemical agents such as enzymes or alpha hydroxy acids	Desquamates the skin Retail: Use twice a week

5   How is nail polish applied properly?

6   Explain the use and benefits of paraffin wax in nail treatments.

7   Choose up to three products from the following list and research them. Find out the properties of the product (What does it do? Why is it different from other products?), the chemical composition of the product (What chemicals is it made of?), and what the effects are of that product on the nails and skin (What will the product do to the nail? Can it have an adverse reaction?). Within your training group, make sure each item on the list is researched, and present yours to the class.

- base coat

- cuticle care

- drier

- exfoliants

- hands and feet soaking products

- masks

- massage mediums

- moisturisers

- nail hardeners

- nail polish removers

- nail polishes

- quick dry

- thinner

- top coat

LO9.3 PROVIDING THE NAIL SERVICE

Always recommend to your clients that they maintain good condition of their bones, muscles and circulatory system. A basic understanding of these skeletal, muscular and circulatory systems helps you to know how to use beauty therapy to enhance the appearance of your clients' skin and nails.

Body systems

Knowledge Evidence

SKELETAL SYSTEM

The bones of the lower arms and legs function in six ways:
- *Support:* The bones hold all other structures in place and maintain the body shape.
- *Blood cell production:* The cells produced in our bones include red blood cells, which transport oxygen throughout the body.
- *Movement:* The bones attach to tendons and ligaments to allow skeletal muscle movements. The joints permit movement.
- *Protection:* The patella protects the knee, the ulna protects the elbow, the carpals protect the wrist and the tarsals protect the ankle.
- *Storage:* The bones store minerals – calcium in the compact bone and iron in the bone marrow.
- *Insulin regulation:* Osteocalcin, a hormone that regulates insulin secretion and sensitivity, is released from the bones. It affects the body's fat deposition and blood glucose levels.

Mobilising joints and stimulating the circulation of blood and lymph to bone tissue during massage improves the health and function of the local tissues of the skeletal system.

MUSCULAR SYSTEM

The muscles of the forearm and hand, lower leg and foot are skeletal muscles. Of the three types of muscles, skeletal muscles are responsible for the movement of the bones.

CIRCULATORY SYSTEM

The circulatory system delivers blood to every cell of the body. The blood supply delivers these key functions to the skin and nails:
- It supplies the cells with nutrients and oxygen and transports waste and carbon dioxide away.
- The blood contains white blood cells that help to fight infection and remove dead cells.

- Platelets help the blood to clot for wound healing.
- Plasma, the fluid that contains the blood components, allows the blood products to be transported to the tissue fluid between the body cells.

For healthy-looking skin and nails, there must be good blood supply. Good circulation can be achieved by a healthy diet, exercise, specialised beauty products, massage and other beauty treatments. Swelling and discolouration around the nails and skin are the result of poor waste removal.

The inadequate circulation that occurs with diabetes can result in poor wound healing. Special client requirements include gentle filing and buffing, not trimming cuticles, taking care when trimming toenails, and spending more time on massage to improve circulation.

Anatomy and physiology of lower arms and hands

Knowledge
Evidence

BONES OF THE HAND AND FOREARM

The wrist consists of eight small carpal bones, which glide over one another to allow movement. These joints are known as condyloid or gliding joints.

There are five metacarpal bones that make up the palm of the hand. The fingers consist of 14 individual phalanges, two in the thumb and three in every finger.

The arm has three long bones: the humerus, the radius and the ulna. Having two bones in the lower arm allows the wrist to rotate. The movement that causes the palm to face downwards is called pronation; and to face the palm upwards is known as supination.

1 Learn the bones in the hand and forearm using the table and diagram below. Identify the bones on a colleague and confirm with your tutor.

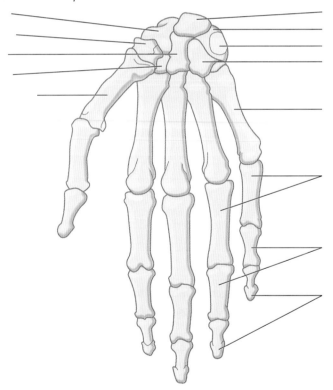

BONE	CHECK WHEN IDENTIFIED
carpal	
metacarpals	
ulna	
radius	
humerus	
phalanges	

MUSCLES OF THE FOREARM AND HAND

The hand and fingers are moved primarily by muscles and tendons in the forearm. The muscles contract, pulling the tendons, which move the fingers, much as a puppet is moved with strings.

The muscles that bend the wrist, drawing it towards the forearm, are known as the flexors (flexor carpi radialis, brachioradialis, palmaris longus and flexor carpi ulnaris). Their opposing muscles, the extensors, straighten the wrist and the hand (extensor digitorum, extensor carpi radialis longus, extensor carpi radialis brevis, extensor carpi ulnaris).

2 Learn the muscles in the hand and forearm using the table and diagram below. Identify the muscles on a colleague by light palpation and confirm with your tutor.

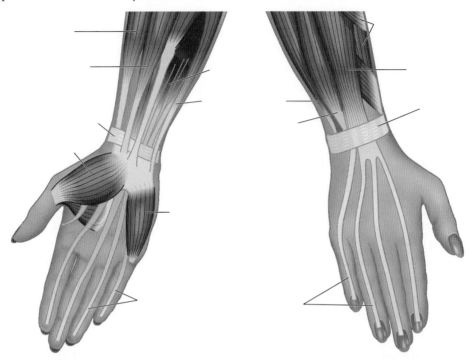

MUSCLE	CHECK WHEN IDENTIFIED	MUSCLE	CHECK WHEN IDENTIFIED	MUSCLE	CHECK WHEN IDENTIFIED
brachio radialis		extensor carpi radialis (longus and brevis)		extensor carpi ulnaris	
extensor digitorum		extensor digitorum tendons		hypothenar muscles	
flexor carpi radialis		flexor carpi ulnaris		flexor digitorum tendons	
palmaris longus		thenar muscles		transverse ligaments	

BLOOD SUPPLY TO THE FOREARM AND HAND

The blood circulates from the heart to the body parts via blood vessels called arteries and returns to the heart via blood vessels called veins.

The arteries

The arm and hand are nourished by a system of arteries that carry oxygen-rich blood to the tissues. You can see the colour of the blood from the capillaries beneath the nail: the colour should be a pale pink underneath the white opacity of the nail cells.

The brachial artery supplies blood to the upper arm. This branches into the ulnar and radial arteries, which supply the forearm and fingers. The radial and ulnar arteries are connected across the palm by a superficial and deep palmar arch. These arteries divide to form the metacarpal and digital arteries, which supply the palm and fingers.

The veins

Veins deliver deoxygenated blood back to the heart. Large veins appear blue at the skin's surface, not simply due to the lack of oxygen in the vessel (deoxygenated blood is maroon) but from the way light is scattered and absorbed by the tissues. Veins often pass through muscles and when the muscle contracts, the vein is squeezed and blood is pushed along. Massage is beneficial for blood flow.

The digital veins drain blood from the fingers. The dorsal arch drains blood from the hands. The cephalic and basilic veins drain blood from the forearm.

ARTERIES OF THE ARM AND HAND

3 Learn the location of the arteries in the arm and hand using the table and diagram below. Identify the arteries on a colleague and confirm with your tutor.

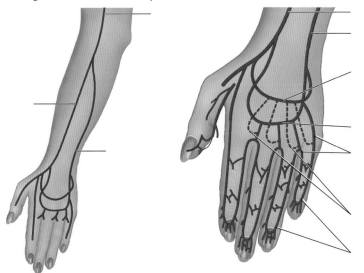

ARTERY	CHECK WHEN IDENTIFIED
brachial artery	
metacarpal arteries	
deep palmar arch	
radial artery (×2)	
digital arteries	
superficial palmar arch	
deep branches of metacarpal arteries	
ulnar artery (×2)	

VEINS OF THE ARM AND HAND

4 Learn the location of the veins in the arm and hand using the table and diagram below. Identify the veins on a colleague and confirm with your tutor.

VEIN	CHECK WHEN IDENTIFIED
right axillary vein	
right clavicle	
right subclavian vein	
right brachial vein	
right median cubital vein	
right basilic vein (×2)	
right cephalic vein (×2)	
right median	

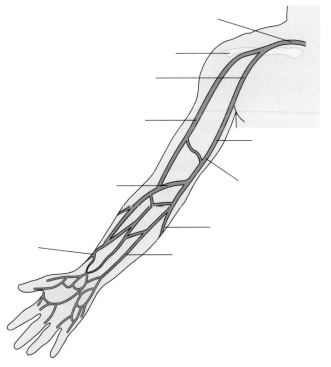

Anatomy and physiology of lower legs and feet

Knowledge Evidence

BONES OF THE FOOT AND LOWER LEG

The foot is made up of seven tarsal (ankle) bones, five metatarsal bones and 14 phalanges (toes). The bones fit together to form arches, which are structured to absorb the impact and preserve balance when we walk, run and jump. The arches of the foot are created by the formation of the bones and joints, and supported by ligaments.

The bones of the lower leg are the tibia, the fibula and the patella. The tibia and fibula have joints with the upper leg (at the knee) and with the foot (at the ankle). The ankle is a hinge joint between the foot and the lower leg. It is composed of the tibia, the fibula and the talus. Both the tibia and the fibula have articulations with the talus; together they allow dorsiflexion and plantarflexion of the foot. When plantarflexed, the foot can rotate (eversion and inversion). Refer to Workbook Chapter 24 for more about bones and joints.

5   If the talus bone is connected to the lower part of the tibia and fibula, does that make it part of the ankle or the knee?

6   Learn the bones in the leg and foot using the table and diagram below. Identify the bones on a colleague and confirm with your tutor.

BONE	CHECK WHEN IDENTIFIED
phalanges	
metatarsals	
tarsals	
calcaneus	
patella	
femur	
tibia	

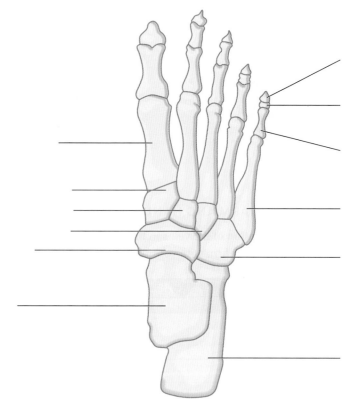

MUSCLES OF THE LOWER LEG AND FOOT

Small muscles in the lower leg are attached to the many small bones in the ankle and foot with several ligaments, making our feet excellent shock absorbers on the ground. As the table below illustrates, numerous small muscles allow a large range of movement at the ankle.

Figure 9.9   Muscles of the arm, hand, leg and foot

MUSCLE	LOCATION	ACTION
Brachioradialis	On the outer (thumb side) of the forearm	Flexes arm at the elbow
Flexor carpi radialis	Middle of the forearm	Flexes and abducts the wrist
Extensor carpi radialis (longus and brevis)	Thumb side of the forearm	Extends and abducts the hand and wrist
Flexor carpi ulnaris	Front of the forearm	Flexes and adducts the wrist joint in towards the body
Extensor carpi ulnaris	Back of the forearm	Extends the wrist and tenses the palm of the hand
Palmaris longus	Middle of the front of the forearm	Flexes the wrist and tenses the palm of the hand
Hypothenar muscles	In the palm of the hand, below the little finger	Flex the little finger and move it outwards and inwards

MUSCLE	LOCATION	ACTION
Thenar muscles	In the palm of the hand, below the thumb	Flex the thumb and move it outwards and inwards
Flexor digitorum (and tendons)	Front of the forearm (and fingers)	Flexes the fingers when contracted
Extensor digitorum (and tendons)	Back of the forearm (and fingers)	Extends the fingers
Gastrocnemius	Calf of the leg	Flexes the knee; plantarflexes the foot
Tibialis anterior	Front of the lower leg	Inverts the foot; dorsiflexes the foot
Extensor digitorum longus	Front of the lower leg	Extends the toes
Soleus	Calf of the leg, under the gastrocnemius (both calf muscles insert at the Achilles tendon to the heel)	Plantarflexes the foot; both calf muscles are used to push off when walking and running
Peroneus longus and peroneus brevis	Attached from the tibia and fibula to the tarsals and metatarsals	Everts the foot; plantar flexes the ankle

7   Learn the location of the muscles in the foot using the table and diagram below. Identify the muscles on a colleague and confirm with your tutor.

MUSCLE	CHECK WHEN IDENTIFIED
flexor digitorum brevis	
abductor digiti mimini	
flexor digitorum tendons	
abductor hallucis	
flexor halluces longum	

BLOOD SUPPLY TO THE LOWER LEG AND FOOT

The lower leg and the foot are nourished by a network of arteries that bring oxygen-rich blood to the tissues. Blood circulation brings warmth to the body, so when circulation is poor, insufficient blood reaches the extremities and they feel cold. Severe lack of circulation to the feet can lead to chilblains.

The arteries

The anterior tibial artery supplies blood to the lower leg and foot. The peroneal artery branches off the posterior tibial artery. At the ankle the anterior tibial artery becomes the dorsalis pedis artery. The posterior tibial artery divides at the ankle to form the medial and lateral plantar arteries. The plantar and dorsalis pedis arteries supply the digital arteries of the toes.

The veins

The digital veins from the toes drain into the plantar and dorsal venous arch. The dorsalis pedis veins drain to the saphenous vein. The following deep veins drain the lower leg: the posterior tibial vein at the back of the leg, the peroneal vein, and the anterior tibial vein at the front of the leg. The deep tibial veins join to form the popliteal vein. Branches of the long and the short saphenous veins can be prone to become varicose veins.

8   Massage movements stimulate blood flow, and the rate, rhythm and repetition of the sequence can be altered for client requirements. If a client has varicose veins, how would you alter the massage sequence? (Select two.)

Performance Evidence

A   Increase stroking movements (effleurage) and reduce deep effleurage movements over the long and short saphenous veins.

B   Reduce kneading (petrissage) over the long and short saphenous veins.

C   Increase kneading (petrissage) over the long and short saphenous veins.

D   Decrease stroking movements (effleurage) and increase deep effleurage movements over the long and short saphenous veins.

ARTERIES OF THE LOWER LEG AND FOOT

9 Learn the location of the arteries in the lower leg and foot using the table and diagram below. Identify the arteries on a colleague and confirm with your tutor.

ARTERY	CHECK WHEN IDENTIFIED
to lateral and medial plantar arteries	
digital arteries (dorsal)	
medial plantar artery	
plantar arch	
popliteal artery (×2)	
calcanean branches of posterior tibial artery	
dorsalis pedis artery	
metatarsal arteries (dorsal)	
plantar digital arteries	
posterior tibial artery (×2)	
calcanean branches of peroneal artery	
lateral plantar artery	
peroneal artery	
plantar metatarsals arteries	
anterior tibial artery (×2)	

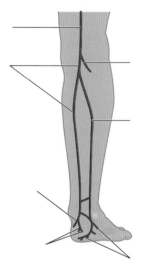
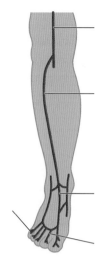
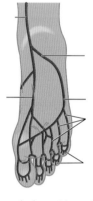

VEINS OF THE FOOT AND LOWER LEG

10 Learn the location of the veins in the lower leg and foot using the table and diagram below. Identify the veins on a colleague and confirm with your tutor.

VEIN	CHECK WHEN IDENTIFIED
femoral vein	
short saphenous vein	
long saphenous vein (×2)	
popliteal vein	
dorsal venous arch	

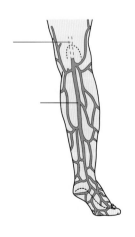

Anterior view   Posterior view

Providing manicure and pedicure treatments

Knowledge Evidence   Performance Evidence

11 List the 'hand-operated' tools required to perform a manicure.

12 What is the difference between reusable and disposable implements?

13 Why is it important for both the nail technician and the client to wash their hands before nail services?

14 What is the difference between a basic manicure and a spa manicure?

15 How is aromatherapy used in manicuring services?

NAIL SHAPING

16 Name the six basic nail shapes.

BUFFING

17 Why would you need to buff a client's nails?

CALLUS REMOVAL

18 During a pedicure/manicure, would you remove calluses before or after exfoliating?

COLOURED NAIL VARNISH APPLICATION AND FRENCH POLISH APPLICATION

19 Give three reasons why a client might want coloured nail polish applied.

20 What is the difference between a traditional application of polish, and a French application?

CUTICLE CARE

21 How would you describe a healthy cuticle?

22 Name two products that could be recommended to a client who needs to care for their cuticles.

EXFOLIATION

23 In what order would you use the exfoliator and the foot spa in the pedicure? Explain why.

MASK

24 A foot mask can contain special ingredients to give specific benefits during the pedicure. Name three possible benefits of the foot mask.

LO9.4 REVIEWING THE SERVICE

Knowledge Evidence

Performance Evidence

1 Review Figure 9.5 in Chapter 9 under 'Conditions and client characteristics' and Figure A.9 in Appendix A Contraindications and restrictions in your textbook. Use the image provided to diagnose and name the disorder shown, describe the salon care you would provide, and the home care you would suggest for each.

	DISORDER NAME	MODIFICATION FOR SALON CARE	HOME CARE
Alamy Stock Photo/RioPatuca			

	DISORDER NAME	MODIFICATION FOR SALON CARE	HOME CARE
Shutterstock.com/Ave Bettum			
Science Source/Biophoto Associates			
Shutterstock.com/Toa55			

Adverse reactions

An adverse reaction occurs when a client has a reaction during the treatment to a product or a service.

Knowledge
Evidence

2   What would you do during a manicure treatment if you noticed the client's skin becoming red around the fingernails when you used a specific product?

_____

_____

Client feedback

Seek client feedback in a range of ways to evaluate the treatment results against expected outcomes. The way you seek feedback should be appropriate to the type of feedback you are looking for and to the type of salon, treatment and client. In the salon, you might seek it verbally, via email or via social networking. In the college, however, the client feedback form can offer a useful tool to source feedback according to your learning needs.

3 Use the form below to obtain client feedback on a nail treatment you have provided.

Client feedback form

As you have just experienced a treatment with a student, please take a minute to complete the below questionnaire, providing your thoughts, observations and impressions of your treatment today. This confidential survey is useful for self-reflection and assists us in monitoring and assessing student progress.

Please answer the questions in accordance with the rating scale: 1 = poor, 5 = exceeding expectations.

	Rating (please tick one)				
	1	2	3	4	5
How well did the student introduce themselves, greet you and make you feel comfortable?					
How well was the treatment area prepared?					
How well did the student communicate and discuss treatment options with you?					
Do you think the products were appropriate to your needs?					
How confident were the student's approaches in performing your treatment?					
How would you rate the student's overall appearance and attitude toward their work?					
Were your timeframe expectations met?					
Please rate the student's overall apparent knowledge and skill in performing your treatment.					
Please rate your level of satisfaction of the treatment performed.					
How confident are you that your health and safety was taken into consideration?					

Please detail any part of your treatment that was unsatisfactory or did not meet your expectations:

_____

Please detail any post-treatment advice, product or further treatment recommendations the student discussed with you today:

_____

Please feel free to make any additional comments:

_____

_____

Thank you for your valuable feedback and assisting our students' learning – we hope to see you for further treatments again soon!

Instructor: _____ Student: _____

Date: ___/___/___ Treatment(s): _____

4 List two other ways you could obtain feedback from the client about their treatment.

Aftercare advice

5 Complete the table below with aftercare advice recommendations for the following items:

	AFTERCARE ADVICE RECOMMENDATIONS
Sun protection	
Gloves	
Filing of nails	
Onychophagy	
High heels	

6 At the conclusion of a treatment, what is the most important thing to do after farewelling the client?

7 Why is it important to record the products used during the treatment?

LO9.5 CLEANING THE TREATMENT AREA

1 What types of linen are likely to be used in a manicure or pedicure treatment?

2 What is the usual protocol for reusable linen following a treatment?

3   Research online to determine how long the following items take to break down if disposed of in a compost soil (imagine an industrial-sized worm farm). You will find a useful guide here: **https://www.sciencelearn.org.nz/ resources/1543-measuring-biodegradability**

ITEM	BREAK-DOWN TIME
Cottonwool	
Disposable towels (made with 100% biodegradable paper or bamboo)	
Glass bottles	
Plastic bags	
Plastic coated cardboard packaging (beauty product packaging would be similar to milk cartons)	

4   Refer to Appendix C Treatment area and equipment in your textbook. Look up the following equipment used during a manicure or pedicure treatment and write the steps needed to clean and store each item correctly.

ITEM	CLEANING PROCEDURE	STORAGE
Disposable pedi paddles Pedicure bowl		
Disposable nail files		
Plastic cuticle pushers Manicure bowl		
Metal cuticle pushers Nail clippers or scissors		

5   Give one example of how you might conserve either product, water or power in the salon during a manicure or pedicure treatment.

_____

_____

CHAPTER 10: NAIL ART AND ADVANCED NAIL ART

LEARNING OBJECTIVES

After completing this chapter, you should be able to:

LO10.1   establish client priorities and design the nail art or advanced nail art based on the client's skin and nail condition, lifestyle considerations and contraindications, including patch test results

LO10.2   prepare the service area sustainably, according to relevant laws and regulations and salon policies and procedures

LO10.3   apply nail art, including hand-painted designs, decals or jewellery

LO10.4   apply advanced nail art for fingers or toes, modifying techniques for two- or three-dimensional nail art decorations such as foils and stencils

LO10.5   review nail art and advanced nail art services, including aftercare and further service planning

LO10.6   clean the nail service area.

INTRODUCTION

Nail art is a niche in the beauty industry reserved for those wishing to use the 'micro-canvas' of the nail to illustrate or express ... whatever! While some designs can be imaginative, mainstream trends can also be subtle styles.

The learning activities in this chapter allow you to demonstrate your knowledge and understanding of the units of competency *SHBBNLS004 Apply nail art* and *SHBBNLS006 Apply advanced nail art*, using the performance and knowledge evidences.

LO10.1 ESTABLISHING CLIENT PRIORITIES

Performance Evidence

Knowledge Evidence

Client record management

1   Fill out a client card for nail art for a client who is aged 24, works as a lab technician, and has generally strong nails. She is after an original design, with some 'bling'.

- Age bracket:

- Sex:

- Occupation:

- Nail analysis:

- Nail design:

Areas requiring special treatment

2   When a client has a skin condition, you need to note on the client card areas that require special treatment. For the following skin and nail conditions, give recommendations for the treatment plan.

   a   Chipped nail on the free edge

   _____

   _____

   b   Pterygium

   _____

When the client has a skin condition that is non-contagious or another contraindication, it is necessary to modify the treatment plan. Nail art design should enhance the nails, but first you need to look at improving the appearance of the hands and nails. Be receptive to the client's needs, as your client may wish to have further treatments, such as aspects of the manicure or pedicure and other beauty services; for example:

- pushing back cuticles
- cutting cuticles or hangnails
- hydrating cuticles or skin
- filing nails to a flattering shape
- beauty services, such as waxing, tanning, spa treatments, etc.

3   Explain the importance of suggesting further services that are relevant to the client.

   _____

   _____

Nails analysis

Knowledge Evidence

4   Find two images of healthy nails, and two of unhealthy nails (either the nails or the skin). Print out the photos and stick them on an A3 piece of paper. For each image, label whether the nails are healthy or not, and make notes around the image pointing to what helped you decide (e.g. near an image of an unhealthy nail you might write that the nail surface has deep ridges, with an arrow pointing to those ridges).

Contraindications

Knowledge Evidence

5   Refer to Appendix A Contraindications and restrictions and Figure 10.2 under 'Contraindications' in your textbook. Based on the images given below of contraindications for nail art treatment, write in the condition, the relevance to nail art, and the action you would take, including referring the client to other practitioners.

CONTRAINDICATION	CONDITION	RELEVANCE TO NAIL ART	ACTION
Shutterstock.com/DD Images			

CONTRAINDICATION	CONDITION	RELEVANCE TO NAIL ART	ACTION
Shutterstock.com/Pan Xunbin			
Shutterstock.com/Hairem			

Design

Knowledge Evidence

6 Complete the following table with the product effects and client suitability for each nail art technique listed in the first column.

NAIL ART TECHNIQUE	PRODUCT EFFECTS	CLIENT SUITABILITY
Decals		
Hand-painted design		
Nail coverings		
Studs		
Diamantes		

NAIL ART TECHNIQUE	PRODUCT EFFECTS	CLIENT SUITABILITY
Charms		
Marble nail art		

7   Activity: Use a sketch pad to sketch different ideas for nail art and advanced nail art. Try to sketch at least one design every day. You will then have over 20 original designs that you can practise with. Though they might not be award-winning designs, you might be surprised at how good they look on the nails, and having something to start with is much easier than having a blank canvas (or nail) with no ideas or inspiration.

COLOURS AND JEWELLERY USED

8   Visit a local salon that offers nail art and ask if you can view their nail art designs. Look for examples of different colour combinations and types of jewellery being used, or of other types of mixed media or freeform designs. If the salon is okay with it, make some notes, and try to come up with three design ideas that you hadn't thought of previously.

Note: some technicians might be protective of their original designs. Make sure you tell them that you are a student who would like to see some examples of different techniques. Respect their decision if they do not want to show you their designs.

9   In what section of the client card would you record the colours and jewellery used in the treatment?

EFFECTS OF CHANGES

10 Design a nail shape and a nail design for a client with overly wide nail beds.

Nail shape:

Nail design:

RANGE AND VARIETY OF NAIL ART

11 There are many different styles and products that can be used in nail art. Complete the following table for each product to explain the:

- effect of the product on the nail art design
- benefit of the product
- disadvantages of the product
- what sort of design the product is best suited to
- the ingredients of the product.

PRODUCT	EFFECT ON NAIL ART DESIGN	BENEFIT OF PRODUCT	DISADVANTAGE OF PRODUCT	TYPE OF DESIGN BEST SUITED TO	PRODUCT INGREDIENTS
Acrylic paint	Bright and long-lasting product to secure nail art structures in place. Acrylic paint can be diluted with water or colours mixed together.	A good range of colours. Durable and can hold nail art structures.			Pigment suspended in an acrylic polymer
Coloured acrylic, gel polish or gel			Needs more time to remove. Nails must be strong to withstand the treatment.		
Nail polish				Freehand designs	

LO10.2 PREPARING THE SERVICE AREA

Performance Evidence

Knowledge Evidence

Sustainability

1   Sustainable practices should be encouraged in all areas of the salon. Think of ways in which water, power or product used can be conserved in the nail art treatment. An example of product conservation is making sure to wash paint brushes correctly so they can be used again and not thrown out.

   a Product

   _____

   b Water

   _____

   c Power

   _____

2   Would you expect a consultant who performs advanced nail art to think differently about the sustainable use of product, water and power than a consultant who only performs more standard nail art applications? Explain your answer.

   _____

   _____

Organisational policies and regulations

Knowledge Evidence

3   Go online to your state/territory and local government's website and find any health and hygiene regulations and requirements that are relevant to nail art services. These might include preparation, client privacy, cleaning, etc.

4   If, after a full consultation and discussion on design, the client is not happy with the final look of the nails, who would you report this to?

_____

_____

_____

TOOLS AND EQUIPMENT

5   Complete the following table of equipment and tools used during nail art treatments. You may need to refer to Appendix C Treatment area and equipment in your textbook or search different products online to ascertain the maintenance instructions from the manufacturer.

EQUIPMENT/TOOL	USE	MAINTENANCE
Tweezers		
Marbiliser/dotting tool		
Nail art brushes		

6   Complete the following crossword.

Nail art tools

Across
3 Used to paint spots
4 They look like gemstones
6 Jewellery may be inserted into it on the artificial free edge

Down
1 Used for a marble effect
2 Can airbrush shapes through it
5 Used to create stripes

HYGIENE, SAFETY AND PRESENTATION

7   You are a therapist who wants to present yourself as best you can for nail art consultations. What is the best way to keep your own fingernails?

_____

_____

_____

8 How could you best present the colours and designs available to clients in the treatment area?

_____

_____

9 Why is ventilation so important? List two reasons.

_____

10 What sort of personal protective equipment could you wear to protect yourself during a treatment?

_____

11 At your training facility or the local nail salon, ask for the details of their air ventilation. Ask if they have specifics, such as the name, the model, and any other details. Go online and find a similar model to learn the specifications.

12 Where in the salon would you keep all safety data sheets? Why is it a good idea for all workers in a salon to know where these are kept and be able to access them easily?

_____

_____

_____

LO10.3 APPLYING NAIL ART

Performance
Evidence

1 Go online to find three product ranges of decals and jewelleries that you could use for nail art designs. Which brand do you like the best? Why do you like that one over the others?

_____

_____

Nail art top coat or sealant

2 True or false: Nail art top coats are usually more durable than normal top coat polishes, to protect the art beneath.

_____

Retouching/repair

3 Retouching and repairing nail art is a service that the salon should also provide. What is the major difference in the treatment that you would provide for these services?

_____

_____

4 In the spaces below, sketch a design with instructions for nail art application. It is best to look online for some inspiration. For example: mixed media foil/mesh nail design or a 'snake skin' look.

stencilling	free form art design
two or more colours	gel or acrylic colour design or art
mixed media	mixed media with a minimum of four colours

LO10.4 APPLYING ADVANCED NAIL ART

Performance
Evidence

Application of sealant

1 How many layers of sealant would you use when rhinestones have been applied to the nail?

Damaged, chipped or broken advanced nail art

2 When repairing damaged, chipped or broken advanced nail art, what about the condition of the cotton tip should you always make sure of?

Retouch

When your client accidentally ruins the nail art during treatment and you have insufficient time left in the allocated appointment time, what do you do? According to the ACCC, you are within your rights to reschedule the client and/or charge for added time if necessary.

If you have completed the client record card properly, retouching a client's nails can be performed quickly by any therapist. Allow 15 minutes per nail.

View some YouTube clips of professional technicians demonstrating retouching ideas and seek tips from your tutor for retouching nails, and take notes on what you would do in the following instances.

3   Mottled nails – the tips of the whole set have been mottled, but the rest of the nails are fine.

4   The client has nicked the surface of a nail polish down to the free edge after leaving the salon.

Range and variety of nail art

Knowledge Evidence

5   What are three products that are specific to advanced nail art?

6   What are the steps for using a stencil and foil on a nail?

7   If using two or more colours, which order would you apply the colours in?

LO10.5 REVIEWING THE SERVICE

Performance Evidence

Aftercare advice

Knowledge Evidence

1   List seven suggestions that you could give to clients to assist them in maintaining their nail art after the treatment.

2   Look in your salon or go online and find three examples of products that you would recommend to a client after receiving their nail art application.

Client feedback

3 Where is the best place to record client feedback?

_____

4 Fill out a client card for a male model, aged 19, who has been happy with French manicures in the past, but is now wanting acrylic nails with a freeform, art-deco style nail art design. He will definitely be wanting it again. (Ensure you record the outcomes of the treatment and client feedback for future reference.)

- Client age bracket:

_____

- Sex:

_____

- Outcome of previous treatments:

_____

- Design:

_____

- Outcome of current treatment:

_____

Adverse reactions

5 What would you say to a client in the following two scenarios?

   a When asking the client if they generally have sensitive skin.

_____

   b Advising the client that the skin around the nail is looking overly red after applying the nail art.

_____

_____

6 Where on the client card would you put any adverse reactions that the client may present with either before, during or after the nail art treatment?

_____

LO10.6 CLEANING THE SERVICE AREA

1 All equipment needs to be properly cleaned and stored following a nail art treatment. List three items that are specific nail art equipment used in the treatment.

_____

_____

_____

2   List three items that are specific to nail art, that you will need to restock after a nail art treatment if they were used.

_____

_____

_____

3   List three items from a nail art or advanced nail art treatment that would need to be disposed of.

_____

_____

_____

4   List two types of linen that might need to be used in a nail art or advanced nail art treatment.

_____

_____

SECTION 4

HAIR TREATMENT AND REMOVAL

11 Waxing services

12 Lash and brow services

13 Eyelash extensions

CHAPTER 11: WAXING SERVICES

LEARNING OBJECTIVES

After completing this chapter, you should be able to:

LO11.1   establish client priorities and design the treatment plan with consultation and skin analysis to determine client requirements

LO11.2   prepare for the waxing service according to salon protocol and relevant health and safety legislative requirements

LO11.3   apply and remove wax to most effectively remove hair with minimal adverse effects

LO11.4   review the service by evaluating it, administering post-treatment care and rescheduling the next treatment

LO11.5   clean the treatment area in preparation for the next service.

INTRODUCTION

Waxing services remove hair temporarily from the face and body. Waxing removes both the visible hair and the root, so the regrowth is of completely new hairs with soft, fine-tapered tips.

Depilatory waxing involves applying wax to the skin and then removing it with minimal damage to the skin. Embedded hairs are removed at the root, and are expected to grow again with the follicle's next hair growth cycle: approximately three to six weeks. It varies according to the body region and the individual. In this chapter, we discuss hot, strip and sugar wax methods.

Tweezing involves grasping the hair near the surface of the skin, and tweezing with the direction of growth. The hair grows again with the next growth cycle. Tweezing is the best way to remove stray hairs after waxing. Threading is a popular alternative to tweezing large areas. For more information about threading, refer to *Milady's Aesthetician's Series: Advanced Hair Removal* by Pamela Hill and Helen Blackmore.

The learning activities in this chapter allow you to demonstrate your knowledge and understanding of the unit of competency *SHBBHRS010 Provide waxing services*, using the performance and knowledge evidences.

LO11.1 ESTABLISHING CLIENT PRIORITIES

Performance Evidence   Knowledge Evidence

Hair growth

Hair follicles grow in cycles throughout one's life (see Figure 11.1). The duration of a cycle varies from person to person and from site to site on the body. Follicles continue to grow even if the body is physically stressed to the point of severe starvation. The largest hairs grow from the deepest embedded follicles, which are also the source of the thickest hair shafts.

Figure 11.1 Rates of hair growth

AVERAGE: MALE AND FEMALE COMBINED	PER DAY
Top of scalp	0.44 mm
Chest	0.40 mm
Chin	0.38 mm
Crown of head	0.35 mm

AVERAGE: MALE AND FEMALE COMBINED	PER DAY
Axilla	0.30 mm
Beard (male)	0.27 mm
Thigh	0.20 mm
Eyebrow	0.16 mm

HAIR GROWTH PATTERNS

1   Investigate the typical hair growth patterns of different parts of the body that might need hair removal. Look online, share information with your classmates, ask at your local salon whether they can share information with you, and even look at your own hair growth pattern. Sketch the body part with arrows showing the direction of hair growth for the following body parts:

  a   front and back of leg

  b   foot and toe

  c   bikini line

  d   underarm

Knowledge Evidence

VARIATIONS IN HAIR GROWTH

Seasonal variation:

- Growth is faster in summer than winter owing to improved blood supply in the warmer months.
- There is no seasonal variation in the tropics.
  Gender differences:
- Female scalp hair is slower growing than male scalp hair.
- Male axilla hair grows faster than female axilla hair.
- Female thigh hair grows faster than male thigh hair.

  Regeneration times can also be quite different for different sites. Figure 11.2 shows the recovery times for plucked hairs.

Figure 11.2   Recovery for plucked hairs

AVERAGE MALE/FEMALE	DAYS
Scalp	129
Axilla	123
Thigh	121
Supra-ear	117
Chin	92
Eyebrow	64

HAIR GROWTH CYCLES

There are three main stages recognised in the cycles of hair growth: anagen, catagen and telogen. These are shown in Figure 11.3 and summarised in Figure 11.4.

Performance Evidence

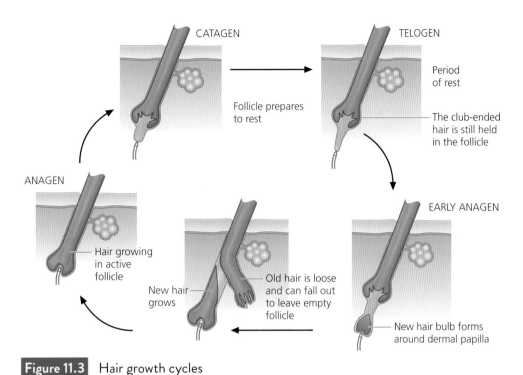

Figure 11.3 Hair growth cycles

Figure 11.4 Stages of hair growth cycles

ANAGEN THE ACTIVE PHASE OF GROWTH	**Early anagen** Downward growth of a follicle from a site (a patch of cells) just below the attachment region of the arrector pili muscle. The new bulb and new tube grow towards the blood vessels that will become the papilla of the hair follicle. At this stage there are two 'zones' of mitosis: • A patch in the follicle tube wall that grows the new tube down into the dermis (this patch is sometimes called 'the bulge' but there isn't really a bulge of cells most of the time.) • A domed crown of actively dividing cells (called the matrix cells) in the swelling of the hair bulb that is beginning to pump out new cells. These will eventually form the hair shaft/inner root sheath (IRS), which rises vertically upward. **Middle anagen** • Full attachment and development of the papilla. • The population of active matrix cells has expanded to several thousand. • Melanocytic cells are active. • Mitosis is rapid and new keratinocytes are streaming off the bulb matrix dome. • This stage may last from months to years. **Late anagen** • Mitosis begins to slow and some reduction in pigmentation may be noted. • The population of matrix cells beginning is to decline. • At the zone of the hair shaft and inner root sheath keratinisation is creeping further downward over the shrinking bulb.
CATAGEN THE REGRESSIVE (RETRACTION) PHASE OF GROWTH	• This stage may only last a few days to a week. • Mitosis has ceased. • The population of matrix cells is reduced to perhaps fewer than 100 non-dividing matrix cells. • These cells eventually become shielded by the dying and hardening cells around the perimeter and over the top of the bulb. • The bulb cells are starting to keratinise and the blood supply (papilla of the hair follicle) detaches and withdraws from the follicle. • The follicle tube starts to shrink and withdraw up inside the vitreous membrane 'sock' towards the surface of the skin, leaving the detached and shrunken papilla in the 'toe' of the sock. • The fully keratinised bulb starts to take on an appearance of a club or 'bottle brush' as the keratinised cells spike outwards. The surviving matrix cells remain sheltered under the keratinised dome of the hair shaft.

TELOGEN THE RESTING PHASE OF GROWTH	• The follicle is now completely inactive, shrunken and shallowly attached. The reason hair looks thinner when waxed is because there are no telogen hairs in the follicles. • The spiky bulb is embedded in the upper dermis and prevents immediate detachment and loss of this 'dormant' follicle's hair shaft. • Over time the attachment will weaken and the hair either be dislodged and fall out or be pushed out by a new hair shaft growing up from underneath from a re-activated bulb. • Telogen phase lasts from days to a couple of months. Telogen hairs are not felt when removed.

Skin Disorders Resource 2015 by Dylan Webb, Victoria University 2015.

2   Complete the following table regarding the growth stages of hair with a sketch of each stage, a definition of each and how each relates to hair removal.

STAGE	SKETCH OF STAGE	DEFINITION	RELATION TO HAIR REMOVAL
Anagen			
Catagen			
Telogen			

Alternative hair removal methods

Knowledge Evidence

Research the various types of hair removal a client could have done (or do themselves) and investigate the advantages and disadvantages of each. Include waxing, shaving, laser, IPL, threading, electrolysis and any others you can think of.

3   Complete the table below, and include what you would say to a client who is asking your professional advice on each option.

HAIR REMOVAL METHOD	ADVANTAGES	DISADVANTAGES	YOUR PROFESSIONAL OPINION
Waxing			
Shaving			
Laser/IPL			

HAIR REMOVAL METHOD	ADVANTAGES	DISADVANTAGES	YOUR PROFESSIONAL OPINION
Threading			
Electrolysis			

Factors affecting hair growth

Knowledge Evidence

4 Give two examples for each of the following:

a hormonal effects on hair growth

_____

_____

b hereditary effects on hair growth

_____

_____

c drugs and chemical effects on hair growth

_____

_____

Hair analysis

5 In the image below, label the features of the hair and hair follicle.

Knowledge Evidence

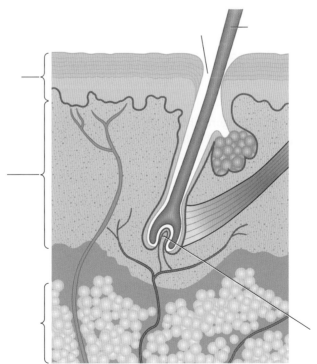

6 Describe the difference between a terminus hair and a vellus hair, and include where on the body you are most likely to find each.

_____

_____

_____

_____

Common hair disorders

Knowledge
Evidence

Figure 11.5 Common hair disorders relevant to hair removal

	DESCRIPTION	INCIDENCE/ EPIDEMIOLOGY	AETIOLOGY (CAUSE)
EXCESSIVE HAIR GROWTH			
Hypertrichosis	Excessive hair growth that does not follow a male pattern of hair growth.	Either congenital or acquired. Very rare.	Congenital – longer silky lanugo hair doesn't shed and become vellus hair but persists throughout life. Acquired – may be due to a serious debilitating disease, an endocrine disorder, certain drugs or anorexia.
Hirsutism	The excess of androgen dependent hair in a male pattern.	Grouped into: Endocrine – (adrenal glands): Cushings syndrome, virilising tumours, congenital adrenal hyperplasia. Pituitary – Acromegaly.	Ovarian – polycystic ovaries, virilising tumors, gonadol dysgenisis. Iatrogenic – due to androgenic drugs. Idiopathic – and due to organ hypersensitivity.
ALOPECIAS			
Traumatic alopecia (trichotillomania)	Hair is constantly pulled or rubbed.	Quite rare, more common in childhood and adolescence. Stress exacerbates the condition.	Patches removed from hair and eyebrows.
Traction alopecia	Hair loss is often just seen around the scalp margins or eyelashes.	Much more common with clients with weak eyelashes due to eyelash extensions.	Due to tight hairstyles, perming and lash extensions.
Cicatricial alopecia	Hair follicles are replaced with scar tissue after injury.	Commonly occurs after injury.	Due to burns or other scarring.

Skin Disorders Resource 2015 by Dylan Webb, Victoria University 2015.

7 For the following hair disorders listed in Figure 11.5 above, suggest a salon treatment and/or referral to an appropriate professional. (See also Appendix B Referrals to professionals.)

a Hypertrichosis

_____

_____

b Hirsutism

_____

_____

c Trichotillomania

_____

_____

d Cicatricial alopecia

_____

_____

Contraindications and scope of practice

8 Refer to Figure 11.7 in your textbook and Appendix A Contraindications and restrictions. For each image of a symptom provided in the table below, complete the name of the contraindication, the relevance to a waxing treatment, and the action you would take.

Knowledge
Evidence

SYMPTOMS	CONTRAINDICATION	RELEVANCE TO WAXING	ACTION
Dreamstime.com/Dmitriy Bryndin			
Shutterstock.com/Faiz Zaki			
Alamy Stock Photo/George Mdivanian			
Shutterstock.com/WEERACHAT			

9 For the following restrictions to waxing, go online and find an image for each one that has hair around it (e.g. a scar on the arm or leg, etc.). Explain for each image where you would be comfortable waxing (use arrows or shading, etc.), where you would need to use tweezers, or where you would avoid completely.

   a Recent scarring

   b Scar tissue

   c Skin trauma

   d Varicose veins

10 What might happen if you did not state a diagnosis of folliculitis on a client?

Relevant medical history and medications

11 When filling out the client card for waxing services and asking about the client's medical history, what would you include if the client offered the following information:

   • Has three children, one of which was a complicated birth

   • Is on the contraceptive pill

   • Gets get neck pain

LO11.2 PREPARING FOR WAXING SERVICE

Knowledge
Evidence

1 Describe three ways you can guard against cross-infection during a waxing treatment.

2 Research your local state or territory's health, hygiene and skin penetration regulations and requirements to find out what you need to do (or not do) in hair removal. Choose one form of hair removal and write a one-page report that summarises the key points.

3 Fill out a client card for a 32-year-old female client who requires a full leg wax, upper lip wax and underarm wax. Her hair is long and fine on her legs, thick and dark under her arms, and fine and short on her lip. She has slight scarring on her knee from a childhood injury.

   Sex:

   Age bracket:

   Restrictions:

   Details:

Hair analysis:

_____

Full leg wax:

_____

Upper lip wax:

_____

Underarm wax:

_____

4   Sometimes when a hair is removed with wax, it bleeds a little bit. What type of hair might this be, and where is this hair most likely to be located? And why does it bleed? What would you do if you saw blood on the wax strip?

_____

_____

_____

5   What is the best stance for a therapist to assume for a waxing treatment?

_____

_____

_____

6   Can you think of three incidents or accidents that can happen during a waxing treatment that should be reported to your manager?

_____

7   Why would you wear an apron during a waxing treatment?

_____

8   What can you do for hair that is too long to wax?

_____

LO11.3 APPLYING AND REMOVING WAX

Performance Evidence   Knowledge Evidence

1   Why can double-dipping the spatula into the wax pot be a source of cross-infection? Explain the routes of infection transmission if that were to happen.

_____

2   What should you keep in mind when choosing a wax product to use in an area prone to perspiration, such as the underarm or lip?

_____

_____

3   How should the arm be placed when removing hair on the ulnar aspect of the forearm?

_____

4 True or false: Skin is the same all over the body; there is no difference between the skin on the face and the skin on the hand. Explain your answer.

5 You should use hot or sugar wax when performing an eyebrow wax unless: (give two reasons)

6 To what length should you trim bikini hair before waxing?

7 When waxing below the knees, what should you ask the client with regards to which areas are to be treated?

8 Always test wax on the wrist before commencement of treatment and after filling or adjusting wax pots. Why is it important to keep hot wax at the right temperature?

9 How full should the wax pot be to maintain wax temperature?

10 Explain what is meant by supporting the skin through a wax treatment.

11 Pressure over the waxed area helps to block the sensation of pain. When the client's hand is used to alleviate pain over raw skin, what personal protective equipment should you offer them?

12 What would you do to try to ensure you have removed all unwanted hair from an area?

13 If there are stray hairs remaining, what can you do? For each of the below, answer true or false:

a Repeatedly spread and remove wax over one area.

b Spread and remove more than twice on sensitive areas such as the bikini line, the face and the underarms.

c Remove any stray hairs using sterilised tweezers.

d In some instances, you can safely wax over the area again (follow the product manufacturer's instructions and it will largely depend on your level of ability).

Products used

Knowledge
Evidence

14 For each of the following ingredient types found in waxes and post-treatment creams, state its
properties and function and give an example of a product that contains it.

INGREDIENT TYPE	PROPERTIES/FUNCTION	PRODUCT EXAMPLE
Emollients and thickeners, e.g. lanolin, paraffin		
Beeswax		
Sugar wax		
Plasticisers		

15 Why is after-wax product so important?

16 Complete the table below by writing the key effect of the application methods of hot wax, strip wax and sugaring.

WAX TYPE	APPLICATION METHOD	KEY EFFECT
Strip wax	Apply with direction of hair growth, remove against direction of hair growth.	
Hot wax	Apply against direction of hair growth, remove against direction of hair growth.	
Sugar wax	Apply against the direction of hair growth, remove with the direction of hair growth.	

LO11.4 REVIEWING WAXING SERVICE AND PROVIDING POST-SERVICE ADVICE

Performance
Evidence

1 When booking a client in for a follow-up treatment, how long would you allow for these treatments?

TREATMENT	TIME SCHEDULED
Bikini line	
Upper-lip wax	
Chin and throat wax	
Lip and chin wax	
Half arm and full arm	
Eyebrow	
Half leg and full leg	

Outcomes of treatment

Performance Evidence   Knowledge Evidence

2 From the following list of adverse reactions, create an information leaflet that could be sent to clients. In the table below, include the name of the reaction, a description of how the client would recognise it, the actions the client should take if they show the symptoms of that reaction, and any other information you think should be communicated.

- Inflammation
- Spotting of skin
- Hive-like reactions
- Ingrown hairs
- Torn skin
- Bruising
- Infections of hair follicle
- Burns
- Bleeding from follicles

ADVERSE REACTION	DESCRIPTION	RECOMMENDED ACTION
Inflammation		
Spotting of skin		
Hive-like reactions		
Ingrown hairs		
Torn skin		
Bruising		
Infections of hair follicle		
Burns		
Bleeding from follicles		

3 If your client has an ingrown hair that would need to be removed with a needle, you should:

   A adhere to salon policies and procedures regarding ingrown hair removal only

   B adhere to state/territory and local health, hygiene and skin penetration regulations and requirements relevant to the provision of waxing services plus salon policies and procedures

   C adhere to federal work health and safety legislation only

   D not treat the client. You you are not qualified to remove an ingrown hair.

4 Locate the skin penetration regulation and state what you must do regarding the disposal of needles.

5 What are the main infections that can be transmitted via skin penetration procedures?

Aftercare recommendations and advice

6 What advice would you give to a client after a waxing treatment to avoid adverse reactions? How might this information best be communicated to the client?

Knowledge
Evidence

7 What is a common active ingredient of hair-retarding products?

8 Research the use of infra-red for clients to use at home to retard hair growth. What is your professional opinion?

Client feedback

9 If you have received client feedback in the past that you left hair behind after a half-leg wax, what might you do to try to remedy that this time?

10 If you want to know whether the client felt the treatment outcomes were as they expected, you will ask them questions such as:

- How does your skin feel?
- How would you describe your skin's appearance now?

You will then offer them some common words to describe their skin.

a Suggest at least three words for your client (to describe their skin after waxing):

_____

b How would you record the treatment results?

_____

Waxing service performed

11 Suggest five reasons why it is important to note the waxing service performed on the client record card.

_____

_____

_____

_____

_____

LO11.5 CLEANING THE TREATMENT AREA

Knowledge Evidence

1 List five single-use items that would be used in the waxing service. Try to come up with a more environmental option for each one.

_____

_____

_____

_____

_____

2 How should the couch be cleaned? What would you need to do for it to be presentable for the next client?

_____

_____

_____

_____

3 Though it is always necessary to use sustainable options where appropriate, there are times when this is not possible. In the case of hot wax, can it be recycled in the salon? Why or why not?

_____

_____

_____

4   To maintain presentation of the treatment area and as part of work health and safety, wax pots should be completely cleaned once per month. Follow the link **https://www.professionalbeauty.com.au/beauty/waxing-massage/how-to-clean-a-waxing-pot/#.YMM-k_kzY2w** for an industry best practice guide to cleaning a wax pot and list the 13 steps here:

CHAPTER 12: LASH AND BROW SERVICES

After completing this chapter, you should be able to:

LO12.1   establish client priorities and develop a treatment plan for the lashes and brows through consultation and a skin and hair analysis

LO12.2   prepare yourself, the client and the treatment area for lash and brow treatments according to salon protocol and legislative requirements

LO12.3   perform chemical treatments, such as tinting or lifting for eyelashes and tinting or henna for eyebrows, and apply post-treatment product

LO12.4   use wax and tweezers to shape eyebrows to suit the client's face

LO12.5   review the service by evaluating and recording the outcome, rebook the service, advise the client on aftercare and make post-care recommendations

LO12.6   clean the treatment area in preparation for the next service.

INTRODUCTION

Lash and brow treatments make the eyes look more vivid and expressive. Tinting, lash lifting and henna are convenient for when make-up is impractical, such as when working out at the gym or swimming, and has the added benefit of lasting for weeks.

The learning activities in this chapter allow you to demonstrate your knowledge and understanding of the unit of competency *SHBBFAS004 Provide lash and brow services*, using the performance and knowledge evidences.

LO12.1 ESTABLISHING CLIENT PRIORITIES

Performance
Evidence

Knowledge
Evidence

Hair growth and analysis

It is important to understand how the hair follicle produces the hair. The follicle is responsible for the growth phases of anagen, catagen and telogen. Review Chapter 11 of your textbook to understand the stages of hair growth and when it is best to schedule hair treatment and removal services.

1   Use the following terms to label the diagram:

- early anagen

- anagen

- catagen

- telogen

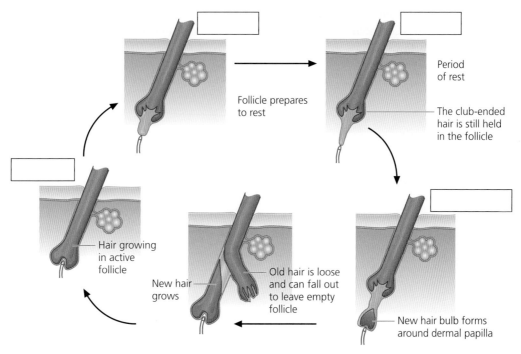

Follicle prepares
to rest

Period
of rest

The club-ended
hair is still held
in the follicle

Hair growing
in active
follicle

New hair
grows

Old hair is loose
and can fall out
to leave empty
follicle

New hair bulb forms
around dermal papilla

2   Though many eyebrows follow a general pattern, there are always variations between individuals. Ask three of your family or friends if you can sketch their eyebrow growth pattern, using the space below. For each one, think about the brow shape you could give them, and any shapes you wouldn't be able to give them. Make notes in the area next to your sketch.

SKETCH OF EYEBROW GROWTH PATTERN	NOTES

3   Refer to the image of simplified cross-section of the skin in Figure 11.3 under 'Hair growth cycle' in your textbook. Using the anatomical terminology in the diagram, explain why you would perform an eyebrow tint before an eyebrow wax.

4 Review Figure 11.2 of your textbook under 'Hair type and density'. Sketch the transitions from short vellus hair, to long and fine vellus hair, to accelerated vellus hair (pigmented and fine), and lastly to terminal hair (long, thick and pigmented) in the space below.

5 Look over Figure 11.3 of your textbook labelled 'The hair in its follicle', where the three main layers of the skin are shown. What are the three layers called? Which ones house the hair and its follicle?

6 The diagram shown in Figure 11.3 is known as a simplified histology diagram. What is a histology diagram and why is it helpful?

7 What can happen if brows are over-tweezed?

8 Using Figure 11.4 from your textbook labelled 'Factors affecting hair growth', give one example of how each of the following factors can affect hair growth:

   a hormonal

   b hereditary

   c drugs and chemicals

Knowledge
Evidence

Eyebrow shape

Place yourself at ease by beginning to practise on an arm or leg, particularly if you are very new to using wax on the face. Get accustomed to controlling the wax on the small spatula and applying with precision at the right temperature to shape the brow.

9   Activity: Draw eyebrows with a kohl eyeliner on a part of the arm or leg where hair growth mimics the direction on the face. Apply the wax, or tweeze as you would on the face. Soon, both student and instructor will be confident that the student is ready to apply wax to the face.

10 Sketch each of the following eyebrow shapes. Go online and find examples of images with people with those shapes.

   • Arched

   • Rounded

   • Oblique

11 The three methods for shaping brows listed below have different benefits. Give a benefit for each one.

   a Waxing

   _____

   b Threading

   _____

   c Tweezing

   _____

Tint and henna

12 Which colour do you think would be most suitable for the eyelashes and eyebrows of the following client types?

   a Dark complexioned

   _____

   b White/grey-haired

   _____

   _____

   c Fair complexioned

   _____

   d Elderly

   _____

   e Red-haired

   _____

13 How would you describe the essential difference between eyebrow tinting and henna to a client?

   _____

Contraindications and scope of practice

Knowledge
Evidence

14 Review Appendix A Contraindications and restrictions, and Figure 12.5 and Figure 12.6 in your textbook. Based on the images below, complete the table with the contraindication, the relevance to an eyebrow shape or tint, and the action you would take if a client had this contraindication.

SYMPTOM	CONTRAINDICATION	RELEVANCE	ACTION
Alamy Stock Photo/Science Photo Library			
Alamy Stock Photo/Mediscan			
iStock.com/Ruslana Chub			

LO12.2 PREPARING FOR A LASH AND BROW SERVICE

Knowledge Evidence

1   Why is it good practice to use a trolley for lash and brow treatment equipment? Give two reasons.

_____

_____

2   How might you present yourself for a lash or brow service in a professional way that also encourages the client to trust your ability?

_____

3   Should you wear gloves during a brow tint procedure? Explain your answer.

_____

_____

4   Look up your local council's health and hygiene regulations and requirements that are relevant to lash and brow services. According to these regulations, are you required to have a water basin in the treatment area?

_____

5 List five procedures that can be implemented in the salon that will help stop the spread of infection during a brow treatment.

_____

_____

_____

_____

_____

Record management and incident reporting

Knowledge Evidence Performance Evidence

6 Fill out a client card for lash and brow services for a male client who needs a brow shaping done. His hair is very thick and dark, his skin tends to get dry, and he has never had a brow shaping done before. He is aged 32, and currently gets regular facials at the salon to combat his dry skin.

- Age bracket:

_____

- Sex:

_____

- Hair analysis:

_____

- Skin analysis:

_____

- Outcomes of previous treatments:

_____

- Current skincare:

_____

- Treatment plan/design:

_____

_____

- Aftercare:

_____

- Homecare:

_____

7 Case study: During an eyelash tint procedure, even though you felt there were no contraindications, and your technique and procedures were correct, you still managed to get tint in the client's eye. The client returned eight hours later to say that after a strong burning sensation, a blood vessel burst and the eye is blood shot. What should you do?

_____

_____

_____

LO12.3 CHEMICALLY TREATING EYELASHES AND EYEBROWS

Products used

Knowledge Evidence

1   When would you use hot wax in an eyebrow wax treatment, and when would you use strip wax?

_____

2   What are the two products that are essential for a permanent tinting treatment?

_____

3   What strength of peroxide is used to activate the tint?

_____

4   When would you mix the product?

_____

5   Activity: In pairs, come up with three reasons to explain why a permanent tint applied to the eyelashes or eyebrows might not have coloured the hair successfully. Discuss your answers as a class group.

_____

_____

_____

_____

_____

_____

6   What are the two types of products used when doing an eyelash lift?

_____

7   How do you select the perfect perming rod for an eyelash lift treatment?

_____

Eyelash and eyebrow tinting and henna

Apply petroleum jelly around the eyebrow to prevent tint or henna staining the skin outside the perimeter of the brow or lash. Some clients like to have the dye on the skin to give the impression of a tattoo for a couple of days.

Performance Evidence

8   Why do you think the dye is not safe to apply on skin?

_____

_____

9   Why is important to remove all eye make-up from the hairs before performing an eyelash or eyebrow tint?

_____

_____

10 What is the purpose of using a cotton tip to check the lashes at the completion of an eyelash tint treatment?

11 When tinting the eyebrows of a client with fair or fine hair, how long should you leave the tint on the eyebrows before removing?

12 Why would exfoliating have an effect on a henna eyebrow treatment?

13 Why is it important for a client to not have a cosmetic tanning treatment in close proximity of receiving a henna treatment for their eyebrows?

Eyelash lift

14 How long does an eyelash lift last?

15 How long should you leave the eyelash lift after the treatment is complete?

Post-treatment products

16 Following an eyelash or eyebrow tinting procedure, what two products should you apply? What should you be aware of with any product you apply?

LO12.4 SHAPING EYEBROWS

Products used

1 Which wax product is best to use for an eyebrow wax?

Knowledge
Evidence

Performance
Evidence

2 When would you use hot wax as opposed to strip wax for an eyebrow wax?

Shaping eyebrows

3 What is the best way to test the temperature of the wax before applying it to the eyebrows?

_____

_____

4 What are the three ways you could remove sticky residue when the eyebrow wax is finished?

_____

_____

_____

5 When tweezing the brows, what is the general order in which you would remove hairs?

_____

_____

LO12.5 REVIEWING THE SERVICE AND PROVIDING POST-SERVICE ADVICE

Performance Evidence

Adverse effects

1 Review Figure 11.21 and Figure 12.13 in your textbook. Complete the table below, by noting down the common names for each adverse reaction, what might have caused it in a brow treatment, and what actions you would take if the client shows the symptoms.

Knowledge Evidence

MEDICAL NAME	COMMON NAME	CAUSE	ACTION
Urticaria			
Folliculitis			

MEDICAL NAME	COMMON NAME	CAUSE	ACTION
Conjunctivitis			
Pustule			

2   If the client is complaining that their eye is stinging while the tint is on, what three steps would you take?

3   Why is spotting on the skin following hair removal a good thing?

Client feedback

4   There are various ways to seek client feedback. In the salon, you might seek it verbally, by showing the client the results with a mirror and asking for their thoughts, or via email and social networking. In the college, however, the client feedback form is a useful tool to obtain feedback according to your learning needs. Use the attached form to obtain client feedback for at least two lash and brow treatments and describe what you have learnt from the client's feedback (both good and bad).

Knowledge
Evidence

Client feedback form

As you have just experienced a treatment with a student, please take a minute to complete the below questionnaire, providing your thoughts, observations and impressions of your treatment today. This confidential survey is useful for self-reflection and assists us in monitoring and assessing student progress.

Please answer the questions in accordance with the rating scale: 1 = poor, 5 = exceeding expectations.

	Rating (please tick one)				
	1	2	3	4	5
How well did the student introduce themselves, greet you and make you feel comfortable?					
How well was the treatment area prepared?					
How well did the student communicate and discuss treatment options with you?					
Do you think the products were appropriate to your needs?					
How confident were the student's approaches in performing your treatment?					
How would you rate the student's overall appearance and attitude toward their work?					
Were your timeframe expectations met?					
Please rate the student's overall apparent knowledge and skill in performing your treatment.					
Please rate your level of satisfaction with the treatment performed.					
How confident are you that your health and safety was taken into consideration?					

Please detail any part of your treatment that was unsatisfactory or did not meet your expectations:

_____

Please detail any post-treatment advice, product or further treatment recommendations the student discussed with you today:

_____

Please feel free to make any additional comments:

_____

_____

Thank you for your valuable feedback and assisting our student's learning – we hope to see you for further treatments again soon!

Instructor: _____ Student: _____

Date: ___/___/___ Treatment(s): _____

Aftercare and post-treatment recommendations

5   Fill in the gaps in the following sentences.

   a   Wear no _____ for at least _____ hours following the eyebrow shaping treatment.

   b   Avoid the use of _____ products as they add to the _____ pH effect of the oxidant, disrupting the skin's protective acid mantle, thereby leaving the skin prone to _____.

   c   Avoid _____ the area immediately following the service. The area is susceptible to infection and fingers may transfer dirt, causing _____.

   d   Advise a client who has a _____ over the eyebrow area that their _____ needs to be washed _____, at least for the first _____ hours.

6   What is the average time that a client will need to book in to maintain their lash tint?

_____

LO12.6 CLEANING THE TREATMENT AREA

1   List three practices in the salon that could be made more sustainable during an eyebrow tint or shape treatment. Think about the use of power, water or product.

_____

_____

_____

2   Mark the following statements as true or false.

   a   Tweezers need to be disinfected at the end of each day.

_____

   b   Items used need to be restocked in preparation for the next client.

_____

   c   Waxing and tweezing terminal hairs contains no risk of blood and bodily fluid transfer and contamination.

_____

3   Define the following and give an example of when you would use each one during a lash or brow service.

   a   Cleaning

_____

_____

   b   Disinfection

_____

_____

   c   Sterilisation

_____

_____

4 Explain how heat and pressure works as a physical disinfecting procedure.

_____

5 True or false: Tweezers should be sterilised after every client. Explain your answer.

_____

_____

6 What sort of reusable linen would be used during a lash or brow treatment?

_____

7 List four items that would need to be disposed of after a lash or brow treatment.

_____

CHAPTER 13: EYELASH EXTENSIONS

LEARNING OBJECTIVES

After completing this chapter, you should be able to:

LO13.1 establish client priorities, design a treatment plan for lash extensions through consultation, prepare the treatment area and perform a patch test

LO13.2 remove damaged, sparse or grown-out lashes safely and hygienically

LO13.3 apply eyelash extensions in a range of styles and to the client's requirements

LO13.4 review the service, provide post-service advice, recommend future services, update the treatment plan and reschedule

LO13.5 clean the treatment area in preparation for the next client.

INTRODUCTION

Eyelash extensions first reached Australia around 2005, originating in Korea and Japan. Their widespread international popularity has seen lash enhancement services offered in niche beauty salons or in dedicated eyelash clinics. It is a lucrative business, and success relies on expertise and professionalism.

The learning activities in this chapter allow you to demonstrate your knowledge and understanding of the unit of competency *SHBBMUP008 Apply eyelash extensions*, using the performance and knowledge evidences.

LO13.1 ESTABLISHING CLIENT PRIORITIES

Performance Evidence

Eyelash extensions create a plumper, fuller and more dramatic look for the lashes that no other eyelash enhancement has yet been able to achieve.

Extension services

FULL-SET APPLICATIONS

1 How many lashes per eye (or what range) is the best for the following requirements?

   a A full set (general)

   _____

   b A full set for a dense upper lash bed

   _____

   c A full set for a sparse upper lash bed

   _____

IN-FILL SERVICES

2 More frequent infills might be required for curly hair. Why?

   _____

REMOVAL SERVICES

3 How should the eye patches sit, and why?

_____

_____

Record management

Performance
Evidence

PATCH TESTING

4 List four of the common adverse reactions you should be wary of when conducting a patch test for eyelash extensions.

_____

_____

PREVIOUS TREATMENTS

5 Damaged eyelashes might be the result of previous services being performed incorrectly. What action would you take in this situation?

_____

_____

_____

_____

CLIENT RECORD CARD

Knowledge
Evidence

6 Fill out a client record card for eyelash extensions for the following client. She is aged 23 and has prominent eyes. She has had a patch test with no adverse effect (negative result). She is after a look to complement her natural look, and is mildly nervous about getting the treatment. She has naturally dark brown hair.

- Age bracket:

_____

- Sex:

_____

- Restrictions:

_____

- Eye analysis:

_____

- Hair analysis:

_____

- Patch test results:

_____

- Treatment design:

_____

Knowledge
Evidence

Contraindications

7 Review Appendix A Contraindications and restrictions and Figure 13.6 in your textbook. In the following table, list the contraindication, the relevance to the eyelash extension treatment, and the action you would take based on the symptom presented.

SYMPTOM	CONTRAINDICATION	RELEVANCE TO TREATMENT	ACTION
Client pulls strands of hair out of scalp during the initial consultation.			
Science Photo Library/Dr P. Marazzi			
Client tells you they wear contact lenses.			
Science Photo Library			

8   If you suspect a client has a medical condition that is relevant to an eyelash extension treatment, what is your first course of action? Why is it important not to diagnose a condition?

_____

_____

_____

Hair growth cycle

Knowledge
Evidence

9   The three stages of hair growth are anagen, catagen and telogen. For each one, explain what the phase is, and how it relates to the growth of the natural eyelash and to eyelash extension treatments.

_____

_____

_____

_____

_____

_____

_____

Benefits and risks of eyelash extensions

Knowledge
Evidence

10   List three examples of benefits and three examples of risks for eyelash extensions in the table below.

BENEFITS	RISKS

LO13.2 PREPARING THE TREATMENT AREA, YOURSELF AND THE CLIENT

Organisational policies and procedures

Knowledge   Performance
Evidence   Evidence

1   What is best practice when presenting yourself for an eyelash extension treatment? Provide answers for your hands, your hair, jewellery and personal protective equipment.

_____

_____

_____

_____

Equipment use and maintenance

2 For each of the following equipment items, explain its relevance to eyelash extension treatments, and describe how you would ensure it is maintained.

EQUIPMENT ITEM	USE	MAINTENANCE
Air blower		
Application tweezers		
Disposable mascara wands		
Isolating tweezers		
Jade stone		
Lint free eye gel patches		
Micro swab/brush		
Medical tape		
Oil-free cleanser		
Silicone pad		
Sponge		
Puff case		
Eyelash extensions remover		
Eyelash extensions		

Regulations and requirements

Knowledge
Evidence

Performance
Evidence

3   Look up your state or territory's health and hygiene regulations and requirements. What can you find that is relevant to the eyelash extension service? How might you ensure your practice area meets the correct guidelines?

_____

_____

4   There are two ways of disposing of biohazardous materials. What are they?

_____

_____

Client complaints

Knowledge
Evidence

5   You apply a full set of lashes to a client, and she finds that she is not happy with the service. What do your salon policies and procedures expect you to do? Explain the steps you should take in following the complaints procedure.

_____

_____

_____

_____

_____

_____

_____

6   After discussing the issue with you as required by the complaints procedure, the client decides she wants the lashes removed. Do you need to report the incident to the manager?

_____

7   The salon's insurer will have specific needs in terms of client health and safety. Why is it important for insurance reasons to document everything that happens when a client makes a complaint?

_____

_____

Remove eyelash extensions

Performance
Evidence

8   List three common reasons why you would need to remove eyelash extensions.

_____

_____

_____

9 Decide whether the following statements are true or false in relation to removing eyelash extensions. If it's a false statement, write how the statement should read.

a You should use a large amount of remover to make sure you get the full extension removed.

_____

_____

b If you are using liquid remover, blot any excess liquid with a paper towel.

_____

_____

c While the bond is still in place, you should pull gently on the eyelash extension with tweezers.

_____

_____

LO13.3 APPLYING EYELASH EXTENSIONS

Performance Evidence

Selection and application of eyelash extensions type

When performing eyelash extensions, it is important to consider a range of factors when deciding what kind of product and technique to use on a client. These factors can be anything from colour, length, the type of curl and the adhesive used. Each client is different, so understanding these factors will help to ensure a positive outcome.

Knowledge Evidence

ELEMENTS OF EXTENSIONS

Knowledge Evidence

1 When planning the fullness of the lash extension to apply to the client, what is the most important factor to consider?

_____

_____

2 How would you write on the client record card that you will be alternating the length of lashes between 12 mm and 14 mm?

_____

3 Complete the following table of thickness options for eyelash extensions with a description of the look and what each is suited to.

THICKNESS	THE LOOK	SUITED TO
0.15 mm		
0.20 mm		
0.25 mm		

4 What colours do lash extensions come in?

_____

CURLS

Knowledge
Evidence

5   In the spaces provided below, sketch the eyelash curls B, C, D and J described in Figure 13.10 under 'Curl' in your textbook. You will find other eyelash curls are defined in the industry. Search online for the eyelash curl types L, V, W and Y curls and sketch them in the table.

B curl	C curl	D curl	J curl
L curl	V curl	W curl	Y curl

6   Go online to find examples of coloured, diamond and glitter eyelash extensions. Record your observations in the table below.

TYPE	OBSERVATIONS
Coloured	
Diamond	
Glitter	

ADHESIVES

Knowledge
Evidence

7   Complete the table below with the effects and benefits of each of the different adhesives used in eyelash extensions.

ADHESIVE TYPE	EFFECTS	BENEFITS
Normal		
Clear		
Coloured		
Sensitive		

Eyelash extension application to suit client eye shape and facial features

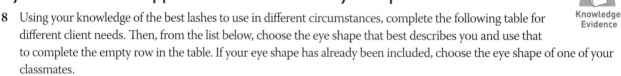
Knowledge
Evidence

8   Using your knowledge of the best lashes to use in different circumstances, complete the following table for different client needs. Then, from the list below, choose the eye shape that best describes you and use that to complete the empty row in the table. If your eye shape has already been included, choose the eye shape of one of your classmates.

- Almond
- Monolid
- Deep-set
- Hooded
- Prominent
- Small

EYE SHAPE	IMPACT ON EYELASH EXTENSION	APPLICATION TECHNIQUE
Round		
Close-set		

Advantages of products and techniques

Knowledge
Evidence

Performance
Evidence

DIFFERENT LASH TYPES

9   Choose the most likely eyelash extension product (either mink, silk or synthetic) for the following:

a   stiff lash

_____

b   feathered look

_____

c   irritant free

_____

d   can lose curl

_____

e   animal product

_____

f   long-lasting

_____

10 If someone is vegan, what might be the best lash option for them?

_____

11 Which lashes of the three types are natural-based products?

_____

DIFFERENT TWEEZERS

Knowledge
Evidence

12 There are many different tweezers, all with their own advantages. Choose three of the following, give their advantages, and sketch how they look in the space provided.

- Curved

- Inverted

- Pointy

- Slanted

- Straight

TWEEZERS	ADVANTAGES	SKETCH

LO13.4 REVIEWING THE SERVICE AND PROVIDING POST-SERVICE ADVICE

Performance
Evidence

Aftercare

Knowledge
Evidence

1 There are many recommendations for homecare that can be made to the client following the eyelash extension treatment. List five of those recommendations. What is the best way to make sure the client goes home with all of the correct advice?

_____

_____

_____

_____

_____

2   The following are products that you could recommend to a client after the eyelash extension treatment. Explain why each one is relevant.

   a   Eyelash conditioner

_____

   b   Coating sealer

_____

   c   Mascara

_____

Client feedback

3   If the client has had an adverse reaction during the treatment, what could you recommend?

_____

Client record card

4   Following a treatment, what would you need to ensure you put onto the client's record card for future services?

_____

_____

_____

_____

_____

_____

_____

Rescheduling

5   Why would you recommend that a client not have eyelash extensions more than four times a year?

_____

_____

6   What is the ideal frequency of the following treatments, and how long should they last?

TYPE OF LASH EXTENSION TREATMENT	FREQUENCY	DURATION
Full set – 40–60 lashes		
Infills		
Full set of specialty lashes (W,Y) – 80–100 lashes		
Infills for specialty lashes		
Lash removal		

LO13.5 CLEANING THE TREATMENT AREA

Knowledge Evidence

1 Eyelash equipment and tools are not currently defined as 'clinical waste' and are unrestricted in Australia. What does this mean in a salon environment?

_____

_____

_____

2 List five items used during an eyelash extensions treatment that you would need to restock after the client has left.

_____

_____

_____

_____

_____

Sustainability

3 List five different ways you could make the eyelash extension treatment in the salon more sustainable.

Knowledge Evidence

_____

_____

_____

_____

_____

SECTION 5

MAKE-UP

14 Design and apply make-up
15 Remedial camouflage make-up
16 Photographic make-up

CHAPTER 14: DESIGN AND APPLY MAKE-UP

After completing this chapter, you should be able to:

LO14.1   establish client priorities and make-up requirements based on client consultation, skin analysis and the occasion

LO14.2   design make-up plan and select products, tools and application techniques to suit the client's skin and facial shape

LO14.3   apply day, evening or special occasion make-up according to the make-up plan, explain application methods to the client throughout treatment and seek feedback

LO14.4   apply strip or individual temporary false eyelashes

LO14.5   provide post-service advice, including homecare and retail product recommendations

LO14.6   clean the service area in preparation for the next treatment.

INTRODUCTION

Make-up services add a special touch to the beauty experience for clients wishing to leave looking their absolute best. Make-up services can be incorporated into a skincare treatment plan or can be a special occasion make-up service.

The learning activities in this chapter allow you to demonstrate your knowledge and understanding of the unit of competency *SHBBMUP002 Design and apply make-up*, using the performance and knowledge evidences.

LO14.1 ESTABLISHING MAKE-UP REQUIREMENTS

Performance Evidence

Contraindications to make-up services

Knowledge Evidence

1   Review Figure 14.1 and Appendix A Contraindications and restrictions in your textbook. Using the images or symptoms listed below, complete the table with the name of the contraindication, the relevance to make-up treatment and the action you would take.

SYMPTOM	CONTRAINDICATION	ACTION
Alamy Stock Photo/Tom Radford		

SYMPTOM	CONTRAINDICATION	ACTION
Client tells you they have minor allergic reactions to some products		
iStock.com/petekarici		
Science Photo Library		

2   If a client has a contraindication such as conjunctivitis or a stye, why should you not proceed with the make-up application?

_____

_____

3   A person with psoriasis or eczema can still receive treatment, but the treatment must be modified in affected areas. Explain what is different about this skin that necessitates a modification to the make-up application.

_____

4   Fill out a make-up services client record card for a mid-20s female who needs her make-up done for her wedding. Her big day will be outside in summer, on the beach, with a reception to be held afterwards in a function room. She has a mild allergic reaction to adulterated lanolin. She has brown eyes with monolids, dark brown/black hair, a heart shaped face, and her skin is not overly oily.

•   Age bracket:

_____

•   Sex:

_____

•   Occasion:

_____

- Outcome of previous treatments/products:

_____

- Skin type:

_____

- Face shape:

_____

- Eye shape:

_____

- Eye colour:

_____

- Hair colour:

_____

- Treatment design:

_____

_____

_____

_____

_____

_____

5 Visit your local beauty salon, your training facility, or look online and ask about the legal and insurance liabilities and responsibilities regarding the application of make-up in the salon or as a mobile make-up artist. Find out the following information and write half a page on your research.

- What are the laws governing the application of make-up?

- What companies insure make-up artists?

- What are the costs involved in being properly insured?

6 There will be different restrictions for various skin treatments such as injectables, intense pulsed light, laser and surgery. What is the best way to be able to advise the client on what the restrictions will be for a make-up treatment?

_____

_____

Skin types and conditions

Performance Evidence   Knowledge Evidence

7 Refer to 'Classify the skin' in Chapter 21 of your textbook and describe the appearance of the following skin types or conditions.

a Oily

_____

_____

b   Dry

_____

_____

c   Diffuse red

_____

_____

8   Refer to 'Common disorders that can be treated' in Chapter 21 of your textbook to answer the following questions.

   a   Which products should you use for couperose skin?

_____

_____

   b   What would you focus on during a consultation for pigmentation disorders?

_____

_____

   c   What can a client with mature skin expect from a daily skincare routine and dedicated treatments? List two of these.

_____

_____

Client needs based on occasion

9   Think about different events for which a client may need specific make-up. Choose three of these and search the Internet for images that display make-up for those events. Make sure you include one day event, one night event, and one for a skin colour, ethnic background or age different from what you are used to. Put the images on a poster with labels and arrows pointing to the lips, eyes, nose, etc. to demonstrate your understanding of the make-up applied.

LO14.2 DESIGNING THE MAKE-UP PLAN AND APPLICATION TECHNIQUES

Performance Evidence

Knowledge Evidence

Client image and occasion

1   A great way to meet new clients, and make more money, is to deliver a make-up lesson or demonstration to a group of friends at a party or other special occasion. Plan a party of your own with some friends. Take photos during the event and present them to your class. Evaluate and report on what went well and what could be done differently next time. Some things to consider:

- Was the make-up right for the audience?

- Did the make-up look different under different light? Were you able to observe it under natural compared to artificial light? What did you observe?

- Were your friends interested in buying the product? Do you think they would have been more/less interested had you not demonstrated the application and communicated the features and benefits of the product?

- Were your friends interested in learning how to apply the product? Was there anyone in the audience with more make-up skill than you? How did/would you deal with this?

- Do you think you could develop a client base by doing this or similar promotional activities? What promotional activities would work best for you?

Colour analysis and design

Make-up is used to enhance and accentuate the facial features to make us appear more attractive, which in turn makes us appear more confident. Make-up is used to create balance in the face, by skilful application of different cosmetic products to reduce or to emphasise facial features.

2   There are three main categories of light that your client's make-up will be viewed in: dusk and dawn, midday and night/evening. For each one, find some images in a magazine that you can use to demonstrate the differences in light, and the differences in make-up application. Create an A4-page display on each category.

COLOUR WHEEL

Colours can be combined with others to create new colours. The colour wheel is a traditional representation of the way colours can be blended. Primary colours and secondary colours are in the centre of the colour wheel.

- Primary colours: red, blue and yellow are the source of all colours. No other colours combined can make these colours.
- Secondary colours: purple, orange and green are the colours made by combining two primary colours.

    Combining a primary and a secondary colour creates a tertiary colour. For example, yellow combined with orange creates yellow-orange.

COLOUR HARMONY AND COLOUR CONTEXT

Colour harmony is the aesthetically pleasing arrangement of colours, and when make-up is applied with harmony, it is smooth, balanced and calming to the eye. The colours need to be matched to the client's personality and characteristics and combined with the right application techniques, product and colour selection. The colours have a contextual arrangement. When colours are put against one another, they work differently. For example, when yellow is placed against blue, it is intense, but against an orange – in another context – it becomes considerably less vivid.

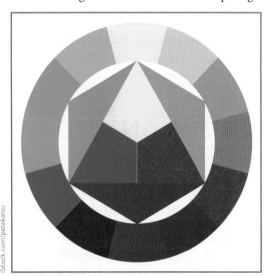

iStock.com/petekarici

Figure 14.1   The colour wheel

- *Analogous colours are two colours that are close to each other on the colour wheel.* Blue or blue-green eye shadow looks great with blue eyes, but two steps farther across, a yellow-green would not enhance the eyes so much, but create a subtle effect. Use analogous colours for understated designs and for bridal, 'everyday' make-up.
- *Complementary colours are opposite each other on the colour wheel.* The use of colours that are opposite each other creates a harmonious contrast, which is appealing in a make-up context. A client with green eyes may select an eye shadow with a purple base. Complementary colours are also used in make-up concealing products. To conceal red tones in the skin, use green to balance.
- *Colours found together in nature.* Application of colours from nature is naturally pleasing to the eye. Imagine using colours that imitate a tree in spring, then in autumn.

WARM VERSUS COOL COLOUR CONTEXT

Make-up colour schemes are broadly described as 'cool', 'warm' or 'neutral', based on colour selection. Match the make-up with the 'temperature' of the natural colouring of the skin, hair and eyes. A person with medium to dark colouring tends to require warmer colours for harmony; a person with fairer, lighter features requires cooler colours. The client with medium-toned features requires a neutral colour selection, which can include both cool and warm colours.

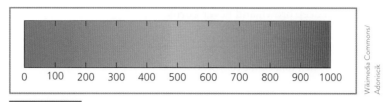

Figure 14.2 Cool to warm mired scale

- Warm colours include dark reds, 'earthy' tones, khaki, plum, burgundy, medium to dark brown.
- Cold colours include bright reds, yellow, coral, pastels, light pink, light blue, light green.
- Neutral colours include teal, navy and mauve, and tints and shades of green, blue and pink.

GREYSCALE

White contains all colours and black contains no colour. The use of tint, tone and shade with colours is the application of black and white to the colour. This technique is used when applying eye shadow to the distinct areas of the eyelids:

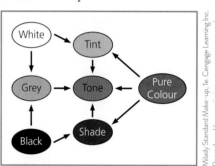

- Tint is the application of white to a colour to make it lighter.
- Tone is the application of grey (both tint and shade) to a colour to reduce its saturation and brilliance.
- Shade is the application of black to a colour to make it darker.
  We apply all three, a 'tint, tone and a shade', when applying eyeshadow.

Figure 14.3 Tint/tone/shade chart

COLOURS IN COSMETICS

3 Which colour will counteract dark purple shades under the eyes? Is that an analogous or a complementary colour?

Knowledge Evidence Performance Evidence

_____

4 Which colour eyeshadow would you choose to complement the eyes – a complementary or an analogous colour?

_____

5 Why are colours found in nature popular in organic ranges?

_____

6 Suggest eye, lip and blush colours for the following client types and state whether you are using a cool or warm palette.

a Medium-tones, hazel (green-brown) eyes

_____

_____

b Fair hair, blue eyes

_____

_____

7 The greyscale is the tint, tone and shade in cosmetic colours. Does shade recede a feature or bring it forward?

_____

8 For the examples given, refer to Figure 22.6 in Chapter 22 to identify the Fitzpatrick skin type for each.

a A client whose ethnicity is Indian

_____

b A client whose skin colour is deep brown and never burns

c A client with pale skin and brown hair and eyebrows

Facial shape and corrective make-up

Knowledge
Evidence

9 Assess the client's face shape and decide the ideal product for contouring. Complete the table below by writing in your own words the corrective steps to bring aesthetic harmony to the face shape.

FEATURE	STRUCTURE ISSUE	CORRECTIVE MAKE-UP
Nose	Nose is too broad	
	Nose is too short	
	Nose is too long	
	There is a bump on the nose	
	There is a hollow along the bridge of the nose	
	Nose is crooked	
Forehead	The forehead is prominent	
	Forehead is shallow	
	Forehead is deep	
Chin	Jaw is too wide	
	Chin is double	
	Chin is prominent	
	Chin is long	
	Chin recedes	
Neck	Neck is thin	
	Neck is thick	

10 Learn to bring aesthetic symmetry to the eyes, and to draw attention to and flatter the eyes according to the client's requirements. Complete the table below by writing, in your own words, the corrective steps you would take for each eye shape.

EYE SHAPE	CORRECTIVE MAKE-UP STEPS
Dark circles	

EYE SHAPE	CORRECTIVE MAKE-UP STEPS
Wide-set eyes	
Close-set eyes	
Round eyes	
Prominent eyes	
Overhanging lids	
Deep-set eyes	
Downward-slanting eyes	
Small eyes	

EYE SHAPE	CORRECTIVE MAKE-UP STEPS
Narrow eyes	
Monolids	

11 Lips are almost never even or symmetrical. Learn to bring aesthetic symmetry to the lips and you can then enhance them according to the client's requirements. Complete the table below by writing, in your own words, the corrective steps you would take for each lip shape.

LIP TYPE	CORRECTIVE MAKE-UP STEPS
Thick lips	
Thick upper or lower lip	
Thin lips	
Small mouth	
Uneven lips	
Lines around the mouth	

Usage of make-up tools

Performance Evidence

Knowledge Evidence

SELECTION, CARE AND INFECTION CONTROL FOR MAKE-UP TOOLS AND EQUIPMENT

12 List two different applicators that can be used to apply foundation. When would you use each one?

_____

_____

_____

13 The experienced make-up artist has personal preferences for specialised brushes and tools. Your own brush kit will evolve with the products you select and as industry innovations develop. Complete the table below by naming the following basic brushes.

BRUSH	NAME	DESCRIPTION
		Powder brushes must have soft bristles to remove excess face powder or to apply specialised powders. The Kabuki brush has dense bristles, which can take previously applied product away.
		The bristles are flat, round-shaped and dense. They may be natural or synthetic. Used to apply and blend liquid or compact foundation for a smooth finish.
		The synthetic fibres bind well to liquid, which enables a flawless application of foundation. It may be used to apply powders, blush and shimmer extremely lightly because the natural hairs bind with the powder.
		To apply facial contouring products to highlight and shade areas of the face.
		A smaller version of the natural hair face powder brush. It is used to apply powder colour to the face and for blending.
		To apply brow powder or soften eyeliner.
		To apply eye shadow, blend and shade.
		To blend powder eye colours and soften harsh lines and colours.
		Most often available in synthetic, and in a range of sizes, it is for the exact placement of concealing product.
Shutterstock.com/frantic00		To remove excess make-up from the brow hair, to add colour, blend eyebrow pencil and groom the brow hair into shape.
		A fine-tipped brush used to apply liquid or powder colour to contour the eyes, creating a precise line.
Shutterstock.com/Michael Kraus		To apply and remove excess mascara and to separate the lashes.

BRUSH	NAME	DESCRIPTION
		To apply lip products and ensure a definite, balanced outline to the lips. The fine tip is for the lip line, the angled lip brush is to enable a controlled application to thin lips, and the diamond-shaped is ideal for even application to the bulk area of the lip.

14 When applying false lashes to a client with straight lashes, what should you always do and why?

15 How often would you sharpen make-up products that require sharpening?

16 Why are spatulas so important in the make-up application service?

17 How would you adequately clean tweezers, scissors and other metal equipment used during a make-up application?

18 How would you deal with a used silicone sponge after a make-up application, compared with a make-up sponge?

19 Go online to research different types of make-up boxes. Find the one that would best suit you as a make-up artist who is just starting out. Remember to take into account the functionality, the size, the look and the price, among other factors. Write down the advantages and disadvantages of one or two for working in the salon, and another one or two for working as a mobile make-up artist.

Product selection

Knowledge Evidence

Performance Evidence

20 If a foundation claims to be 'anti-ageing', what ingredient will it likely contain? What does this ingredient do to help with ageing skin?

21 Listed in the table below are some common chemicals that go into make-up products and can cause adverse reactions. Research the chemicals and fill in the gaps, giving a product that the chemical is found in, what that chemical is used for and the adverse reaction it can cause, if any. Then find three other chemicals found in products around the salon, and research what each chemical does, what products it can be found in, and whether it can cause any adverse reactions.

CHEMICAL NAME	PRODUCTS	IT'S USE OR ORIGIN	ADVERSE REACTIONS
Bismuthoxychloride	Mineral make-up	Gives a matt or shimmery appearance, white pigment	
Butyl stearate	Eye make-up, lipsticks		Comedogenic
Ethyl hexyl palmitate	Many make-up and skincare products		Is a palm oil extract (not sustainable) and an irritant
Adulterated lanolin	Skincare products, lip products	Adulterated by wool that has not been washed before extracting the lanolin	
Lanolin BP grade	Skincare products, lip products	Sheep sebum extracted from the wool	
Octinoxate	Sunscreens and lip balms		An irritant

22 For each of the following make-up products, go online and find a product that you would likely use as a make-up technician. Give the advantages and disadvantages of your chosen product.

PRODUCT	BRAND/NAME OF PRODUCT	ADVANTAGES	DISADVANTAGES
Blushes			
Cleansers			
Concealers			
Eye shadows			
Eyeliners			
False lashes			
Liquid and solid foundations			
Lip gloss			
Highlighters			
Lip liners			
Lipsticks			
Mascaras			
Pencils			
Powders			
Pre make-up stabilisers or primers			
Skincare			

Health and hygiene regulatory requirements

Knowledge Evidence

23 A client insists they bring in their dog, which is not a seeing eye dog, into the salon. They are not happy with you when you refuse the dog entry. What do you do?

24 If the client mentioned above were to complain about the incident, what is important for you or your manager to do, both for the salon's records and for legal and insurance reasons?

25 Go to the website of your state/territory or local government work health and safety guidelines and write a half-page report on any health and hygiene regulatory requirements relevant to make-up services. Find out whether you are required to:

- wear an apron
- wash your hands in between each client
- sanitise your brushes, and how
- cleanse the client's skin.

LO14.3 PROVIDING MAKE-UP SERVICES

Performance Evidence

1 What are the four things that should be remembered to ensure you make a good impression as a professional make-up artist?

Applying make-up

Knowledge Evidence

2 Though you may have a general sequence that you use to apply make-up for a certain occasion, what are two reasons you may have to alter your normal technique?

3 The following make-up looks differ from each other in various ways. For each one, list two features of the make-up look which distinguish it from the others.

a Day make-up

b Evening make-up

c Special occasion make-up

d Make-up for mature skin

4 The following article gives six ways to create a winged eyeliner effect: **https://www.stylecraze.com/articles/create-perfect-winged-eyeliner-spoon.** If you are starting out, you might need some tricks to help you create the perfect winged eyeliner; however, some of these tricks will affect your professional image. Explain this statement.

Client feedback

5 What is one way you could seek feedback from the client to know if they are happy with the make-up application?

LO14.4 APPLYING FALSE EYELASHES

Performance
Evidence

1 What are the three different types of temporary false lashes?

2 How would you explain to a client the difference between false eyelashes and eyelash extensions? Refer to Chapter 13 Eyelash extensions to assist.

3 What would be the ideal way to patch test for eyelash adhesive?

4 Explain why the following factors matter when choosing false eyelashes for a client.

a The client's age

b Sparse lashes

c The natural eyelash colour

5 Next to each image in the table below, write, in your own words, the corrective steps you would take when applying artificial lashes for each eye shape.

EYE SHAPE	CORRECTIVE STEPS WHEN APPLYING ARTIFICIAL LASHES
Small eyes	
Close-set eyes	
Wide-set eyes	
Downward-slanting eyes	
Round eyes	
Deep-set eyes	
Overhanging lids	

6 Why would a salon never recurl strip lashes to be reused?

LO14.5 PROVIDING POST-SERVICE ADVICE

Aftercare advice

1 List three tips you would provide to the client on maintaining the make-up application.

Knowledge Evidence

2 What advice would you give a client about removing make-up with a cleanser?

3 It is important to know the products you are using so that you have the potential to make a retail sale if the client is impressed. Where are two places that you could learn about a product?

LO14.6 CLEANING THE SERVICE AREA

Knowledge Evidence

1 When restocking equipment and products between clients, it is important to know what needs to be restocked and when, to ensure a smooth transition. List five equipment items and five products you would need to restock at the end of a make-up treatment.

2 List five ways you could make your make-up service more sustainable.

CHAPTER 15: REMEDIAL CAMOUFLAGE MAKE-UP

LEARNING OBJECTIVES

After completing this chapter, you should be able to:

LO15.1 establish client make-up requirements, identify skin type and conditions and match to camouflage products

LO15.2 design a make-up plan based on consultation and skin analysis, and select products and equipment based on client requirements

LO15.3 apply remedial camouflage make-up to a range of skin conditions according to the treatment plan, seek client feedback and adjust application as required

LO15.4 review the camouflage make-up service by discussing outcomes, recommending products, demonstrating techniques and providing homecare advice

LO15.5 clean the treatment area for the next service.

INTRODUCTION

Remedial camouflage make-up is a technique using specialised make-up to conceal skin irregularities and blend the area with the surrounding skin. Camouflage make-up is specifically designed to be both natural-looking and hard-wearing for everyday use. Camouflage products are a unique formulation for fragile skin, and they have better coverage and more durability than other make-up.

The learning activities in this chapter allow you to demonstrate your knowledge and understanding of the unit of competency *SHBBMUP004 Design and apply remedial camouflage make-up*, using the performance and knowledge evidences.

LO15.1 ESTABLISHING REMEDIAL CAMOUFLAGE MAKE-UP REQUIREMENTS

Performance Evidence

1 There are many reasons a client may need or desire remedial camouflage make-up. List three of these.

_____

_____

_____

Areas and conditions requiring camouflage make-up techniques

2 What are the four groups of conditions that may be camouflaged? Give an example of each one.

_____

_____

_____

_____

DIFFERENCES IN SKIN APPEARANCE THAT MAY REQUIRE REMEDIAL CAMOUFLAGE MAKE-UP

Knowledge Evidence

3   To identify when remedial camouflage make-up is required, and what procedures and products to use, the make-up artist must first identify the different conditions that a client may present with. Complete the following table with the features that would help you recognise each condition. You may need to also refer to Appendix A Contraindications and restrictions.

CONDITION	IDENTIFYING FEATURES
Chloasma	
Hypertrophic and keloid scars	
Leucoderma and vitiligo	
Moles or pigmented naevi (i.e. not cancerous)	
Psoriasis	
Rosacea	

4   Which of the following cannot be classified a birthmark?

- Strawberry naevus
- Port wine stain
- Mole
- Scar tissue

5   Research the difference between a capillary naevus, strawberry naevus and spider naevus and answer 'no' or 'yes' to complete the following table.

	CAPILLARY NAEVUS	STRAWBERRY NAEVUS	SPIDER NAEVUS
Present at birth			
Develops shortly after birth			
Over-proliferation of cells that line blood vessels			
Malformed blood vessels			
Flat			
Deeper vessels involved (red overlies a bluish hue)			
Capillaries broken or stretched, radiating from a single distended venule in the centre			

6   Can all tattoo colours be camouflaged?

_____

7   Why is it important to know what the skin type of the client is before planning their cosmetic camouflage? Use a skin type to describe why this is important.

_____

_____

_____

Client requirements and concerns

Performance Evidence   Knowledge Evidence

8   What is the golden ratio? How does it relate to facial shapes and make-up application?

_____

_____

_____

_____

9   Where the client is wearing their remedial camouflage make-up will change the application because there will be differences in the lighting. What would you do in the following lighting situations and why? Fill in the table.

LIGHTING SITUATION	REASONING FOR MODIFICATION	ACTION
Outdoors, overcast		
Poor lighting		
Low natural light in the treatment area		

Contraindications to make-up services

Knowledge
Evidence

10 Complete the following table of contraindications that restrict or prevent a remedial camouflage make-up treatment. Review each image or description and provide the condition's name, its relevance to the remedial make-up treatment and the action you would take.

SYMPTOM	CONTRAINDICATION	RELEVANCE TO REMEDIAL CAMOUFLAGE MAKE-UP	ACTION
Shutterstock.com/VLADGRIN			
Dreamstime.com/Dmitry Bryndin			

11 List three bacterial, fungal, parasitic and viral infections, and three abnormal skin or skin disorders relevant to cosmetic camouflage.

_____

_____

_____

_____

_____

_____

12 A client presents with what seems like an undiagnosed tumour on the area they would like to have the remedial make-up treatment applied. It can be easy to accidentally state what you suspect you see to the client; however, this is not within the scope of your practice. Write a script of what you might say in such a situation, so that you are prepared.

_____

_____

_____

RECORD MANAGEMENT

13 Fill out a client record card for cosmetic camouflage make-up for the following client: she has a strawberry birthmark on her forehead, dry skin, and skin on the lesion that looks fragile in parts but is not broken.

- Sex:

_____

- Contraindications and restrictions:

_____

- Skin analysis:

_____

- Products and techniques used:

_____

_____

- Aftercare:

_____

LO15.2 DESIGNING THE MAKE-UP PLAN

Colour analysis and design

Refer to the section 'Colour analysis and design' in Workbook Chapter 14 and answer the following questions.

1   Give example colours for each of the following:

   a   primary colours

_____

   b   secondary colours

_____

   c   tertiary colours

_____

2   Define the following:

   a   tint

_____

   b   tone

_____

   c   shade

_____

3   Using the camouflage of a deep green tattoo as an example, how would you apply the following colour theories to your make-up design?

   a   Greyscale

_____

_____

   b Knowledge of the colour wheel (i.e. analogous and complementary colours)

4 It is important to know how the make-up products will look under different light sources. What is a rule of thumb to use if the client will be photographed with a flash?

Products and application techniques

Knowledge Evidence

5 There are different ways that specific make-up products can be applied in order to conceal the condition the client has presented with. Using your knowledge of highlighting, shading and contouring, go online and find four images of conditions that could have remedial camouflage make-up applied to them. Stick them on poster paper, and draw lines to the areas you would focus on, describing the techniques you would use.

6 If a product range claims the following, what would you expect to see:

   a the range has a full set of colours

   b the range has products to camouflage hypotrophic and hypertrophic scarring and growths

INGREDIENTS AND PIGMENTS

Knowledge Evidence

7 Some ingredients and pigments will have different effects on the skin when applied. For example, some ingredients have reflective qualities, so are not useful when the client will be photographed with flash photography. Research the three chemical ingredients or pigments in the table below and note down the results of your research, including the products they are used in, their effect on the skin, and alternative products that could be used.

INGREDIENT	PRODUCT IT IS USED IN	EFFECT ON THE SKIN	ALTERNATIVE OR ADDITIONAL PRODUCTS (IF ANY)
Zinc oxide			
Mica			
Rubber			

Complying with health and hygiene regulations and requirements

Knowledge Evidence

8 How does the privacy legislation relate to the remedial camouflage make-up treatment?

9 If your salon does not require you to wear gloves, what are some general rules you should follow?

10 Go online and research the relevant state/territory and local health and hygiene regulations and requirements for the remedial camouflage make-up service. Write a half-page report detailing items specific to your local area.

11 What sort of personal protective equipment is needed for the therapist when applying remedial camouflage make-up? Why is no more needed?

12 Imagine you had a client with a severe telangiectasia, and you didn't want to treat the skin but the client insisted. It looked OK, so you proceeded to complete the client record card and obtained signed consent. When stippling over the blood vessel, it bled and the client became very distressed. What are the implications and what can you do? Consider:

a client record card

b incident reporting

c insurance and liabilities

LO15.3 APPLYING REMEDIAL CAMOUFLAGE MAKE-UP

Performance
Evidence

1 Why is it important to drape the camouflage make-up client?

2 What would guide your decisions about the initial tones to use to camouflage the following skin irregularities:

a a deep red mark

b a dark brown mark

c a lighter mark

3   True or false: The airbrush only needs one pass over an area to have it sufficiently covered.

Client feedback

4   What two steps should you follow when obtaining feedback from the client?

5   If the client is dissatisfied with the make-up application, how could you use the opportunity to show your skills as a make-up technician?

LO15.4 REVIEWING THE SERVICE

1   When demonstrating techniques, you should advise the client that the make-up will hold better if the skin is in good condition. How does the client maintain their skin in a good condition?

2   During the demonstration, the client may tell you that they don't think they could replicate this make-up on their own. How could you respond?

3   How often would you recommend scheduling future appointments for a review of the make-up application technique?

Product recommendations

4   If the client is to be receiving regular cosmetic camouflage treatments, what products might you recommend in relation to:

a   cleansing

b   moisturising

Aftercare advice

Knowledge
Evidence

5 It is always a good idea to recommend the products that you have used to the client for home use. List five pieces of advice that you could give the client when discussing the best use of the product out of the salon.

_____

_____

_____

_____

_____

LO15.5 CLEANING THE TREATMENT AREA

Knowledge
Evidence

1 How would you clean and dry brushes between clients, and at the end of each day?

_____

_____

2 List five things that could be done in a more sustainable way throughout the remedial camouflage make-up treatment.

_____

_____

_____

_____

_____

3 Consider the following disposable items used in the remedial camouflage make-up service, and describe how each one should be disposed of correctly.

ITEM	METHOD OF DISPOSAL
Make-up wedges	
Mascara wands	
Blood-stained tissues	

CHAPTER 16: PHOTOGRAPHIC MAKE-UP

After completing this chapter, you should be able to:

LO16.1   analyse the context of the shoot and work with the creative team in a range of photographic formats
LO16.2   establish make-up requirements by cleansing the skin to determine contraindications, skin characteristics and facial features
LO16.3   design the make-up according to the design brief or preshoot
LO16.4   apply photographic make-up according to the design brief and photographic make-up plan
LO16.5   clean the service area in preparation for the next treatment.

INTRODUCTION

Photographic make-up is for beauty therapists and make-up artists working in a creative sphere. The artist can work in any location where make-up services are needed: in a beauty salon, as part of a creative team at a fashion show or film set, or in a make-up studio. The make-up artist is expected to make collaborative, informed and creative decisions about which make-up products and application techniques are appropriate for the photographic situation.

The learning activities in this chapter allow you to demonstrate your knowledge and understanding of the unit of competency *SHBBMUP003 Design and apply make-up for photography*, using the performance and knowledge evidences.

LO16.1 ANALYSING PHOTOGRAPHY CONTEXT

Performance Evidence

1   For the following roles in a photo shoot, define what they would be responsible for:

a   photographer

b   make-up artist

c   stylist

d   model

e   art department

2   When you check the brief, and consult with the client, which one of these is not necessary?

A   You understand team roles for additional work.

B   You understand the model's budget and time constraints.

C   The use of resources conforms to the design plan.

D   You have factored costs for product and your time appropriately.

Production environments

Knowledge
Evidence

3 Depending on the photography context, the lighting, and so the make-up needs, will change. Fill in the table below to show your understanding of lighting and its effects for each context.

CONTEXT	LIGHTING
Interior	
Exterior	
Studio	

Make-up services for print outcomes

Performance
Evidence

Knowledge
Evidence

PHOTOGRAPHIC MAKE-UP PRINCIPLES

4 What is meant by a monochromatic shade range?

5 Which colours are best to use in black-and-white images to show up as black? Which ones are best if a natural look is required?

6 In digital colour photographs, what are the effects of the high resolution image?

7 Describe how you would appropriately consider the following lighting conditions by describing the light and the effect on make-up:

a a dawn photo shoot in hot weather

b a midday photo shoot in overcast weather

The most important cosmetic ingredients that have an effect in photographic make-up are:

- reflective SPF sunscreens
- mineral make-up with titanium dioxide (reflective SPF).

8 Explain the effects of each of these products on the photographic make-up.

LO16.2 ESTABLISHING MAKE-UP REQUIREMENTS

1 Explain what is meant by a test shoot and a paid shoot.

Performance
Evidence

Contraindications and scope of practice

2 If you suspect the client has herpes simplex (cold sores), what would your course of action be? Refer to Figure 14.1 in Chapter 14 to help answer this question.

Knowledge
Evidence

APPEARANCE OF COMMON SKIN TYPES AND CONDITIONS

3 Draw a line to match the following contraindications with their relationship to photographic make-up.

RELATIONSHIP TO PHOTOGRAPHIC MAKE-UP
Microbes cause contagious skin and eye diseases.
Skin that can look or feel different, for example it may be swollen, red or itchy.
If make-up product collects behind it, it is difficult to remove, and serious eye damage can occur.
Treatments that make changes to the skin, affecting the photographic make-up design plan.

CONTRAINDICATION
Contact lenses
Recent skin treatments such as injectables, intense pulsed light, laser and surgery
Bacterial, fungal, parasitic and viral infections
Abnormal skin

4 Which of the following describes oily skin?

A Shiny appearance, finger slips over the surface

B Matt appearance, finger glides over the surface

C Shiny appearance, finger drags over the surface

5   List at least four ways you can adjust the photographic make-up design for a client with oily skin.

_____

_____

_____

_____

6   Which of the following describes dry skin?

   A   Flaky skin, finger drags over the surface

   B   Open pores, finger drags over the surface

   C   Flaky skin, finger glides over the surface

_____

7   List at least four ways you can adjust the photographic make-up design for a client with dry skin.

_____

_____

_____

_____

Facial shape

Refer to Chapter 14 Design and apply make-up for the following questions.

8   What are the seven facial shape categories for make-up application?

_____

9   Working with a classmate, determine the following:

   •   What is their face shape?

_____

   •   What features do they have that helped you determine that?

_____

   •   What corrective make-up steps would you take for them?

_____

_____

LO16.3 DESIGNING THE MAKE-UP PLAN

Performance
Evidence

1   In a client record card for photographic make-up design, where do you write the following details:

   a   The photo shoot is to be held in a studio as advertising for a hair product.

_____

   b   The client needs the photographs to be black and white.

_____

2 Where on the design sheet do you record areas requiring correction or camouflage?

3 If the photographer is doing a portrait with the light coming in from one side of the face, is it necessary to complete the make-up on the other side with as much detail? Choose your answer from the following options.

  A No, only one side needs to be considered.

  B Yes, but not with as much detail.

  C Make-up needs to be balanced because you never know what will happen in the shoot and they might want to shoot from a different angle.

4 Where do you get the colour design concept from?

5 Suppose the brief contained dramatic considerations – they want the make-up dark and moody. Which of the following does that mean?

  A They are not using as much lighting; use lighter make-up.

  B They are not using as much lighting; go stronger with make-up.

  C It is light and bright; use lighter make-up.

6 You are doing a portrait photographic make-up and there is a lighting angle change. What are some adjustments you need to make for the following three issues?

  a Tan lines are showing

  b The client's redness in the nostril became more obvious

  c A strong jawline in the model's face shape became more obvious

7 Suggest some possible continuity issues for when you are working on a film, and what can you do to avoid them.

Effect of changes

8 Powder is used more often for photographic make-up because:

  A shine directs a lot of lighting on the face.

  B it helps the make-up to last all day.

9 True or false: Does a bronzer with an ashy undertone create a shadow?

10 Is outside lighting generally cool or warm?

11 You should always follow the client's brief when deciding how to interpret a theme for the day. Come up with some colour tones for products to approach photographic make-up with the following themes.

a   Christmas theme

_____

b   Winter day theme

_____

c   Beach theme

_____

d   Dark and moody theme

_____

LO16.4 APPLYING MAKE-UP FOR PHOTOGRAPHY

Performance Evidence

1   Distinguish the slight differences between the different types of make-up by collecting some images of models in the following contexts. Write two sentences to describe the features of each look.

CONTEXT	FEATURES (1 OR 2 SENTENCES)
Business	
Wedding	
Fashion	
Commercial	
Catwalk	

Make-up applications for different conditions

Performance Evidence

2   True or false: You need to consult with the photographer in regards to lighting.

_____

3   True or false: Make-up needs to be stronger for natural light.

_____

4 True or false: Backdrop colours need to be considered in the studio.

5 If you were doing a dewy, glossy look on the eyelid, would this be the first or the last thing you do?

6 Justify your answer to question 5 above, and state whether you would need to maintain the look throughout the photo shoot.

Organisational policies relevant to make-up services

Knowledge Evidence

7 What is one way you can present yourself as organised and ready when arriving to a photo shoot?

8 If your photo shoot is on location, what is the best practice for dealing with waste disposal?

9 Who might you report to if an incident occurred on location? Who might liable if you don't report the incident?

10 List at least three items that you should advise the client/model about if it is their first time on a shoot.

LO16.5 CLEANING TOOLS AND EQUIPMENT

Knowledge Evidence

1 While on a photo shoot, is it better to use disposable or reusable equipment? Why?

2 What is the correct way to remove all cosmetic products from their containers? Why?

Sustainability

Knowledge Evidence

3 True or false: You can reuse metal spatulas and palettes as a more sustainable way of preventing unnecessary waste.

4   When thinking about the sustainable use, conservation and disposal of products, it's important to know what's in them. Follow the link to answer the questions below regarding cosmetics and chemicals: **https://www.science. org.au/curious/people-medicine/chemistry-cosmetics**

    a   Use the interactive to figure out what the average amount of chemicals are in the following products:

PRODUCTS	AVERAGE AMOUNT OF TOTAL CHEMICALS
Eye liner	
Sunscreen	
Lipstick	
Foundation	

    b   Think about the products you use daily, then work with the interactive to figure out how many chemicals on average you use in your typical day.

SECTION 6

BODY TREATMENTS AND SERVICES

17 Cosmetic tanning

18 Provide body massages

19 Aromatherapy massages

20 Aromatic plant oil blends

CHAPTER 17: COSMETIC TANNING

After completing this chapter, you should be able to:

LO17.1 establish client priorities with thorough consultation, skin analysis and patch testing

LO17.2 select and prepare tanning products and equipment, and prepare yourself, the treatment area and the client for comfort and according to work health and safety (WHS) legislative requirements

LO17.3 use spray tanning equipment to apply cosmetic tan according to salon protocol and the treatment plan

LO17.4 review the service, providing aftercare and homecare recommendations and a future treatment plan

LO17.5 clean the treatment area in preparation for the next client.

INTRODUCTION

Cosmetic tanning products have been in Australia and New Zealand since 1960 when Coppertone® released its first cosmetic tanning product QT®, or 'Quick Tanning Lotion'. Cosmetic tanning is a temporary tan that gives the skin a healthy, bronzed appearance without the harmful effects of UV radiation.

The learning activities in this chapter allow you to demonstrate your knowledge and understanding of the unit of competency *SHBBBOS007 Apply cosmetic tanning products*, using the performance and knowledge evidences.

LO17.1 ESTABLISHING CLIENT PRIORITIES

Performance Evidence

Knowledge Evidence

Contraindications and scope of practice

1   Use the image or description provided in the table below to identify the contraindication, explain its relevance to the cosmetic tanning service, and give the action you would take if a client had it. Remember to include when you would refer clients to a medical practitioner.

SYMPTOM	CONTRAINDICATION	RELEVANCE TO COSMETIC TANNING	ACTION
Alamy Stock Photo/WENN Ltd			

SYMPTOM	CONTRAINDICATION	RELEVANCE TO COSMETIC TANNING	ACTION
iStock.com/petekarici			
Client says they are short of breath and have very mild asthma			
Alamy Stock Photo/Hercules Robinson			

PIGMENTATION DISORDERS

Knowledge Evidence

2 What are the two types of pigmentation disorders that affect the cosmetic tanning application? How do they affect it? Where would you put this information on the client card?

_____

_____

_____

Recognising skin reactions to patch tests

Performance Evidence

3 Patch testing for product allergies is very important. Refer to Appendix A Contraindications and restrictions and identify which of the following should be considered a positive (i.e. bad) reaction to a patch test? (Hint: You may select more than one.)

A Streaking

B Patchiness

C Anaphylaxis

D Erythema, rash and swelling

E No reaction

Client record card

Knowledge
Evidence

4 Fill out a client record card for body treatments for a male client. He is a model and cannot have tan lines. He has had cosmetic tanning before, and had a slight redness from the product, so is trying your salon because you use an organic line of products. He is in his mid-20s, has naturally pale, dry skin.

• Age bracket:

_____

• Gender:

_____

• Outcomes of previous treatments:

_____

• Skin analysis:

_____

• Patch test:

_____

_____

• Modifications to treatment:

_____

_____

_____

LO17.2 PREPARING TO APPLY COSMETIC TANNING PRODUCTS

Knowledge
Evidence

1 Go online to research the laws in your state or territory around who is considered a minor, or underage, in relation to cosmetic tanning. Write a one-page report, including the answers to the following questions:

   a What are the restrictions?

   b What must a parent or guardian consent to, and must they be present?

   c Would you prefer they stayed through a treatment with a minor? Why?

   d Who is liable if the guardian stays through the treatment, without consent?

2 Answer the following questions in relation to preparing the client for the cosmetic tanning treatment.

   a How should the client stand when receiving the treatment?

_____

   b What should they be wearing?

_____

   c How long do you leave the skin to dry?

_____

3 Is it appropriate for the therapist to greet the client wearing their full personal protective equipment (PPE) for the cosmetic tanning treatment? Why?

_____

_____

_____

Colour choice

4 What are the general rules to use when choosing how many shades darker to go with the tanning solution?

Knowledge Evidence

_____

_____

_____

5 In groups, role play a situation to suggest modifications to the treatment procedure, where the:

   a client is underweight or overweight

   b client prefers a darker tan

   c client is particularly tall or petite

   d client is over 40 years old.

6 Fill in the table below with the best tanning solution option, basing your decision on skin type alone. The first row has been done for you.

FITZPATRICK LIGHT/ DARK	NORMAL	COOL (OLIVE, ASHEN)	WARM (PINK, YELLOW, ORANGE)
Type I	e.g. Caramel, olive 8% DHA	e.g. Caramel, olive 8–10% DHA	e.g. Caramel, violet 8–10% DHA
Type II			
Type III			
Type IV			
Type V			
Type VI			

7   Research at least two professional cosmetic tanning products used in the industry, one of which must be certified as organic. The product companies don't often tell you any disadvantages, so you will have to find other sources; for example:

- online blogs

- reviews

- industry magazines

- newspapers

- industry mentors

- salons.

Complete the table below by identifying the advantages and disadvantages of the brands.

COSMETIC TAN BRAND	ADVANTAGES	DISADVANTAGES

Procedures followed and products used

Knowledge Evidence

8   Although there is a range of different tanning products, they all act in a similar way to create a tanned look. Describe the reaction that takes place to create the tanned look.

_____

_____

_____

9 The two main active ingredients in most tanning products used in the salon are *dihydroxyacetone* (DHA), 1–16%, and *erythrulose*, 1–3%. In the table below, list everything you know about each one, including where it comes from, the tone it creates, and so on.

DIHYDROXYACETONE	ERYTHRULOSE

10 List four other additives that can be found in tanning products in the salon, and what their functions are.

_____

_____

_____

_____

_____

_____

_____

_____

_____

Health and hygiene

Knowledge Evidence   Performance Evidence

11 Why do the therapists always wear black and an apron?

_____

12 Why should you always have your hair tied back and wear minimal jewellery when performing a tanning treatment?

_____

13 Where are the following items disposed of?

a Disposable underwear

_____

b Disposable cap

_____

c Cottonwool

_____

14 If a client has an anaphylactic reaction to the tanning product, what should you do first? To whom do you report it?

_____

PPE

15 What are the PPE items that should be worn by you and the client during the cosmetic tanning treatment?

_____

_____

_____

_____

_____

LO17.3 APPLYING PRODUCT WITH THE SPRAY GUN

Exfoliation and preparation

1 The most popular method of exfoliation prior to the tanning service is the exfoliation mitt or the more luxurious dry brush method, because softening the skin immediately before treatment could interfere with the tan application. List the steps of the dry brush method.

_____

_____

_____

_____

_____

_____

_____

_____

_____

_____

_____

_____

Spray gun settings

Knowledge
Evidence

2 What is a pressure gauge?

_____

3 What is PSI, and how does it relate to spray gun settings?

_____

_____

_____

4 The following table lists issues that might occur with a spray gun when applying cosmetic tan. Complete the table by providing possible solutions for each issue.

ISSUE	POSSIBLE SOLUTION
Too much spray	
Leaking fluid cup	
Pressure low/no spray	

Even colour

Knowledge
Evidence

5 For an even colour, tanning enhancers are used. Tanning correctors are also useful to ensure the colour stays even between treatments. Describe how and when to use the tanning corrector.

_____

_____

6 How can you monitor your technique during application to ensure an even colour?

_____

_____

_____

_____

_____

_____

LO17.4 REVIEWING THE SERVICE

Performance
Evidence

Service outcomes

1 When reviewing the service, there might be some results that are not as the client expected. Explain why the following results may have occurred.

a Too light

b Patchy areas

c Green colour

2 If a client shows results that are unsatisfactory, what are the three things that you should make sure to record on their client record card?

Aftercare advice

Performance
Evidence   Knowledge
Evidence

3 When can the client bathe or shower after a tanning treatment?

4 Does the client need to wear sunscreen after treatments? Explain your answer.

5 Provide a description for each of the products listed below. Then visit a local salon that provides cosmetic tanning services, and write down the brand and name of the product it sells that falls into that category.

PRODUCT CATEGORY	DESCRIPTION	BRAND AND NAME OF PRODUCT
Exfoliator		
Exfoliating mitt		

PRODUCT CATEGORY	DESCRIPTION	BRAND AND NAME OF PRODUCT
Tan application mitt		
Tan extender		
Tan corrector		
Bronzer		
Full cosmetic tan mousse/gel/lotion/cream/spray/wipes		

6   How long would you advise the client to leave it before their next tanning appointment? Why?

_____

_____

_____

_____

Adverse reactions

Knowledge Evidence

7   Complete the following table by describing the appearance of each adverse reaction, and the action you would take if a client had that reaction during a cosmetic tanning treatment.

ADVERSE EFFECT	APPEARANCE	ACTION
Coughing		

ADVERSE EFFECT	APPEARANCE	ACTION
Watery eyes		
Fainting		

Follow-up advice

Knowledge Evidence

8　It is important that the client prepares themselves as best they can before the cosmetic tanning service. List five pieces of advice you could give the client before their appointment.

_____

_____

_____

_____

_____

ENHANCERS

9　Tanning enhancers bring out the best in the tan. They can extend the length of the tan, make it look more natural and/or add shimmer. Complete the following table to describe the ingredients, benefits and use of tanning enhancers as you would explain them to the client.

Knowledge Evidence

TANNING ENHANCER	USE	BENEFITS AND INGREDIENTS
Extender		

TANNING ENHANCER	USE	BENEFITS AND INGREDIENTS
Bronzer		
Full cosmetic tan		

Future treatment plan

10 How regularly should you schedule appointments for a client? What constraints might affect this?

LO17.5 CLEANING THE TREATMENT AREA

Knowledge
Evidence

Write notes on the steps the manufacturer has advised you to take to maintain the spray gun. There are specific instructions to ensure the pipes or nozzle don't get blocked, spray irregularly or leak. The parts are expensive to replace and can cost the salon a lot of money.

1 How often is the spray gun to be cleaned?

2 Describe the cleaning and reassembly procedure for your spray gun.

Associated risks

3 List some of the risks associated with a malfunction of the compressor.

Knowledge
Evidence

4 What risk is associated with overspray of product?

_____

5 If the salon and tanning equipment are not cleaned appropriately, what risks does that pose to both therapists and clients?

_____

_____

Sustainability

Knowledge
Evidence

6 When choosing a sustainable product to use in cosmetic tanning, what sort of things would you be looking for?

_____

7 Review the Chapter 7 text regarding disposal of items in the salon, and go online to check the regulations on disposal of waste from the salon. List everything you can find that would relate to the cosmetic tanning service.

8 List 10 different ways you could make the cosmetic tanning treatment more sustainable. Think about how water, energy and electricity are used, along with the disposal of single-use items and the cleaning of reusable items.

_____

_____

_____

_____

_____

_____

_____

_____

_____

_____

CHAPTER 18: PROVIDE BODY MASSAGES

After completing this chapter, you should be able to:

LO18.1   establish client priorities with a thorough consultation and body analysis

LO18.2   design the massage treatment according to the client's requirements, considering duration, frequency and cost

LO18.3   prepare for a body massage according to salon protocol and work health and safety (WHS) legislative requirements

LO18.4   provide a body massage treatment to the client's requirements, adjusting the massage movements, routine and massage medium to suit

LO18.5   review the massage, plan and schedule future treatments and provide body care homecare advice

LO18.6   clean the treatment area in preparation for the next client.

INTRODUCTION

The body massage typically used in the beauty salon is commonly called 'Swedish' massage, and is a medium-pressured massage that works the body tissues *with* the flow of the blood circulation against the resistance of the underlying bones and muscles. Body massages in beauty salon and spa treatments promote relaxation and wellbeing, and improve the condition of the skin, body and mind.

The learning activities in this chapter allow you to demonstrate your knowledge and understanding of the unit of competency *SHBBBOS008 Provide body massages*, using the performance and knowledge evidences.

LO18.1 ESTABLISHING CLIENT PRIORITIES

Performance Evidence

Knowledge Evidence

Effects and benefits of body massage on anatomy and physiology of body systems

Here you will learn the effects and benefits of massage and understand the limitations of the Swedish massage techniques in beauty therapy.

Figure 18.1   Effects and benefits of massage on the body systems

EFFECTS	BENEFITS
ARTICULAR AND SKELETAL	
Massage improves joint mobility due to warmth in the area.	Helps to soothe and prevent symptoms of osteoarthritis.
Swedish massage can reduce pressure on muscles attached to stressed bones.	Help to prevent shin splints (fasciitis).
CIRCULATORY AND LYMPHATIC	
The blood flow in the arteries is increased by effleurage and petrissage manipulations, and the vessels, by reflex, dilate (vasodilation), which results in an erythema reaction to the skin. Massage improves the gaseous exchange between the capillaries, the interstitial fluid and the body cells.	The arteries and arterioles aid the transportation of oxygen and nutrients to the cells. Venous return eliminates waste products, toxins and carbon dioxide from the cells, which gives the skin good colour and tone.

EFFECTS	BENEFITS
They lymph vessels on the surface of the skin deliver excess fluid and waste products, larger particles in particular, to be further processed and delivered to the venous circulation. The lymphatic system is also our immune system.	Very light pressured massage helps to move the lymph to improve skin function and appearance as well as enhance overall body wellness.
ENDOCRINE	
Massage stimulates the parasympathetic nervous system, which can have a positive effect on hormonal imbalances in the body. Be sure to have medical approval before using massage to aid in relaxation when a person has an endocrine disorder.	Symptoms from conditions that are affected by irregular glandular secretions such as polycystic ovarian syndrome, hyperthyroidism or menopause can be eased through massage. Endocrine disorders are often exacerbated by stress, which occurs as a hyperactivity of the sympathetic nervous system.
INTEGUMENTARY	
The vasodilation action creates warmth in the skin, increasing nutrients and oxygen, the action of fibroblasts and the output of the sudoriferous glands. Fibroblasts produce collagen, elastin and the gel matrix in the dermis. Eccrine glands help to remove the waste products and urea from the body more efficiently.	Massage improves every function of the skin. Stimulation of the sebaceous glands by general massage manipulations produces more sebum, which helps to soften the skin and make it more resistant to infection due to its antibacterial property. Frictions used in the massage routine will increase activity in the stratum germanitivum and aid desquamation of the dead skin cells, leaving the skin smoother and softer. Sensory nerve endings can either be soothed or irritated depending on the massage manipulation used; for example, effleurage or tapotement (percussion). The effect on subcutaneous tissue is primarily on the heating and mobilisation of the fat.
MUSCULAR	
The increased blood circulation feeds the muscle tissue, bringing extra oxygen and nutrients and aiding the removal of waste products in the venous return. The increased production of heat created by vasodilation produces a warming effect and the skin's surface temperature is raised.	Muscles function better, improving tone, elasticity and extensibility. With various techniques, muscles that are tense and shortened can be relaxed and stretched; weakened muscles can increase in tone.
NERVOUS	
The peripheral nerve endings can be either soothed or stimulated, depending on the massage movements performed.	Vigorous manipulations can have a stimulating effect – tissues and organs can be influenced to work more efficiently. Slow rhythmic manipulations can induce mind relaxation and sleep. Massage can also have a soothing effect on the nerves when effleurage and vibrations are performed, causing temporary pain relief. The parasympathetic division of the autonomic nervous system is stimulated, which has a calming effect on the body.
RESPIRATORY	
Percussion movements have the most direct effect on the lung tissue.	Massage to the muscles of respiration, the intercostal muscles, aids in respiration and can be good for clients with asthma. It can deepen the breath by inducing mind relaxation through stimulating the parasympathetic nervous system.

1   Refer to Figure 18.9 in your textbook and the above information to complete the following table to identify the benefits and effects of massage and which body system is affected (the first one has been done for you). Describe the benefit in your own words.

EFFECT	BENEFIT	PRIMARY BODY SYSTEM AFFECTED
Increases blood circulation	e.g. Nourishes and oxygenates body cells and systems. Warms the skin and muscle tissues.	Circulatory system

EFFECT	BENEFIT	PRIMARY BODY SYSTEM AFFECTED
Stimulates superficial and deep layers of the skin		
Aids desquamation – the natural shedding of dead skin cells		
Increases sebaceous secretions of the skin		
Increases lymphatic flow		
Relaxes tense and contracted muscles		
Stimulates the nerve endings		
Softens and breaks down localised fatty deposits		
Breaks up poorly cross-linked collagen fibres		

EFFECT	BENEFIT	PRIMARY BODY SYSTEM AFFECTED
Improves lymphatic and venous circulation		
Invigorates and energises, clears the mind		
Calms and relaxes the mind		
Uplifting to the psyche, temporarily gives the body a sense of wellbeing, improving confidence, and lifts depression		
Percussion movements have the most direct effect on the lung tissue		
Massage improves joint mobility due to warmth in the area		

Contraindications and conditions

Knowledge
Evidence

2   Research the following body system disorders that might restrict or prevent you from performing a massage treatment on a client.

    a   Identify the system of the body each disorder affects:

        i   endocrine system

        ii   circulatory system

        iii   blood-borne/circulatory system

        iv   muscular system

        v   nervous system

        vi   skeletal/articular system

b   Describe the complications of the disorder that might restrict or prevent you from performing a treatment.

c   Suggest a treatment modification or an appropriate referral to a medical or complementary professional that might assist the client.

You may use the content in Chapter 18 of your textbook, Appendix A Contraindications and restrictions and Appendix B Referrals to appropriate professionals as a guide.

BODY SYSTEM DISORDER	BODY SYSTEM	COMPLICATIONS THAT MIGHT RESTRICT OR PREVENT YOU FROM PERFORMING A MASSAGE TREATMENT	ACTION (REFERRAL OR MODIFICATION TO TREATMENT)
Heart diseases: e.g. high blood pressure (hypertension)			
Menopause (HRT)			
Sprains			
Strain			

BODY SYSTEM DISORDER	BODY SYSTEM	COMPLICATIONS THAT MIGHT RESTRICT OR PREVENT YOU FROM PERFORMING A MASSAGE TREATMENT	ACTION (REFERRAL OR MODIFICATION TO TREATMENT)
Low blood pressure (fainting due to vasovagal response)			
Skin trauma (risk of cross-infection – hepatitis, AIDs)			

3   Refer to Appendix A Contraindications and restrictions in your textbook and research the appearance and characteristics of the skin diseases and disorders listed below. Use images/sketches and written descriptions to complete the following tables.

CONTRAINDICATION	APPEARANCE AND CHARACTERISTICS
Bacterial, fungal, parasitic and viral infections	
Cancer: squamous cell carcinoma	

CONTRAINDICATION	APPEARANCE AND CHARACTERISTICS
Suspicious undiagnosed skin growths	
Clients under the influence of alcohol or drugs	
Symptoms of infectious disease	
Localised fractures	
Pain	

CONTRAINDICATION	APPEARANCE AND CHARACTERISTICS	RELEVANCE TO BODY MASSAGES
Allergies		Client may react with erythema, oedema, rash or anaphylaxis to products used in treatment.
Areas exhibiting loss of tactile sensation		Pressure and touch receptors sense the firmness of the massage. Pain and heat receptors sense when a product is burning or stinging. Heat receptors sense when something is too hot, e.g. hot towel.
Inflammation and swellings		Compressing the tissues will burst dilated capillaries and can cause bruising. Massage increases circulation and draws fluid to the area.
Lumps and tissue changes such as skin trauma		Lumps and tissue changes can arise for many reasons, such as: • trauma and injury • cancer (benign or malignant) • congenital or acquired • infectious • inflammatory (acute or chronic) • hormonal (e.g. glandular).
Phlebitis and deep vein thrombosis		Venous problems can lead to vein inflammation and skin ulceration. Massage could force a blood clot to move through the bloodstream, causing a blockage elsewhere, such as the heart muscle, the brain or the lungs.
Rashes		The skin is a barrier, so open skin will absorb products and cause cross-infection in the salon.
Recent scar tissue		For aesthetically appealing scar tissue, the wound needs to be left alone for an initial period. Wound healing takes 2 years to fully complete, and can range from 3 to 6 months for the wound to be safe for massage.
Severe oedema		Test the skin for fluid retention, and if it is a medical/clincal oedema, Swedish massage can make it worse.

LIMITATIONS TO BODY MASSAGE FROM SKIN DISORDERS

Knowledge Evidence

4   Refer to Appendix A Contraindications and restrictions in your textbook. List the appearance of and the limitations to body massage for the following skin disorders and diseases.

SKIN DISORDER	APPEARANCE	LIMITATION TO BODY MASSAGE
Dermatitis and eczema		
Ichthyosis		

SKIN DISORDER	APPEARANCE	LIMITATION TO BODY MASSAGE
Lupus erythematosus		
Scleroderma		
Skin tumours		
Urticaria		

5   What are at least three things you should always list on the client record card when recording outcomes of current treatment?

_____

_____

SCOPE OF PRACTICE

6   Rehabilitation and treatment for clients with conditions such as osteoporosis are not within the realm of beauty therapy. If a client presents with what you suspect is osteoporosis, they might eventually see an endocrinologist. Would you refer this client to an endocrinologist? Explain your answer.

Knowledge
Evidence

_____

_____

7   A client tells you they have a recurring issue with their right shoulder and neck area, and that sometimes a massage has helped the pain. What might you do in this situation?

_____

_____

Body analysis

8   What are the main things you are observing in a body analysis?

_____

Through a consultation, a client may tell you they are wanting massage in order to slim down. Massage alone cannot mobilise fat for slimming. Healthy nutrition and muscle movement through cardiovascular exercise is necessary to move the fat through the blood and lymph circulation.

9   List the three different types of body fat and their characteristics.

_____

_____

_____

10 What kind of fat is cellulite? Which beauty therapy treatments would you recommend for cellulite?

11 When measuring muscle tone, what does good and poor muscle tone feel like?

12 What is the average adult BMI? What is it considered if it's above that?

LO18.2 DESIGNING THE MASSAGE TREATMENT

Performance Evidence

Effects and benefits of massage movements

Knowledge Evidence

1 List the effects and benefits of the following massage movements.

a Effleurage

b Petrissage

c Frictions

d Tapotement

_____

_____

_____

_____

_____

e Vibrations

_____

_____

_____

_____

2 Explain how the following terms relate to massage technique.

a Repetition

_____

b Rhythm

_____

c Variation

_____

3 How can body massage treatment aid clients who have skeletal or articular issues?

_____

_____

4 Complete the following table by listing the massage movements recommended for each client reaction/outcome.

CLIENT REACTION/OUTCOME	VARIATION
Acne and congestion	
Oedema	

CLIENT REACTION/OUTCOME	VARIATION
Pregnancy	
Male	
Sensitive/ticklish areas	

5   Use the correct anatomy and physiology terminology to describe the following massage movements.

   a   Move your hands from the upper back down to the lower back.

   _____

   b   Move from the heel to the gluteal area with a superficial stroking effleurage.

   _____

   c   Stroke up the torso, over, and around the shoulders and up the neck.

   _____

Suitability for client needs

6   Complete the following sentences.

   a   Clients with budgetary and t_____ constraints can be recommended shorter massages that are less frequent.

   b   Determining the outcomes of previous treatments indicates what went well and what did not go well, and helps you to decide on the treatment p_____.

   c   Massage is not suited to clients with some p_____ attributes, such as excessively thin skin.

Using correct anatomy and physiology terminology

7   On the client record card, where and how would you document that the client has lateral left and right lower back pain?

   _____

   _____

Treatment duration, frequency and cost

8   How long would you spend massaging the following specific areas during a full body massage?

AREA	TIME (MINUTES)
Arms, each	
Neck/chest	
Abdomen	
Legs, each	
Buttocks	
Back	
Scalp	
Face	

9   How long is a full body massage usually scheduled for? How long for a back massage?

10 When discussing the cost of the treatment, should you remind the client that they will need to be able to afford the full treatment course over the time of the treatment course? Or would you only discuss the cost of an individual treatment?

Areas to treat and expected outcomes

11 When designing a treatment, you must take into account both the physiological and psychological effects and benefits of massage to adequately suit the client's needs. List three effects and three benefits of each below.

CLIENT NEEDS	EFFECT	BENEFIT
Physiological needs		
Psychological needs		

LO18.3 PREPARING FOR BODY MASSAGE TREATMENTS

Performance Evidence

Products, equipment and treatment area

1 For each of the following massage mediums, describe its use in the massage treatment.

PRODUCT	NAME	USE
Massage cream Alamy Stock Photo/kubala	Massage cream	
Alamy Stock Photo/Pick and Mix Images	Massage oil	
iStock.com/ThitareeSarmkasat	Purified talc powder	
Shutterstock.com/Vladimir Gjorgiev	Water-based emulsion	
iStock.com/Sergey Nazarov	Massage gel	

2 What are three items of equipment that you will need to have prepared before the body massage treatment?

_____

_____

_____

3 What temperature should you set the room at? Why is this important?

_____

_____

4 List three things you can do to ensure the treatment area is presented in an appealing way for a body massage treatment.

_____

_____

_____

Presenting yourself

5 List 10 ways your college or salon expects you to professionally present yourself for massage treatments.

Knowledge Evidence

_____

_____

_____

_____

_____

_____

_____

_____

_____

_____

_____

Health, hygiene and safety

6 What is RSI and how does it relate to a massage treatment?

_____

_____

7 If you were to develop RSI, what are the first best steps to deal with it?

_____

8 Feet are considered one of the more unhygienic areas of the body. Give an example of how a salon might clean the client's feet in preparation for a treatment.

_____

9 It is important to only gather information about a client that is relevant to the treatment that is being discussed. How would you respond if, upon asking after a client's weight prior to a body massage treatment, they became upset and claimed you were discriminating against them based on their size?

_____

_____

_____

Legal responsibilities

10 Up to what age is someone considered a minor when it comes to massage treatments? If someone that age or younger wanted a massage treatment, what would you need to do first?

_____

_____

_____

11 When do you need to gain a signed client consent form to perform a body massage treatment?

_____

_____

12 What does a client consent form do in terms of liability?

_____

_____

13 If a client states that they require a person of a particular gender to perform their massage treatment for cultural or religious reasons, what should your reaction be?

_____

_____

_____

LO18.4 PROVIDING BODY MASSAGES

1 What are the four tapotement massage movements?

_____

_____

_____

_____

2 Why is it necessary to know the client's weight and height before performing tapotement movements in a body massage treatment?

_____

_____

While providing the massage

3 For each of the following massage movements, describe how they are performed.

MASSAGE MOVEMENT	HOW IT IS PERFORMED
Effleurage	
Friction	

MASSAGE MOVEMENT	HOW IT IS PERFORMED
Petrissage	
Tapotement	
Vibration	

4   While massaging the arm and hands:

    a   When rotating the forearm, how do you steady the client's elbow?

    b   What does circumduct mean? What do you circumduct when massaging the arm and hand?

5   While massaging the neck and chest:

    a   What is a trapezius lift?

    b   How do you apply straight traction to the cervical spine?

6   While massaging the abdomen, how do you drape the client for modesty?

7   Refer to 'Step-by-step: Abdominal and chest massage' in Chapter 18 of your textbook. In which direction do you massage the abdomen? Why do you think it is appropriate to massage in this direction?

8   While massaging the legs and feet:

    a   What is the first massage movement you would perform on the foot?

    b   When massaging the back of the leg, what could you do if the client suffers from poor circulation or cellulite on the upper lateral or posterior aspect of the thigh?

9   What is the first step of the back massage?

10 What is the name for the full body routine described in the textbook?

11 Sometimes the client requirements mean that you will need to use a different type of massage. List three other types of massage you could learn.

LO18.5 REVIEWING MASSAGE AND PROVIDING BODY CARE ADVICE

Performance Evidence

Adverse reactions

Knowledge Evidence

1 You might be able to assess adverse reactions by observing the client. How else would you know if a client has experienced an adverse reaction?

---

2 Complete the missing sections in the table below to show your understanding of adverse reactions in a body massage treatment.

ADVERSE REACTION	PHYSIOLOGY IN RELATION TO BODY MASSAGE	REMEDIAL ACTION
Allergies	Professional massage mediums tend to be derived from high quality nuts and seeds. Have a range of hypoallergenic and non-comedogenic massage mediums on hand.	
Bruising		Prevent bruising by using correct application of massage technique to the client's requirements, i.e. correct pressure and depth during petrissage and tapotement. If bruising occurs, apply cool compress lightly over area.
Dizziness	Vasodilation causes blood pressure to lower, and with less blood in the brain the head generally feels faint and dizzy. Other reasons for dizziness include toxin release, nervous energy, systemic illness, dehydration, low blood sugar and vertigo.	
Emotional release		Manage the emotional release in a professional manner. With integrity, create a safe, supportive environment so that the client feels it is acceptable to express emotions in your presence.

ADVERSE REACTION	PHYSIOLOGY IN RELATION TO BODY MASSAGE	REMEDIAL ACTION
Erythema	Vasodilation causes a normal redness in the skin known as erythema.	
Joint sounds		'Cracking joints' can give some people a nice feeling; if there is pain associated with a joint sound, refer the client to a GP for diagnosis and specialist treatment. Arthritic joints will make a 'grinding' noise and should be referred for pain management.
Headaches	Continual pain in the head is known as a headache. A range of factors contribute to a headache, such as muscle tension, toxic build-up and work/life stresses. The change in blood flow and nerve stimulation during massage can trigger a headache.	
Muscle spasms		Gently stretch the muscle. Refer the client to a GP if the spasm remains. More serious causes for muscle spasms include dystonia, types of arthritis and diseases associated with the spinal discs.
Pain or discomfort	There should be no cause for pain or discomfort in a Swedish massage. Unexplained pain could be due to skin damage or disorder.	
Blemishes and breakout		Always purchase high quality professional products. Stock a bland mineral oil with no additives for clients with congested or reactive skin that tends to break out after massage.

Future treatments

3   How often would you recommend a full body massage in order to maintain the effects?

_____

4   What is the maximum frequency that a client should receive a full body massage?

_____

5 What are the treatment times you would schedule for the following body massages?

TREATMENT	TIME SCHEDULED (MINUTES)
Full body massage	
Back massage	
Add-on scalp massage	
Add-on back, arm, foot massage	

Post-treatment lifestyle and product advice/recommended future treatments

Knowledge
Evidence

6 What future treatment recommendations would you make to a client for the following specific treatment aims?

a Fat reduction

_____

_____

_____

_____

b Anti-ageing

_____

_____

c Stress relief

_____

_____

7 Clients will benefit from relevant advice regarding specific lifestyle factors. Complete the table below with advice on lifestyle changes that would be appropriate for each client. You also need to write in the lifestyle factor that best matches each client from the list below into the second column of the table:

- alcohol consumption
- exercise routine
- hobbies
- tobacco consumption
- type of employment.

CLIENT	LIFESTYLE FACTOR	EXPLAIN LIFESTYLE CHANGE WITH RELEVANCE TO CLIENT CONDITION AND BODY MASSAGE
Woman suffering adult acne from work stress (Consider: when the skin is not functioning properly, skin disorders like adult acne can take over.)		

CLIENT	LIFESTYLE FACTOR	EXPLAIN LIFESTYLE CHANGE WITH RELEVANCE TO CLIENT CONDITION AND BODY MASSAGE
Lonely client at home with small children suffering from lacklustre skin and dark circles		
Smoker with skin lacking in colour, slow circulation and ageing quickly		
Woman in her 30s concerned with weight loss		

LO18.6 CLEANING THE TREATMENT AREA

Knowledge Evidence

1 What are the items of linen that will need to be cleaned after a body massage treatment?

_____

_____

_____

2 Which products will you need to ensure are adequately restocked before the next client arrives?

_____

_____

_____

3 What will you need to take into consideration when deciding to use disposable linen or reusable linen?

_____

_____

4 How might you choose a more sustainable product to use during a body massage treatment?

_____

_____

CHAPTER 19: AROMATHERAPY MASSAGES

After completing this chapter, you should be able to:

LO19.1   establish client priorities with thorough consultation, including skin and body analysis

LO19.2   design and recommend the aromatherapy massage treatment with modifications adapted to the client's requirements and discuss the benefits, potential adverse effects and treatment sequence

LO19.3   prepare for aromatherapy massage treatment following protocol, sustainable work practices and relevant health and safety legislative requirements

LO19.4   provide the aromatherapy massage according to the client's requirements

LO19.5   review the treatment and provide aftercare, homecare advice and a future treatment plan

LO19.6   clean and clear the treatment area in preparation for the next treatment.

INTRODUCTION

Aromatherapy massage is a specialised treatment that delivers aromatic plant oils to promote relaxation, rejuvenation and harmony. The specific effect of the massage is determined by the expertly selected and blended aromatic plant oils, combined with specific aromatherapy massage manual techniques and inhalation for optimum absorption into the body.

The learning activities in this chapter allow you to demonstrate your knowledge and understanding of the unit of competency *SHBBBOS010 Provide aromatherapy massages*, using the performance and knowledge evidences.

LO19.1 ESTABLISHING CLIENT PRIORITIES

Performance
Evidence

Knowledge
Evidence

Capabilities of aromatherapy

1   For the following client requirements for aromatherapy massage, list *at least* two lifestyle cause examples, and *at least* one treatment priority.

CLIENT REQUIREMENT	LIFESTYLE CAUSE EXAMPLES	TREATMENT PRIORITY
Skin healing (dermatitis, psoriasis, etc.)		
Relaxation		
Sleep disturbances		
Muscular tension		

Contraindications and conditions

2   For the following symptoms, identify the contraindication, give the relevance to aromatherapy massage, and the action you would take if a client had it.

Knowledge
Evidence

SYMPTOM	CONTRAINDICATION	RELEVANCE TO AROMATHERAPY MASSAGE	ACTION
Chronic sharp pain			
 Alamy Stock Photo/Mediscan Client has had epilation in the past few days			
Heart disease			

3   For each of the following restrictions, identify the body system affected, and explain how you would modify the treatment plan if the client had the condition.

CONDITION	BODY SYSTEM	MODIFICATION TO TREATMENT PLAN
Skin trauma		
Dizziness		
Fractured wrist		
Area exhibiting a loss of tactile sensation		

Relevant medical history and medication

Knowledge Evidence

4 If a client tells you they are currently taking antibiotics, how would you modify the aromatherapy treatment to accommodate this?

_____

_____

_____

5 Complete the following table for skin disorders and diseases. It is important to understand when aromatherapy can assist in easing a client's skin disorder or disease, and be able to recognise when you should refer the client to a medical professional.

SKIN DISORDER OR DISEASE	AROMATHERAPY TREATMENT	REFERRAL
Benign neoplasms and hyperplasias		
Dermatitis and eczema		
Ichthyosis		
Lupus erythematosus		
Scleroderma		
Skin tumours		
Urticaria		

6 A female client tells you they are transgender. What does this mean? What is the relevance to aromatherapy massage treatments?

_____

_____

_____

_____

_____

_____

Client record management

7   Fill in the client record card for aromatherapy for the following client: a 48-year-old male who is suffering from headaches due to stress. He has a letter from his GP to say that he has no underlying medical issues, and needs to take measures to relax his muscles and his mind.

- Age bracket:

_____

- Gender:

_____

- Contraindications:

_____

- Restrictions:

_____

- Treatment objectives:

_____

- Modifications to treatments:

_____

8   Conduct consultations with four people to develop aromatherapy massage treatment plans that address a range of client needs. Record the details on the aromatherapy client record card. Take into consideration:

- contraindications
- client's main concern and aromatic preferences
- skin and body biology
- variations to the plan; e.g. massage techniques, oils used, areas treated, duration, etc.

LO19.2 DESIGNING AND RECOMMENDING AROMATHERAPY MASSAGE TREATMENT

Performance Evidence

Effects and benefits of products used and blended oils

1   Visit your local supplier of skincare (e.g. salon, chemist, etc.), and look for products that have aromatherapy blended oils in their ingredients. Most products will use the benefits of the plant oils they include to market their product as being superior to those of their competitors. Find three such products and determine whether they have successfully identified the benefits of using that particular plant oil. If you don't agree with the product's claim, write what you think it *should* claim.

Knowledge Evidence

PRODUCT AND BRAND	PLANT OIL USED	WHAT IT CLAIMS	WHAT IT SHOULD CLAIM

2 In the table below, identify the image of the plant by giving the name of the oil and its plant of origin (the scientific name). List the details for each one, including any information on how the oil is extracted and what its benefits are.

PLANT	PLANT SCIENTIFIC NAME	NAME OF OIL	DETAILS
 Shutterstock.com/Madlen			
 Shutterstock.com/Elena Ray			

PLANT	PLANT SCIENTIFIC NAME	NAME OF OIL	DETAILS
Shutterstock.com/Skyprayer 2005			
iStock.com/slallison			
iStock.com/HansJoachim			

3 List an aromatherapy oil that could be used for the following client objectives.

   a Reducing depression

_____

   b Boosting immunity

_____

   c Skin hydration

_____

4 Why are carrier oils used?

_____

_____

5 Explain the following purposes of carrier oils.

   a Dilution

_____

_____

b Percutaneous absorption

_____

_____

_____

c Longer-lasting effects

_____

_____

_____

d Massage medium

_____

_____

_____

6 For each of the following carrier oils, give the plant of origin (scientific name), its aroma, texture, colour and use.

CARRIER OIL	PLANT OF ORIGIN	AROMA	TEXTURE	COLOUR	USE
Apricot kernel oil					
Sunflower oil					
Avocado oil					
Coconut (fractionated)					

7 In the table below, write in the oils used in aromatherapy treatments and the effect the aromatherapy treatment can have on the various body systems listed. The first row has been done for you.

BODY SYSTEM	AROMATIC PLANT OIL	EFFECT
Circulatory	Camphor	Stimulates circulation
Digestive		
Integumentary		
Lymphatic		
Muscular		
Nervous		
Respiratory		
Skeletal and articular		

8   What do you think the dangers are of purchasing oils that are 'pre-mixed'? That is, the oils are pre-blended in the carrier oil and ready for use to your client's requirements? Consider a person with atopic skin.

_____

_____

_____

_____

_____

Massage routine

9   Using the space provided, place the body parts listed in the first column in the order you would usually massage in an aromatherapy massage, from first to last. Put a star next to the body parts you would give the client the option of skipping.

BODY PART	CORRECT ORDER
chest and shoulders feet abdomen front of the legs back back of the legs face, neck and scalp arms and hands gluteals	

HOW AROMATHERAPY CAN AFFECT THE SKIN

Knowledge
Evidence

10 Explain why aromatic plant oils can penetrate the skin, but carrier oils cannot.

_____

_____

11 The skin on the body parts listed below is either highly permeable or not as permeable. Complete the table by writing either 'high' or 'low' in the permeability column for each body part.

BODY PART	PERMEABILITY
Trunk	
Soles of feet	
Scalp	
Legs	
Forearms	
Abdomen	
Palm of hands	
Forehead	

12 Aromatherapy massage is a very light-pressured massage. It is quite common for clients who don't like a gentle-pressured massage, because it is ticklish or irritating, to insist on a firm-pressured massage despite your recommendation that gentle pressure enhances the penetration of oils. Why should we respect a person's absolute threshold for the sensation of pressure?

_____

Treatment duration, areas treated and areas not treated/treatment objectives/ recommended future treatments

13 Draw a line to match up the treatment need with the correct duration and frequency of aromatherapy massage.

TREATMENT NEED		CORRECT DURATION
Relaxation		75 minutes with herbal compresses. Weekly.
Improving sleeping patterns		1 hour. Weekly until skin gets better.
Relieving muscle tension		75 minutes. Weekly with a view to treating every 4–6 weeks for maintenance.
Resolving skin problems		1 hour. Every 3–4 weeks.

14 There are many reasons why a client might require an aromatherapy massage. What are the four client objectives listed in your textbook? Go online and research aromatherapy massage to find and list some more. Can you get to 10 different treatment objectives?

LO19.3 PREPARING FOR AROMATHERAPY MASSAGE TREATMENT

Knowledge Evidence

1 Go online and find example images of how *not* to present yourself for an aromatherapy massage. For each, post it to the discussion board or paste it to a separate sheet of paper, and describe what you think is unacceptable about the therapist shown.

2 Fill in the missing words to complete the sentences. Each should give an accurate statement of how the treatment area should be prepared.

a Set the room temperature to between _____ and 21 °C.

b If the client is receiving the treatment while sitting on a treatment chair or stool, _____ the client and have _____ ready to add support and _____ throughout treatment.

c The treatment area should induce _____ and client comfort; lighting should be _____ and decor subtle and _____ biased. Sound levels should be _____ for relaxation.

Health and safety

3 Go online to find your state/territory and local health and hygiene regulations and requirements relevant to aromatherapy massages. Is there any legislation that requires you to maintain quality oils to prevent damage to client health? Write an A4-page report on your findings. Consider:

- work health and safety
- incident reporting
- who is liable if there is damage to health.

LO19.4 PROVIDING AROMATHERAPY MASSAGE TREATMENT

Performance Evidence

Benefits of different massage movements

Knowledge Evidence

1 Each massage movement and technique has its own benefits. Describe each below and why you would choose it.

a Effleurage

_____

b Friction

_____

c Petrissage

_____

2 Fill in the missing words to complete the following three sentences.

a When you need to remove some movements from your sequence to accommodate a client's requirements, you need to adjust the _____ of other movements to make up time.

b If the client had slackened, elderly skin, you would need to adjust the _____ of the stroke to ensure there was no dragging.

c When the client requires a more relaxing movement, you need to _____ the rhythm of the movement.

3 For aromatherapy massage, it is important that you know the acupressure/tsubo points in various parts of the body to access energy channels. In pairs, locate the acupressure/tsubo points on the arms, face and neck.

Suitability of aromatherapy

4 The factors that can affect the suitability of treatments for a client's needs are listed below. In small groups, brainstorm and write in some ways you could still recommend the aromatherapy massage treatment if these factors are brought up by the client.

Knowledge Evidence

FACTOR	POSSIBLE RECOMMENDATIONS
Budgetary and time constraints	
Outcome of previous treatments	
Physical attributes	

LO19.5 REVIEWING MASSAGE TREATMENT AND PROVIDING POST-TREATMENT ADVICE

Knowledge Evidence

Client feedback

1   What are two ways you could obtain client feedback at the end of an aromatherapy massage?

2   How often are you required to record the treatment review?

Outcomes of previous and current treatment

Knowledge Evidence

3   If a client were to tell you that the product they used following their last aromatherapy massage made their skin react, where on the client record card for body massage would you place this information?

4   Suggest one health and safety implication of not providing post-treatment advice to the client.

5   Refer to the section 'Potential adverse effects' in Chapter 18 of your textbook. Complete the missing information in the 'Physiology' and 'Remedial action' columns in the table below.

ADVERSE REACTION	PHYSIOLOGY	REMEDIAL ACTION
Allergies	Professional massage mediums tend to be derived from high quality nuts and seeds. Have a range of hypoallergenic and non-comedogenic massage mediums on hand.	
Bruising		Prevent bruising by using correct application of massage technique to the client's requirements, e.g. correct pressure and depth during petrissage and tapotement. If bruising occurs, apply cool compress lightly over area.
Dizziness	Vasodilation causes blood pressure to lower, and with less blood in the brain the head generally feels faint and dizzy. Other reasons for dizziness include toxin release, nervous energy, systemic illness, dehydration, low blood sugar and vertigo.	

ADVERSE REACTION	PHYSIOLOGY	REMEDIAL ACTION
Emotional release		Manage the emotional release in a professional manner. With integrity, create a safe, supportive environment so that the client feels it is acceptable to express emotions in your presence.
Erythema	Vasodilation causes a normal redness in the skin known as erythema.	
Joint sounds		'Cracking joints' can give some people a nice feeling; if there is pain associated with a joint sound, refer the client to a GP for diagnosis and specialist treatment. Arthritic joints will make a 'grinding' noise and should be referred for pain management.
Headaches	Continual pain in the head is known as a headache. A range of factors contribute to a headache, such as muscle tension, toxic build-up and work/life stresses. The change in blood flow and nerve stimulation during massage can trigger a headache.	
Muscle spasms		Gently stretch the muscle. Refer the client to a GP if the spasm remains. More serious causes for muscle spasms include dystonia, types of arthritis and diseases associated with the spinal discs.
Pain or discomfort	There should be no cause for pain or discomfort in a Swedish massage. Unexplained pain could be due to skin damage or disorder.	
Blemishes and breakout		Always purchase high quality professional products. Stock a bland mineral oil with no additives for clients with congested or reactive skin that tends to break out after massage.

Post-treatment lifestyle and product advice

Knowledge Evidence

6 When discussing which products the client uses, or has used, what are some specific details you might ask the client for? List at least three.

_____

_____

_____

_____

_____

_____

7 Bathing can be a positive lifestyle change for a client's wellbeing. Why? How would you recommend the client prepares a bath with aromatherapy oils?

_____

_____

_____

_____

Aftercare advice

Knowledge Evidence

8 What are the four general pieces of aftercare advice you should give to every client?

_____

_____

_____

_____

9 Visit your local salon or spa and ask to see its range of retail products. Based on what is available for purchase by clients, write down what you would recommend for yourself following an aromatherapy massage treatment.

_____

_____

LO19.6 CLEANING THE TREATMENT AREA

Knowledge Evidence

1 List five things that need to be cleaned following an aromatherapy treatment, either straight after the treatment or at the end of the day.

_____

_____

_____

_____

2 List the types of linen that can be used in the aromatherapy massage treatment.

---

3 What do you do with biohazardous waste? What sort of items should you be particularly careful of when performing the aromatherapy massage?

Storing oils

4 What are the risks of not storing aromatic plant oils correctly?

Knowledge
Evidence

---

5 List seven ways you must store aromatic plant oils to ensure their longevity. Some things to keep in mind are the labelling, the decanting and resealing, the protection from light, and the temperature.

Sustainability

6 List eight ways you could make your aromatherapy massage treatment, salon or spa more sustainable in the ways that power, water or products are used.

Knowledge
Evidence

_____

_____

_____

_____

_____

_____

_____

_____

CHAPTER 20: AROMATIC PLANT OIL BLENDS

INTRODUCTION

Aromatic plant oils are used in aromatherapy massage, spa and other salon treatments and cosmetics. Learning about aromatic plant oils and how they blend extends your knowledge and understanding of cosmetic ingredients and their effect on the skin, mind and body.

The learning activities in this chapter allow you to demonstrate your knowledge and understanding of the unit of competency *SHBBCCS006 Prepare personalised aromatic plant oil blends for beauty treatments*, using the performance and knowledge evidences.

LO20.1 ESTABLISHING CLIENT PRIORITIES

Performance
Evidence

Knowledge
Evidence

Contraindications and scope of practice

1 For each of the following contraindications, list the relevance to aromatherapy massage and the action required if a client has it.

CONTRAINDICATION	RELEVANCE TO AROMATHERAPY MASSAGE	ACTION REQUIRED
Heart disease		
Bacterial infection		
Recent scar tissue in the area		

CONTRAINDICATION	RELEVANCE TO AROMATHERAPY MASSAGE	ACTION REQUIRED
Deep pain with no logical explanation		
Recent chemical peel		

2   Give an example of a plant oil that should be avoided for the following conditions or client types:

a   pregnancy

b   epilepsy

c   atopic client

d   irritable skin

3   What do the essential oils of wormwood, rue, horseradish and thuja all have in common?

4   Give an example of a complementary therapist who will treat a client's condition with the appropriate aromatic plant oils.

5   What is the scope of an aromatherapist for a client presenting with a skin disorder, according to the Therapeutic Goods Administration?

Client needs

6   A client will always enter a salon with an idea of what they are able to do, as well as other factors that might affect the treatment plan. For each of the following factors, suggest three constraints that a client might present to you when booking in for a treatment that includes aromatic plant oils.

Knowledge
Evidence

a   Budgetary and time constraints

b   Outcomes of previous treatments

---

c   Physical attributes

---

7   Complete the table with an example aromatic plant oil and the effects for each of the client concerns.

CLIENT CONCERN	AROMATIC PLANT OIL	EFFECTS
Relaxation		
Stress reduction		
Skincare: acne vulgaris		

LO20.2 DESIGNING OIL BLENDS

Performance Evidence   Knowledge Evidence

Carrier oils

1   What are the four main reasons carrier oil is used?

---

2   Complete the table by filling in the missing details each carrier oil, including the plant of origin, aroma, texture, colour and use.

NAME OF OIL	PLANT OF ORIGIN	DETAILS
Kukui (Hawaiian nut oil)		

NAME OF OIL	PLANT OF ORIGIN	DETAILS
Avocado oil		
Shea butter		
Evening primrose oil		

3   The better carrier oils are polyunsaturated vegetable oils, which combine well with the aromatic plant oils in aromatherapy. Do an online search to learn the difference between polyunsaturated, monounsaturated and saturated fats. Briefly describe these differences.

_____

_____

4   Why might a carrier oil be used as an additive in an aromatic plant oil blend?

_____

5   When could you use a cream or similar other medium as a carrier in an aromatic plant oil blend?

_____

Application methods

6   Complete the table with the effects and benefits of aromatic plant oil application methods.

Knowledge
Evidence

APPLICATION METHOD	EFFECTS AND BENEFITS
Compresses	
Facial and body massage	
Hand and foot massage	
Hydrosols	
Poultices	
Spa treatments	
Vaporisations	

Characteristics and properties of aromatic plant oils

7   Select one of the aromatic plant oils listed below. Design an information pamphlet that could be shared with the class to describe the following aspects of the plant oil:

Knowledge Evidence

- properties

- profiles

- plant family

- botanical and common names

- effects and benefits

- toxic effects

- contraindications.

Remember to focus on presentation for your pamphlet with colour, headings, images and clear information.

Basil	Bermagot	Chamomile	Citrus (orange)	Citrus (grapefruit)
Citrus (lemon)	Citrus (lime)	Citrus (mandarin)	Citrus (tangerine)	Clary sage
Cypress	Eucalyptus	Frankincense	Geranium	Juniper
Lavender	Rose	Rosemary	Sandalwood	Tea tree

OLFACTORY SENSE

8   What does the olfactory sense do?

Knowledge Evidence

9 Organoleptic properties can be separated into four odour categories. What are they?

10 Give an example of an odour source for the following scents that a client might say is their preference:

    a minty

    b woody

    c citrus

    d earthy

ORGANIC CHEMISTRY OF AROMATIC PLANT OILS

Knowledge
Evidence

11 Use this article to answer the following questions: **https://www.khanacademy.org/science/ ap-biology/chemistry-of-life/elements-of-life/a/carbon-and-hydrocarbons**

    a Why is carbon a good element to use as a backbone in molecules?

    b What are the types of bonds that carbon makes with other carbon molecules?

    c How many other atoms can carbon bond with?

    d What is a hydrocarbon?

12 Use this article to answer the following questions: **https://www.khanacademy.org/science/ap-biology/ chemistry-of-life/elements-of-life/a/functional-groups**

    a What is a functional group and why is it important to understanding aromatic plant oils?

b Draw the structure and identify the properties (polar, non-polar acidic or basic) of the common functional groups in biology.

FUNCTIONAL GROUP	STRUCTURE	PROPERTIES
Hydroxyl		
Methyl		
Carbonyl		
Carboxyl		
Amino		
Phosphate		
Sulfydryl		

13 It is vital that the terminology of chemistry is used in a precise and accurate manner so that mistakes are not made. Use a blank sheet of paper to write down all the chemistry terms that you do not understand, and find the definition of each one in a dictionary. You might have a few words, or many. This is not about how many you have, but about improving your scientific vocabulary.

14 Use this article to answer the following questions:
https://basicmedicalkey.com/families-of-compounds-that-occur-in-essential-oils/

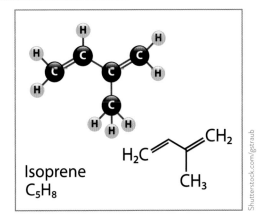

Isoprene
C_5H_8

a What does it mean that isoprene is an aliphatic molecule?

_____

b How does a terpene form a hydrocarbon chain?

_____

c How does a terpenoid differ from a terpene?

_____

d How does a monoterpene differ from a terpene?

_____

e How many isoprene units are in a sesquiterpene molecule?

_____

f If monoterpenol is a monoterpene molecule that has an alcohol functional group attached, how would you describe a sesquiterpenol?

_____

Prepare profiles and plant information

15 Go online to search for the chemotypes of these commonly used aromatic plant oils and list them here.   Performance Evidence

    a Thyme

_____

    b Basil

_____

    c Sage

_____

16 When using aromatic plant oils, it is important to know the botanical name of a plant. What is the difference between a botanical name and a common name of the plant ingredient?

_____

_____

_____

17 Draw a line to match up the following botanical names with their common names.

BOTANICAL NAME
Lavandula angustifolia
Ocimum basilicum
Citrus sinensis
Citrus reticulate
Eucalyptus globulus
Santalum album

COMMON NAME
Sweet orange
Sandalwood
Lavender
Eucalyptus
Mandarin
Sweet basil

18 List the six different methods of extracting aromatic plant oils.

_____

_____

19 Complete the following sentences about plant families and synonyms of aromatic plant oils.

    a Chamomile is from the _____ plant family

    b Antiseptic, diuretic and regenerative are all characteristics of the _____ family

    c An example of an essential oil from the Cupressaceae plant family is _____.

LO20.3 SETTING UP FOR BLENDING

Knowledge
Evidence

1   Why should you only purchase aromatic plant oils in bulk if they are definitely going to be used?

_____

2   Why is it important to always use aromatic plant oils according to the manufacturer's instructions?

_____

_____

3   Why is it important to always store aromatic plant oils in a closed container?

_____

4   Why is it good practice to have the treatment area separate from the blending area?

_____

_____

5   Some aromatic plant oils can have an adverse reaction on a client's health, even if they are only smelling it.
    If a client has an adverse reaction to a scent, and you are going to treat them in a space where that scent has been
    used, what could you do to try to mitigate the client's adverse reaction?

_____

_____

Organisational policies and procedures

Knowledge
Evidence

6   Go online and research your local, state or territory guidelines in regards to aromatic plant oils. Write
    a half-page report on the preparation and use of aromatic plant oils, including purchase, storage and
    appropriate usage.

7   Using the client record card for aromatherapy plant oil blend, fill out the appropriate sections for the
    following client:

    •   female, early 30s

    •   has a history of muscle tension in her upper back, neck and shoulder area

    •   has had aromatherapy massage using aromatic oil blends in the past, but is looking to change the oil blend

    •   prefers citrus blends.

HEALTH, SAFETY AND HYGIENE

Knowledge
Evidence

8   What does the *Therapeutic Goods Act 1989* regulate with regard to aromatic plant oils?

_____

9   If a client develops a skin reaction after the use of a new aromatic oil blend, what steps should you take to manage
    the incident?

_____

_____

10 In order to protect a client's clothing from oils used in the salon, what should you do before a treatment that uses aromatic plant oil blends?

_____

_____

_____

11 List three things that a therapist should do in order to maintain a high level of personal hygiene and presentation in the salon.

_____

_____

_____

LO20.4 PREPARING AROMATIC OIL BLENDS

Performance Evidence   Knowledge Evidence

1   Why is ventilation important in a salon when using aromatic oil blends?

_____

_____

2   In terms of work health and safety, why is it a good idea to have your aromatic oils already blended before the beginning of the treatment, if possible?

_____

_____

_____

3   If your dropper delivers 20 drops to the millilitre, and you need to make a 2% solution, how many drops would you add to 100 mL of carrier oil?

_____

4   What three things should be done to minimise the deterioration of the oils and blends?

_____

_____

_____

LO20.5 PROVIDING TREATMENT USING AROMATIC OIL BLENDS

Performance Evidence

1   Give three examples of treatments when an aromatic plant oil could be used.

_____

_____

_____

2   Complete the following table by explaining each of the application methods.

APPLICATION METHOD	EXPLANATION
Compress/Thai herbal balls	
Facial, body, hand and foot massage	
Hydrosols	
Poultices	
Spa treatments	
Vaporisations	

LO20.6 REVIEWING TREATMENT AND PROVIDING POST-TREATMENT ADVICE

1   When discussing a second treatment for a client, why is it important to know about their initial treatment?

2   What is a reasonable interval between treatments for the client to gain the benefits of the aromatherapy treatment?

Adverse reactions

3   List four common types of adverse reactions that can occur following an aromatherapy treatment.

4 The client might experience a healing crisis during or after a treatment. What does this mean?

_____

_____

Client record card

5 Where on the client record card for aromatherapy plant oil blend would you put the following information?

Performance
Evidence

a Contraindications

_____

Performance
Evidence

_____

b Ingredients and ratios used

_____

c Delivery method

_____

d Medical history or medications

_____

_____

e Post-treatment aromatic oil

_____

f Lifestyle advice

_____

g Treatment objectives

_____

h Recommend future treatments

_____

Post-treatment care and lifestyle advice

Knowledge
Evidence

6 Why is it a good idea for clients to have a bath at home after an aromatherapy treatment?

_____

7 What is vaporisation in regards to aromatherapy treatments? Give an example of a method of vaporisation.

_____

_____

_____

LO20.7 CLEANING THE TREATMENT AREA

Knowledge
Evidence

1 List ten things that need cleaning after an aromatherapy treatment. These can be either from the treatment, or from blending the aromatic oil blend.

2 For each of the following, explain why it is important to remember when storing aromatic plant oils.

a Keep the lids on.

b Don't shake the bottles.

c Never store a small quantity of aromatic plant oil in a large bottle.

3 What should all aromatic plant oil bottles be labelled with?

4 All local councils have different standards about what they can recycle. Go to your local council website to see whether you are able to recycle aromatherapy oil bottles in your area.

5 Choose one of the following and write a brief explanation that demonstrates your understanding of sustainable conservation of:

- product
- water
- power.

SECTION 7

FACIAL TREATMENTS

21 Provide facial treatments and skincare recommendations

22 Specialised facial treatments

CHAPTER 21: PROVIDE FACIAL TREATMENTS AND SKINCARE RECOMMENDATIONS

LEARNING OBJECTIVES

After completing this chapter, you should be able to:

LO21.1 establish client requirements with a skincare consultation and skin analysis

LO21.2 design the treatment plan, recommend facial treatments and record modifications on the client record card

LO21.3 prepare for the treatment according to salon protocol and relevant health and safety legislative requirements

LO21.4 cleanse, exfoliate and perform extractions as required

LO21.5 perform facial massage with an appropriate massage medium and to the client's requirements

LO21.6 apply specialised facial products, such as masks and post-treatment needs

LO21.7 review the facial treatment, evaluate and record outcomes and update the treatment plan

LO21.8 recommend a post-treatment skincare regimen

LO21.9 clean the treatment area in preparation for the next treatment.

INTRODUCTION

Facials are focused, non-invasive skin treatments that are customised to suit the skin's needs. Beauty salons offer different types of facial treatments and product lines so that you can design a treatment and home skincare regime to match the client's objectives and skin characteristics. The overall aim is to maintain and improve the health and appearance of the skin.

The learning activities in this chapter allow you to demonstrate your knowledge and understanding of the unit of competency *SHBBFAS005 Provide facial treatments and skin care recommendations*, using the performance and knowledge evidences.

LO21.1 ESTABLISHING CLIENT PRIORITIES

Performance Evidence

Knowledge Evidence

Effects and benefits

1 Look at your college or salon's professional products and use your product knowledge to recommend a facial sequence for two different professional product ranges for a client with oily but dehydrated skin. Complete the tables below with the stage of the facial (e.g. cleanse, tone, etc.), the name of the product used for that stage, its active ingredient and its effects and benefits.

FACIAL SEQUENCE 1, PROFESSIONAL PRODUCT RANGE:

STAGE IN FACIAL	PRODUCT	ACTIVE INGREDIENT	EFFECT AND BENEFIT

STAGE IN FACIAL	PRODUCT	ACTIVE INGREDIENT	EFFECT AND BENEFIT

FACIAL SEQUENCE 2, PROFESSIONAL PRODUCT RANGE:

STAGE IN FACIAL	PRODUCT	ACTIVE INGREDIENT	EFFECT AND BENEFIT

2 You would normally steam a client's skin if they have an oily, dehydrated skin. What is the recommend steaming duration and how far away from the client should you hold the steamer?

3 Marine extracts can have certain contraindications, so you need to be well informed about any products that contain them and any client allergies. If your product contains algae, for example, and the client is allergic to crustaceans only, can you proceed with product application?

4 If you have a seaweed extract as one of your ingredients, what would you need to know about a client before you proceed with treatment?

Contraindications

5 Review Appendix A Contraindications and restrictions as well as Figure 21.11 and Figure 21.13 in your textbook. Complete the following table of contraindications that restrict or prevent a facial treatment. Use the image and description given to list the name of the contraindication, its relevance to the facial treatment and the action you would take if a client had it.

Knowledge
Evidence

SYMPTOM	CONTRAINDICATION	RELEVANCE TO FACIAL TREATMENT	ACTION
Science Photo Library/CID - ISM Red, patches of skin containing papules. Typically located in sun-sensitive areas, such as on the face, scalp, arms, legs or trunk. Can become raised, elongated or ring-shaped and scaly, similar to psoriasis. Patches can leave hyperpigmentation.			
Shutterstock.com/lpen Erythema with flat or raised, round, whitish skin wheals. Chronic urticaria occurs in a generalised area and can last for minutes or hours. In some cases, the lesions can cause intense burning or itching. Acute hives happen after insect bites.			
Shutterstock.com/Faiz Zaki A bacterial infection concentrated to one follicle. Boils tend to create a tender lump and pustule. Multiple boils in one skin area may be folliculitis. Multiple heads on one boil is a carbuncle. People naturally carry *Staphylococcus aureus* bacteria; some have more than others. Damaged follicles and certain body areas are more prone to boils.			

SYMPTOM	CONTRAINDICATION	RELEVANCE TO FACIAL TREATMENT	ACTION
Alamy Stock Photo/Science Photo Library Red, inflamed open and closed comedones, papules and pustules.			

Scope of practice

Knowledge Evidence

6 What grades of acne are treatable in the beauty salon?

7 How do you identify acne vulgaris that must be referred to a dermatologist?

8 What can a complementary health practitioner do for a person with acne?

9 Inflammation with acne vulgaris occurs due to many reactions that take place in the body. It involves the hormones of the endocrine system and the immune system responses. Both internal and external conditions can cause erythema. List six other causes of erythema.

10 Research dermatological advice on extracting acne pustules and cysts. One such weblink is: **https://www. sinclairdermatology.com.au/clinic/acne-acne-scarring.** Why do you think you should not extract (or pick at) an acne pustule or cyst?

11 It's important you know the names of some common medications that can restrict or prevent a facial treatment. What are two names of retinoids?

12 If a client has a medical condition that will not clear, name one of two things they might do.

LO21.2 DESIGNING AND RECOMMENDING FACIAL TREATMENTS

Performance
Evidence

1 Is a dry skin devoid of oil or water? Explain your answer.

2 Is oily skin often thicker?

3 Which part of the skin is considered the T-zone?

4 How would you adjust the treatment plan for an oily T-zone with exfoliation and mask?

5 When planning a treatment, what is the skin most sensitive to, and so what should you avoid/use?

6 The age at which a person has 'mature' skin depends on the client. How old do you think your client should be before you can tell them they have mature skin?

7 How would you recognise a diffuse red skin type?

Explaining the treatment

Performance
Evidence

8 Your client has open and closed comedones and a few pustules, but you can't extract them as they wish – the skin is too tight. What do you say to the client?

_____

_____

_____

_____

9 A client has come to you with diffuse red skin, but it is not warm. What can you recommend for her?

_____

_____

_____

_____

10 Your client with mature skin has presented with acne scarring and wants it reduced. You know that your product company has products that say that they improve the appearance of wrinkles and scars. What do you tell the client?

_____

_____

_____

_____

Suitability of facial

Knowledge
Evidence

11 It is important to consider factors that are likely to affect the client's suitability for a product when selling skincare. It can also affect your treatment plan. When is it appropriate to exclude an area (i.e. back of neck, face, décolletage, neck, shoulders) when the client is not suited to the treatment during a facial sequence? Suggest four factors and give an example of a client situation for each.

_____

_____

_____

_____

_____

_____

_____

_____

LO21.3 PREPARING FOR FACIAL TREATMENTS

Knowledge
Evidence

1 How would you position the trolley in the treatment area for a facial treatment?

2 At what stage of the treatment should you put on personal protective equipment (gloves and mask, as appropriate)?

3 List the types of linen most likely to be used in a facial treatment.

4 Which machines should definitely be checked when performing a safety check on the electrical equipment used in the facial treatment room?

5 If there is a faulty machine, who would you report it to?

6 When working with cleaning chemicals, it is important to be able to access the safety information easily. Work health and safety laws in Australia insist that safety data sheets (SDSs) are completed and stored safely for all potentially hazardous substances and dangerous goods. This ensures that the chemicals are handled and used as they ought to be. It is the responsibility of the manufacturer or importer of the product to prepare and provide the SDS. Locate a SDS and list at least 10 items of information the SDS should include

Client details

Legislation as well as workplace procedures specify how you are to record the information regarding the use of cosmetic chemicals in the salon. Always keep a record of the products sold to your clients. The client card should be filled in for clients for product purchases as well as treatments.

7   List all pieces of relevant information that should be on the client record card.

_____

_____

_____

_____

_____

_____

_____

_____

_____

_____

_____

LO21.4 CLEANSING AND EXFOLIATING THE SKIN AND PERFORMING EXTRACTIONS

1   What is a superficial cleanse and what is a deep cleanse? When would you perform each?

_____

_____

_____

_____

_____

_____

2   Fill in the missing words in the below sentences.

   a   The natural physical process of losing dead _____ cells from the stratum corneum layer of the
       _____ is called _____.

   b   _____ is a salon technique used to accelerate this process. It is normally carried out after the
       skin has been _____ and _____.

3   How often does skin regenerate according to age?

   a   Teenage skin

   _____

   b   Person in their 40s

   _____

   c   Person in their 80s

   _____

4   How should the client be positioned when performing a steam treatment?

   _____

5   What is the purpose of skin steaming?

   _____

   _____

6   How does an electric steamer work?

   _____

7   Why would you use towels for skin steaming? How hot should they be?

   _____

   _____

8   What is the common name for a closed comedone and an open comedone?

   _____

9   Milia are small white bumps on the skin, which appear when keratin or old skin flakes get trapped under the skin.
    What is the main difference between milia and closed comedones?

   _____

   _____

10 When extracting a milium or closed comedone, in which direction do you hold the needle?

   _____

LO21.5 PROVIDING FACIAL MASSAGE

Performance
Evidence

1 Draw a line to match the following images of facial massage movements to their name, and include
the effects of each.

FACIAL MASSAGE MOVEMENT	NAME OF FACIAL MASSAGE MOVEMENT
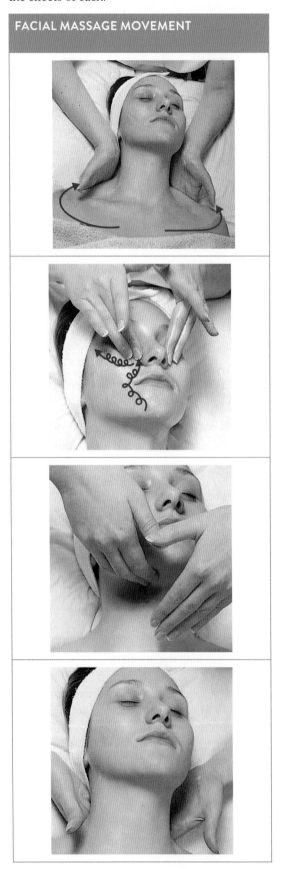	Tapotement
	Vibration
	Effleurage
	Petrissage

2   How would you adapt the repetition, rhythm and variation for the following skin conditions?

   a   Mature, dry and dehydrated skin

_____

_____

   b   Acneic skin

_____

_____

   c   Sluggish skin

_____

_____

3   What is the ideal number of times you should perform effleurage to the neck, shoulders, décolletage, and up to the base of the skull?

_____

4   What is the correct method for finger kneading to the shoulders? Include names of muscles.

_____

5   When performing vibrations on the neck, are your hands flat or cupped?

_____

6   True or false: When you are massaging the neck with your hands cupped, you should slide your hands up the side of the neck, across the jawline, and down the right side of the neck, then reverse.

_____

7   Which finger is the best to use when circling the inner and outer eyes?

_____

8   What is the correct procedure for rolling and pinching the cheek area? How many times do you repeat after the initial motion?

_____

_____

LO21.6 APPLYING SPECIALISED PRODUCTS

Knowledge Evidence

Clients are increasingly well-informed on aspects of cosmetic chemistry and, as a beauty therapist, it is important to always be one step ahead. As new technologies in product ingredients arise, you need to decide whether your product line best serves your clientele. In order to keep up with such an innovative and fast-growing industry, it is essential to have a good understanding of how cosmetic ingredients affect the skin. That is, to know about the chemical and physical reaction that takes place so that the product can have a beneficial effect on the skin.

Basic concepts of cosmetic chemistry

ATOMS

An atom is the smallest part of a chemical element. An element cannot be broken down into a simpler substance. Examples of elements include oxygen, carbon and hydrogen. Atoms consist of a nucleus containing neutrons (no charge) and protons (positive charge). Orbiting the nucleus are electrons (negative charge).

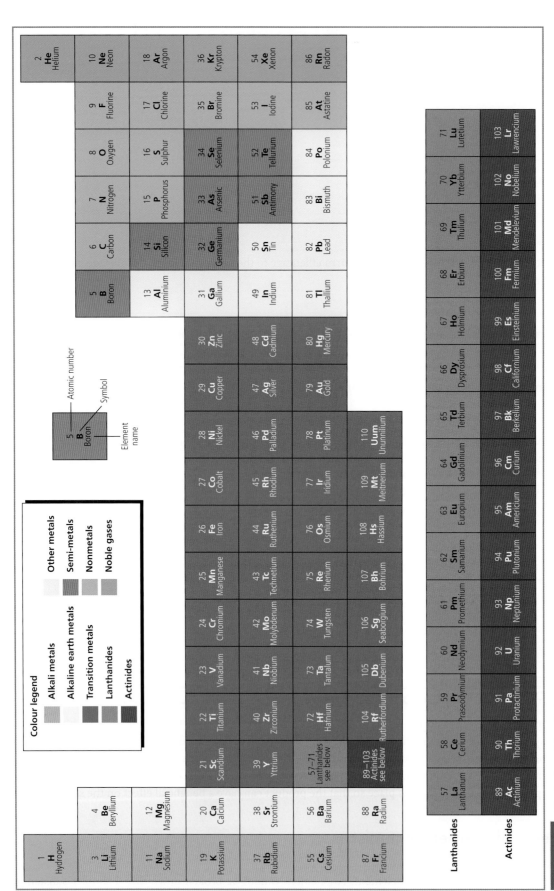

Figure 21.1   Periodic table of elements

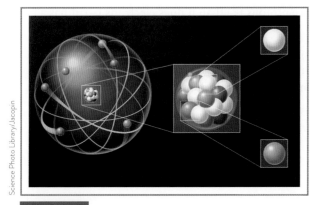

Science Photo Library/Jacopin

Figure 21.2 Atomic structure

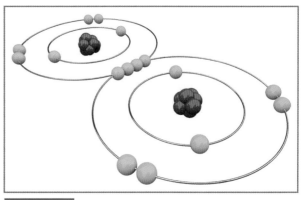

shutterstock.com/Qoncept

Figure 21.3 Atomic structure of oxygen, as O_2

The periodic table of elements (see Figure 21.1) was originally based on defining elements according to their typical reactions and atomic weight. It is organised by increasing atomic number. The atomic number of an element is the number of protons in the element's nucleus. In neutrally charged atoms, the atomic number also reflects the number of electrons in the element. Some elements may exist in forms (isotopes) with different atomic weights, due to differing numbers of neutrons. For example, oxygen's atomic number is 8, its atomic weight is 16 and therefore it has 8 protons and usually 8 neutrons. When oxygen is neutral, it has 8 electrons. When atoms are bonded with other atoms, the number of electrons can change.

As Figure 21.3 indicates, there are orbits around the nucleus. These are known as 'shells'. Each atom in its ground state fills the electron shells from the inner shells outwards, and each shell has a distinctive maximum number of electrons. For oxygen to have a full outer shell it seeks to gain two electrons. Greatest stability is associated with an even number of electrons and filled shells.

1 Draw your own diagram of an atom with its outer shells. Then, with the aid of the periodic table of elements, add the electrons for the element carbon.

BONDS

Bonds are the forces that hold atoms together, and are divided into three types: covalent bonds, hydrogen bonds and ionic bonds. Covalent bonds occur when electrons are shared between atoms. They are stronger than ionic bonds and are not broken by solvents. Water is covalently bonded. Hydrogen bonds occur between hydrogen atoms and other molecules. In water molecules (H_2O), hydrogen forms a weak bond with the oxygen from another water molecule. It is a relatively weak bond but can lead to relatively stable associations. The last type of bond is ionic, which occurs between atoms of opposite electric charge. An atom is said to be charged when it does not have the same number of electrons as protons. The charged atom or molecule is called an *ion*. When an ion has more electrons (−) than protons (+), it has

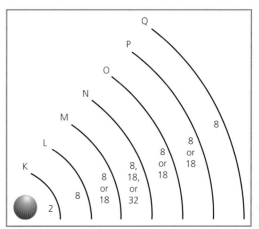

http://openticle.com

Figure 21.4 Maximum number of electrons that a shell can hold

a negative charge and is called an *anion*. When the ion has more protons (+) than it does electrons (−), it has a net positive charge, and we call it a *cation*.

2   Draw the electron configuration structure for the elements hydrogen and oxygen. Now, in order to make hydroxide (OH⁻), you must bond one oxygen atom and one hydrogen atom. The electrons must be shared. With the help of your teacher, draw the atoms bonded together. On your own, try to draw the covalent bonding of hydrochloric acid (HCl).

3   Draw the electron configuration for the elements sodium and fluorine. Now, with the help of your teacher, draw them in their bonded form to make sodium fluoride, which is found in toothpaste.

4   What is hydrogen bonding?

5   What is a cation?

6   What is the strongest type of bond?

   A   Hydrogen bond

   B   Ionic bond

   C   Covalent bond

7   What is characteristic of the type of bond you identified in Question 6?

Physical and chemical change

PHYSICAL CHANGE

Matter is everywhere. It is made up of elements and compounds and can be found in various states. The three states of matter that concern us in the beauty room are solid, liquid and gas. A compound or element can change from one state to another and still remain the same substance. For example, water can be solid (ice), liquid (droplets) and gas (vapour). The change in the state or shape of a substance without a change in chemical structure is a physical change.

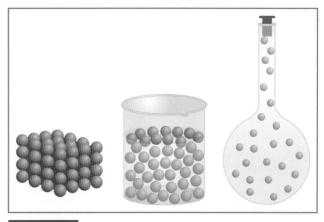

CHEMICAL CHANGE

Figure 21.5 Solid, liquid and gas particles

A chemical change occurs as a result of a change in the bonding of atoms, ions or molecules of a substance. For example, when a beauty product is not kept in a cool, dark place and airtight, you may see a change in the colour of the substance. This often happens because the chemicals have oxidised on exposure to air and light and have changed to another, potentially harmful, substance.

Chemical and physical changes occur on the skin when beauty products are applied. The natural chemicals found on our skin interact with the products for a desirable effect.

A *chemical reaction* must take place for chemical change to occur. A chemical reaction is something that happens when two molecules interact. Energy triggers (or *catalyses*) the event, which brings about a chemical change. For example, in a beauty treatment, the therapist may use heat, such as with a steamer, to increase the intensity of the product on the skin. This may increase the number of chemical reactions occurring on a molecular level or may assist the absorption of the products through the epidermis.

8 Oxidation here means the addition of oxygen to a substance causing chemical change, although it has a more general meaning to chemists. What type of physical change can occur to a product that has oxidised?

Acids and bases

THE pH SCALE

The term pH is an abbreviation for 'potential hydrogen'. The scale is a measure of how acidic or alkaline a substance is. A pH of 7 is neutral if we are considering solutions in water at room temperature. A substance with a pH of more than 7 is a base (alkali) and is an acid if it has a pH of less than 7. The scale is logarithmic; that is, a change of one unit corresponds to a 10-fold change, so that a pH of 4 is 10 times more acidic than a pH of 5, and 100 times more acidic than a pH of 6, and so on.

Figure 21.6 The pH scale set as a table

SCALE	pH	SUBSTANCE
Acidic	1	Battery acid
	2	Hydrochloric acid (stomach acid), lemon juice, vinegar
	3	Vitamin C (L-ascorbic acid, pH 3.5–2.2)
	4	Cosmetic AHAs and BHAs (pH 3.5–4.5), sebum (pH 4.5–6)
	5	Normal skin acid mantle (pH 4.6–6.8 on average), toners, fake tan
	6	Moisturiser (pH 5–6), saliva, urine
	7	Water (neutral), detergent-based cleansers (pH 5–7), vitamin A (retinol)

SCALE	pH	SUBSTANCE
	8	Sea water
	9	Soapy cleansers for oily skin, deodorant
	10	Ordinary soap (pH 9.5–11)
	11	Eyelash perm solution
	12	Bleach used for cleaning
	13	Sodium hydroxide
Alkaline	14	Drain cleaners

As in the pH scale above, the more corrosive chemicals are either strongly acid or strongly alkaline. The pH measures dealt with in the practice of beauty are mostly between 4 and 7. Using products that are too acid or alkaline can be irritating to the skin and, in extreme cases, can lead to skin burns.

Figure 21.7 Properties of acids and bases

ACIDS	BASES
Taste sour	Taste soapy
React with metals to produce hydrogen gas	Feel slippery
Produce H^+ ions in aqueous solution (water)	Produce OH^- ions in aqueous solutions (water)
Turn blue litmus to red	Turn red litmus to blue
Conduct electricity	Conduct electricity
Examples: stomach acid (HCl), lemon juice, vinegar	Examples: soaps, ammonia, cleaning 'lye' (KOH or NaOH)

ACID MANTLE

The acid mantle of the skin is the balance of sebum sweat and lipids, which ideally should be around pH 4.5–6.2. Sebum is more acidic, so an oily skin will have a lower pH than a dry skin.

NEUTRALISATION AND BUFFERING

Making a product for the skin requires the addition of chemicals that moderate the pH. Cosmetic chemists use neutralisers and buffer chemicals to increase, decrease or stabilise the pH as needed. Buffering allows the preparation to resist changes in its pH.

Neutralisation occurs when the pH is brought to 7 (neutral) via a chemical reaction between an acid and a base. For example, when acids (hydrogen ions H+) and bases (hydroxide ions OH–) combine, the result is water. In a facial treatment, this reaction is encouraged with electricity to 'rebalance' the skin's pH in galvanic desincrustation treatments. The product manufacturer may have a 'neutraliser' to deactivate the peel and settle the skin after an acid peel. When neutralisation occurs on the skin it can cause a tingling sensation.

The pH of the skin cannot be 'rebalanced' permanently. The skin's pH will return to normal after a short period of time. The pH of the skin varies with sweat and sebaceous secretions, and skin microflora. pH can change with the season or environment and depends largely on a person's genetics, age, diet and sex. The pH is different on different parts of the body.

Buffering is necessary to ensure a product's pH is ideal for the skin type and condition. A buffer solution includes a weak acid and a weak alkali that will resist change in the pH when an ingredient is added. For example, a moisturiser with '2% glycolic acid' contains only 40 per cent as much glycolic acid as a moisturiser with '5% glycolic acid'. A corresponding reduction in added ammonia will maintain the required pH. Buffers present in beauty products include glycolic acid/ammonium hydroxide, and citric acid/sodium citrate.

THE EFFECT OF pH ON HEALTH

Strong acids and strong bases can be poisonous or corrosive to the human body. Examples of these are shown in the table of the pH scale above. Strong acids or alkalis will burn by damaging the cells and structural proteins of the tissue. The most sensitive body tissues are the eyes, the lungs and the inner lining of the nose.

| Figure 21.8 | First aid for acids and bases |

AREA AFFECTED	FIRST AID IMMEDIATE ACTION
Digestive system	If acidic, drink milk, raw egg, or an antacid. If alkali, drink citrus juice (e.g. orange or lemon) or vinegar. Seek medical attention in either case.
Respiratory system	Prevent further inhalation. Seek medical attention.
Skin	Flood with water, and if acid, add baking soda.
Eyes	Wash with water for 15 minutes.

Balancing pH levels within the body is imperative to health. Substances throughout the body require a specific pH to allow normal chemical processes to occur. Examples are the following:

- Stomach acid (pH 1–2): this is contains hydrochloric acid, HCl, which helps to break down food by activating the enzyme pepsinogen. When stomach acid leaks into the oesophagus or small intestine it is known as 'reflux', which burns the tissues due to the strong acid pH. The pancreas produces an alkali, sodium bicarbonate, neutralising stomach acid as the food leaves the stomach and, as a consequence, the pH gradually comes to a level of 4–5.
- Blood (pH 7.35–7.45): the pH of blood is by necessity slightly alkaline. The blood oxygen levels are very sensitive to the pH level, in what is known as the Bohr effect. An increase in pH causes haemoglobin to bind to oxygen, while a slightly low pH causes haemoglobin to release oxygen as carbon dioxide, which migrates from the tissues of the body to the blood.

9 Is sodium hydroxide an acid or a base? What is its pH?

10 What is the pH of water?

11 What does a buffer do in a cosmetic?

12 What is neutralisation?

Organic chemistry

Organic chemistry relates to all living things. It is the study of all chemicals that are based on carbon atom chains. Organic compounds are carbon structures but may equally contain other elements such as hydrogen, oxygen and nitrogen.

Carbon needs to gain four electrons to fill its outer shell, and therefore can make four different bonds. It can make single, double or triple bonds; that is, one or more electrons can be used to form a bond to another atom.

HYDROCARBONS

Hydrocarbons are made only of carbon chains bonded to hydrogen, which makes them unreactive, although the smaller volatile forms are usually extremely flammable. The mineral hydrocarbons, such as paraffin and petrolatum (extracted from crude oil), are good at smoothing and sealing the surface of the skin, but do not combine with the chemicals in the skin. They are inexpensive, and since they do not react readily with oxygen (oxidise), they have a longer shelf life.

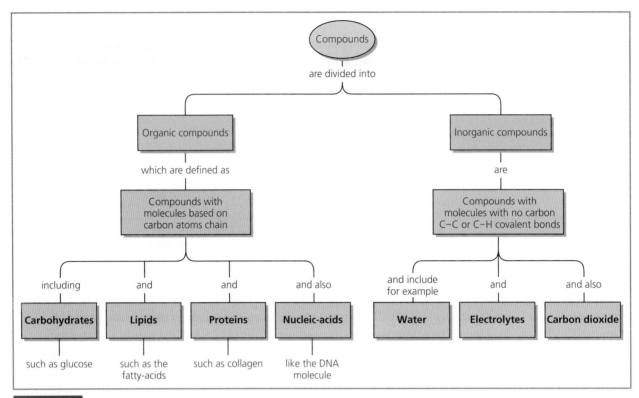

Figure 21.9 Compounds

Note: Products may label themselves 'organic', but this has nothing to do with organic chemistry. It implies that the product has been produced in a biodynamic way, when really it is simply carbon-based. This particular product labelling claim is accountable to the *Australian Certified Organic Standard 2013.*

FUNCTIONAL GROUPS

Functional groups are those parts of molecules that contain characteristic groups of atoms that have recognisable activity, such as acids and bases. Functional groups are responsible for the chemical reactions a compound may undergo. These groups may affect a product's ingredients in binding (physically rather than chemically) with each other and with the skin. Every functional group has a distinctive chemical configuration. Some are listed in Figure 21.10 along with examples of their use in cosmetics.

Figure 21.10 Functional groups and examples of use in cosmetics

FUNCTIONAL GROUP	EXAMPLE OF USES IN COSMETICS
Alcohols	Anti-foaming agents, astringents, surfactants, agents that prevent thickening
Aldehydes	Fragrances, preservatives
Amides	Emulsifiers, surfactants, cleansers
Amines	Foaming agents, emulsifiers, surfactants
Carboxylic acids (fatty acids)	Emollients, thickening agents, soaps
Esters	Emollients, fragrances (fruity)
Ketones	Fragrances, essential oils, solvents (e.g. acetone)

SATURATED AND UNSATURATED BONDS

A saturated compound contains only single bonds between the atoms. Unsaturated compounds contain one or more double or triple bonds between carbon atoms and this is measured by the amount of hydrogen needed to convert these to single bonds.

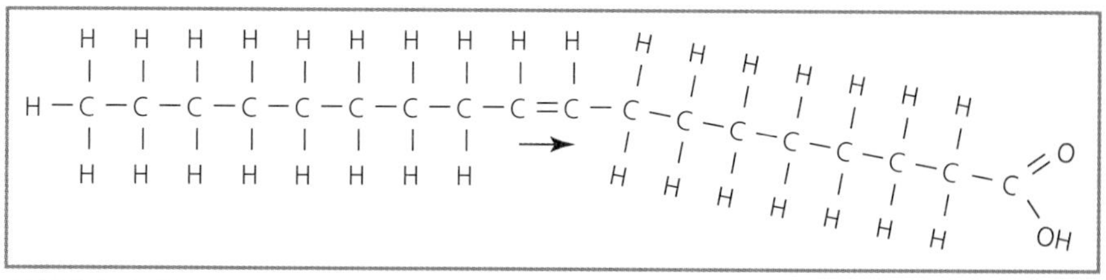

Figure 21.11 Unsaturated compound: oleic acid (found in olive oil)

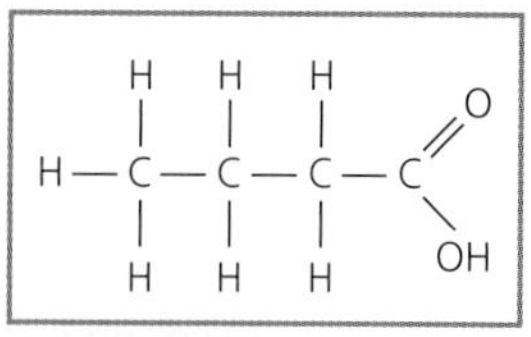

Figure 21.12

Saturated compound: butyric acid (found in butter)

Saturated fats have long flexible carbon chains, which can readily align with each other into stable arrangements. Unsaturation reduces this flexibility by introducing relatively rigid regions and destabilises the chain alignments. As a result, unsaturated compounds melt at lower temperatures than their saturated equivalents.

However, the more unsaturated double or triple bonds in a chemical, the more susceptible it will be to rancidity as these are prone to oxidation. Monounsaturated fats have one double bond, polyunsaturated fats have more than one double bond. In cosmetics, an ingredient that is polyunsaturated is more fluid than a monounsaturated ingredient. Cosmetic chemists can add all types of ingredients to the formulation to prevent rancidity, such as an antioxidant to prevent oxidation.

13 Which of the following best describes organic chemistry?

    A It is chemistry derived only from plant-based origins.

    B It is chemistry that relates to all living things.

    C It must be certified organic to be organic chemistry.

    D All of the above are true.

14 Which of the following is a feature of hydrocarbon chemicals in beauty?

    A They have a longer shelf life (don't oxidise quickly).

    B They don't tend to react/combine with the skin chemicals.

    C They are often cheap products.

    D All of the above are true.

15 a What is a functional group?

    b Give an example of a functional group with a cosmetic that might contain an ingredient with that functional group.

16 What is the difference between an unsaturated and a saturated fat?

Elastin

Interwoven among the bundles of collagen is a network of elastic fibres that give the skin its elastic properties. Elastin is a fibrous protein that makes up 0.6–2.1 per cent of the dry weight of the skin (compared to 72% dry weight for collagen).

STRUCTURE OF ELASTIN FIBRES

Elastin fibres consist of two parts:

1 fibre membrane envelope
2 internal microfibrils and shapeless matrix gel.

In the embryo, new fibres are composed almost entirely of microfibrillar protein coat. Later in development, the proportion of internal elastin and matrix increases. More than 90 per cent of a fully developed fibre is elastin matrix.

The basic molecular unit is a linear polypeptide composed of fibrillin with prevalent amino acids: one-third glycine, unevenly distributed with analine, valine and some hydroxyproline. Molecules called *desmosine*, an amino acid that occurs only in elastin fibres (not to be confused with desmosomes of the cell membrane), link elastin fibres together making a supramolecular structure of elastin, which is insoluble.

ROLE OF ELASTIN

Elastin has a complementary role with collagen. It is able to restore a normal fibrous array after deformation by external mechanical forces, due to the elastic rebound nature of fibres. The structural arrangement of the fibres is probably far more complex than has been shown. All histological (microscope) slides of skin that have been specifically stained to show elastin fibres always produce a random arrangement of fibres going in many directions. This is quite possibly an artefact of slide sectioning (cutting) and staining. The true pattern may never be fully known because the elastin fibres recoil and shrink during slide preparation.

TURNOVER OF ELASTIN

Elastin is also synthesised by fibroblasts. It is speculated that elastin synthesis follows a process similar to that of collagen synthesis, whereby a soluble precursor (pro-elastin) is synthesised within the cell, which then passes into the extracellular space, where the telopeptides are removed (tropoelastin) and then structural changes due to cross-linking occur (final elastin molecule). How the actual fibre envelope with its internal mix of microfibrils and gel matrix is organised is still a mystery.

Although the metabolic turnover of connective tissue protein is slow, a portion of elastin is continuously degraded and replaced by newly synthesised fibres. The enzymes responsible for degradation are called *elastases*, such as *stromelysin* produced by fibroblasts.

CHANGES WITH AGE

There appears to be a change in the balance between breakdown and synthesis with increasing age. It appears that it is the sub-epidermal plexus (in the papillary dermis) that shows a marked decrease in elastin with age, rather than the deeper (reticular) dermal layers. The papillary dermis typically has fine vertical fibres attached to the basement membrane and threading down into the lower reticular dermis.

The reticular dermis has coarser fibres which intertwine. UVA radiation damage to the upper dermis usually damages the fine vertical fibres and this may cause the out-folding of the upper skin (remember the basement membrane shows significant changes with ageing) resulting in the patterns of skin wrinkles, also known as *rhytids*. As women have a thinner epidermis and dermis than men, the extent of damage and alteration of protein fibres like elastin become more obvious at an earlier age. Certain elastases in the body have been shown to increase in activity with increasing age, thus, perhaps, causing this change in balance. Elastin fibres in the dermis typically degrade with chronological ageing. However, in photo-aged skin there is a massive increase in the elastin produced by UV-stimulated fibroblasts. It is a thicker, formless elastin that appears somewhat like a lump of cooked spaghetti on a plate.

Adapted from Dylan Webb's work in Victoria University's *Skin Science Student Resource 2015.*

17 What are rhytids?

_____

18 Why are women more affected by UVA damage to the elastic fibres in the papillary layer than men?

_____

Water and the skin

Water plays a critical role in the structure and function of the skin. All living cells contain a high proportion of water (about 70–80%) and this is also true for the living cells of the dermis and epidermis (stratum germinativum to the stratum lucidum), but not the dead cells of the stratum corneum.

Water in the dermis is also present in high proportions in the matrix of the dermis where glycosaminoglycans (GAGs) bind and hold onto large quantities of water (see later section on glycosaminoglycans and the dermis).

In the stratum corneum, the water content is considerably lower than in living cells. At maximum hydration, the water content of the stratum corneum is about 30 per cent but when superficially dehydrated it can be as low as 10 per cent. The beauty therapist uses moisturising treatments to maximise the water content of the stratum corneum (but probably only the upper cells). At 30 per cent, the keratin proteins are softened and are quite flexible but at 10 per cent they are dry and toughened and do not bend or flex easily – a bit like the difference between wet spaghetti and dry spaghetti. Wet keratin also 'plumps up' rather nicely, making 'fine lines and wrinkles disappear'. This is how moisturising treatments work.

Moisturising treatments cannot, of course, increase the water content of living cells. The homeostatic mechanisms controlling the body's water content are the major factors involved in maintaining the water content of living cells. This involves kidneys, blood and lymph systems and hormones. Any significant reduction of the water content of these cells would cause serious cell damage and death. So, dehydration of the deeper epidermis and dermis would only occur when the whole body is significantly dehydrated; for example, from hyperhidrosis, strenuous exercise on a hot day, sickness involving vomiting and/or diarrhoea. A person who has died of dehydration certainly has not dried out from the outside in.

WATER AND THE STRATUM CORNEUM

Water in the stratum corneum can evaporate. The rate of evaporation of water from the skin surface is influenced by a variety of features:

1 environmental:

- temperature: higher temperatures mean higher evaporation

- relative humidity: drier air causes higher evaporation

- wind velocity: faster drying on a windy day

2 physiological:

- health and wellness: stress and disease interfere with skin growth

- integrity of the skin: ability to produce the *epidermal lipid barrier*

- diet: full intake of nutrients, vitamins and mineral to grow good skin

- genetics; for example, dermatitis, psoriasis and eczema

3 pharmacological:

- drug use; for example, hydrocortisone, retinoids, Warfarin®

- hormone treatments; for example, oral contraceptives, hormone replacement therapy (HRT)

- toxic chemicals; for example, alcohol, caffeine, nicotine

4 cosmetic:

- products; for example, scrubs, moisturisers

- exposure to fragrances

- treatment effects from exfoliations and peels, UV exposure and massages.

All of the above affect the texture, feel and appearance of the stratum corneum. (Remember that this is only a very, very thin layer on top of the very thin epidermis.) It is therefore important to understand something of the nature of water and the skin if you are going to be able to assess causes of skin conditions and effects of skin treatments and products. The biggest market in cosmetics is skin products that increase the 'moisture content' of so-called 'dry' skin.

WATER MOLECULES AND HYDROGEN BONDING

As you no doubt already know, each and every molecule of water is polar – it has a partial negative charge on the oxygen and partial positive charge on the hydrogen atoms. This means each pole can and does form weak associations with some other atom of the opposite charge. Thus, water molecules can link up with each other.

Water molecules can also link up with charged atoms on other molecules, especially 'nice, fat' slightly negative (electron rich) oxygen, nitrogen or sulphur atoms sticking out from larger molecules.

Molecules in the skin that can form hydrogen bonds with water may be of two types:

1   Small and water soluble: these are humectant substances that can be washed from skin (if the skin is pre-washed with acetone). The collection of such substances is called natural moisturising factor (NMF) and includes:

- sodium pyrrolidone carboxylic acid (NaPCA)
- lactic acid
- sodium lactate
- potassium lactate
- urea

2   Larger, insoluble molecules: insoluble and larger molecules in the skin that can form hydrogen bonds with water. These include:

- keratin
- involucrin
- keratohyalin/filaggrin.

These hydrogen bonds bind water molecules to substances in the stratum corneum and are very weak – about 10 per cent of the strength of a covalent bond. But if there are enough of them you have an effective water 'trap'. Water in the skin is considered to be either *free water* or *bound water*.

FREE WATER VERSUS BOUND WATER

Free water is highly mobile and easily available when the need for water molecules is called upon. It:

- acts as a solvent for most of the body's chemicals
- has an important role in osmosis and diffusion
- can move by capillary action (creeping) along membrane surfaces and through microfine channels within the stratum corneum.

Bound water is not highly mobile and more difficult to break away (like it is handcuffed in place) from the attachment because of the hydrogen bonding. It:

- plays a vital role in the structure of proteins by linking across two molecules. This causes proteins to swell ('plump') as the water molecules muscle their way between two protein molecules
- acts as a reservoir of water to be accessed only when free water has been exhausted.

MEASUREMENT OF WATER IN THE SKIN

There are two main types of measurement related to water and the skin that can be made experimentally:

1   total water content – measuring how much water there is in the skin
2   the rate of movement of water through the skin (TEWL).

The two measurements are closely related. Just how it is measured is critical to understanding the moisture content and movement of water molecules in the skin. Both cannot be measured at the same time with any degree of precision or accuracy.

CONCLUSIONS ON WATER AND THE STRATUM CORNEUM

You might well ask: If scientists can't work out how to measure water, let alone what is happening, how can I figure it out? But there *are* some useful 'truths' beginning to emerge:

1   In the stratum corneum at low water content (10%), the water molecules are tightly bound. That is, they are all held as bound water.
2   At high water content (30%), water molecules in the stratum corneum act as bulk, liquid water. The skin absorbs moisture very poorly from the air up to 70 per cent relative humidity (RH) ('dry' air). At over 90 per cent RH there is dramatic absorption of moisture from the air.
3   As RH increases, TEWL is reduced. That is, the skin becomes saturated.
4   As temperature increases, TEWL dramatically increases as the water molecules in the stratum corneum become more 'active' with higher kinetic energy.
5   TEWL varies considerably at different anatomical sites.

6   Thick skin is not more resistant to water loss. In fact, the opposite occurs. Finger and toenails lose water faster than skin. TEWL is lower for abdominal skin than for callused skin on the hands or feet.

7   TEWL of highly hydrated, 'plumped up' skin may be higher than for untreated dry skin (i.e. there is an increase from 10 micrometres [μm] to 40 μm thickness of the stratum corneum) because it has many more free water molecules to lose.

8   Epidermal lipids are crucial to the skin barrier properties and rate of TEWL. This may account for the difference in TEWL between abdominal stratum corneum and volar skin.

As the temperature of the skin increases, the number of binding sites for water molecules in the stratum corneum increases because the molecular structure of the proteins of the stratum corneum 'open up'. This may be a way of preventing too much TEWL with higher temperatures.

Therefore, the stratum corneum appears to be equipped with a mechanism that attempts to reduce the impact of excessive loss of water through TEWL when water content is getting low and environmental conditions seem to favour even more rapid loss of precious water molecules. This 'opening up' means more proteins grab water molecules that are drifting free and hang onto them, or risk going very dry and hard and possibly leading to epidermal cracking and fracturing.

Physiologically speaking, this means that the cells of the stratum germinativum must be up to the task of producing daughter cells (keratinocytes) capable of producing the right type of protein (mainly the upper stratum corneum type of keratins K1/K10) and the right combination of the highly specific epidermal lipids (multi-laminate ceramide composites). Should something go wrong with these cells and/or their ability to grow and differentiate, then the very real possibility of a defective stratum corneum arises.

The reality is that the vast majority of Australians have already done quite a lot of damage to their skin before they are out of their teens. They will then go on in life producing defective skin that will exhibit a variety of symptoms of sun damage and photo-ageing.

One of the symptoms of photo-aged skin, and in some people a related (usually inherited) disorder, is called *xerotic* skin.

XEROSIS (DRY SKIN)

Xerosis, the term used to describe the 'dry' or rough quality of skin that occurs in all age groups, is more common among the elderly. The condition may be generalised, but is usually prominent on the lower legs and is exacerbated by a low humidity (low RH) environment classically found in dry, overheated rooms and during dry, cold weather in the winter months.

Xerosis reflects minor abnormalities in the epidermal differentiation and maturation processes. Lipogenesis and keratinogenesis are damaged, which results in an irregular orientation at the surface for the stratum corneum. The corneocytes don't lie flat. They are tilted upward a fraction so that these tips dry and curl, giving the skin a white reflective surface and a scratchy feel.

The initial assumption that the disorder resulted from a lack of water in the skin has been disproved by several investigators. Water content in the stratum corneum of lower abdominal skin has been reported to increase slightly, although not statistically, from a mean value of 58 per cent in the fourth decade to 63 per cent in the ninth decade; while the rate of TEWL across the stratum corneum as tested with in-vivo experimentation appears constant with age.

Once xerosis is present, frequent regular use of a topical emollient makes the skin more attractive and more comfortable. However, this doesn't get at the root cause of xerotic skin, which seems to lie in the stratum germinativum.

Be aware that the mechanism action of emollients and even the origins of xerosis is so poorly understood by most physicians and virtually all patients that it is no wonder product promotional material so misrepresents the condition and treatment expectations.

SOLVING DRY SKIN PROBLEMS

There are two possible solutions to superficial dehydration (usually limited to the upper regions of the stratum corneum):

1   Temporary solution: daily use of humectant or occludent moisturiser and regular skin (surface) treatments. We are creatures of habit and this solution suits most individuals. We may even enjoy the daily cosmetic rituals that reinforce a positive attitude that comes with looking our best and making a difference.

2   Permanent solution: destruction of the 'altered' stratum germinativum cells with the growth of (hopefully) more normal healthy cells. This could involve:

   • chemical skin peels (phenols, TCA, enzymes, AHAs)

   • laser resurfacing

- dermabrasion

- therapeutic drugs (retinoids).

Some of these treatments are not for the faint hearted and should only be done following expert advice. The aim is to eliminate the underlying cause of this condition lurking in the abnormal keratinocytes in the stratum germinativum. It is not a cosmetic solution. It is a therapeutic one. The outcomes are not always successful.

LIFESTYLE ADVICE

Drinking six glasses of water will not make your skin wetter. Remember homeostasis: intra- and intercellular water concentrations are regulated by kidney function. However, if you drink more water you are drinking less coffee, tea, soft drink and alcohol and this reduces the toxic load your liver has to process and your body functions much better – so you can grow better and healthier skin.

A real contributor to healthy skin is the positive attitude toward better health that makes your brain release more 'feel good' hormones and growth factors.

19 Why is it not necessary to always 'drink more water'?

Glycosaminoglycans

Glycosaminoglycans (GAGs) are a type of sugar molecule; they were formerly called mucopolysaccharides. They form the major part of the matrix of the dermis. They are gel-like in structure with extremely high water-binding ability (humectant). Humectants are important in the skin because they bind to water and keep collagen and elastin supple. In the dermis, the main cell producing GAG is the fibroblast. GAGs are also present on the cell membranes and in the intercellular spaces of the epidermis. Here it is produced by the keratinocytes. There are two major types:

1 Pure GAG: not bound to protein molecules

- hyaluronic acid (HA)

2 Proteoglycans: GAGs bound to a core protein

- dermatan sulphate (DS)

- chondroitan

- chondroitan 4 sulphate

- chondroitan 6 sulphate

- keratan sulphate

- heparin (found in the blood).

GLYCOSAMINOGLYCAN STRUCTURE

As basic units, all GAGs are complex carbohydrates consisting of disaccharides (double sugars):

1 hexosamine sugar (6 carbon/nitrogen containing sugar: either glucosamine or galactosamine)
2 acid sugar (either glucuronic or iduronic acid).
The double sugar units alternate along the length of the molecule.

POLYMER STRUCTURE

These large polymers may have between 15 and 5000 disaccharide units in long chains. They form proteoglycans when a special linking unit at one end attaches the polysaccharide to a special core protein. Many GAGs attach to the protein core. Unlike proteins, GAGs do not appear to adopt tertiary or quaternary levels of structure and remain as linear or twisted shapes.

This means they can be long, floppy and physically unstable. These chains may be from several thousand MW (molecular weight unit) to several hundred thousand (as a comparison, a water molecule is 18 MW units), which means these molecules are extremely large.

All GAGs are anionic in charge; that is, they carry negative charge; this is important to know when doing electrical treatments.

GAGs form complex arrangements with proteins often called super-molecules or 'feather-shaped' structures, allowing them to bond to many, many molecules of water. In these structures, many GAG molecules are attached to a backbone protein, the 'core protein'.

One such proteoglycan molecule could consist of approximately 100 chondroitan, 6 sulphate chains and 30–60 keratan sulphate chains attached to a core protein. They may have a MW size of a couple of million MW units.

In the dermis, many hundreds of these proteoglycans may be attached to a HA molecule, making a huge supermolecule. The attachments of the proteoglycans to the HA are by very weak hydrogen bonding, thus these structures are difficult to extract in one piece.

FUNCTIONS OF GAGS

GAGs have the following functions:

1   Stabilise the matrix: GAGs hold cells and fibrous proteins in place as 'setting' gel.
2   Provide resistance to compression: the gel softens the impact.
3   Viscoelastic rebound: when the skin is distorted by force, it will rebound into the previous shape. The gel has a certain 'shape memory'. (The inside of your eyeball is packed with glass-clear HA, which maintains the shape of the eyeballs after you squeeze or rub them.)
4   Morphology: GAGs fill out tissue to give form (the earlobes are filled with GAGs).
5   Effect osmotic changes and material transport: diffusion and osmosis are altered in a gel material. White blood cells (WBC), especially macrophages, 'swim' through the dermis and sometimes drag themselves along fibres of collagen. They ease their journey by liquefying the gel in front of themselves and re-gelling it behind them as they go. This is achieved by controlling the release and absorption of salts across their cell membranes.
6   Lubrication: gel is slippery. This is especially important with the dermatan sulphate coatings around collagen fibres. This effectively 'greases' them up and allows them to slide over one another. Cell membranes also have gel sugar coatings (e.g. syndecan, a proteoglycan is involved in the organisation of the cytoskeleton).
7   Hydration: bound water can be reclaimed as free liquid water in times of stress.

Adapted from Dylan Webb's work in Victoria University's *Skin Science Student Resource 2015*.

20 What is the role of dermatan sulphate in the dermis?

_____

21 How would you describe, in simple terms, the structure of a glycosaminoglycan structure?

_____

22 List one true glycosaminoglycan and three proteoglycans that are found in the dermis.

_____

_____

_____

_____

23 Which of the following is *not* a function of GAGs in the skin?

   A   Lubrication

   B   Provide resistance to compression

   C   Hydration

   D   Waterproofing

   E   Morphology

   F   Stabilise the dermal matrix

Percutaneous penetration of cosmetic ingredients

The degree of absorption of cosmetic products into the skin depends on how the chemicals in the product will mix with the chemicals in the skin. It helps to understand how the skin works as a barrier and then to know the ways cosmetic chemists are able to work the product into the skin.

THE SKIN AS A BARRIER

The skin is a very good external barrier, designed to protect against moisture loss and foreign invasion. More specifically, the tough corneocytes of the stratum corneum and intercellular lipids that form a liquid crystalline lamellar structure made of lipids and ceramides help prevent TEWL. As the cells differentiate from the stratum granulosum, where keratohyalin becomes insoluble keratin, the lamellar granules move to the outside of the cell, forming the intercellular matrix. The keratinocytes and the intercellular matrix of the epidermis resemble bricks and mortar (see Figure 21.14) – the cells are the bricks and the intercellular lipids are the cement mortar. Just as TEWL occurs, products can penetrate this barrier.

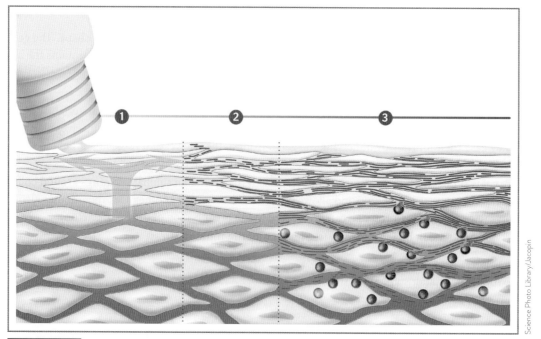

Science Photo Library/Jacopin

Figure 21.13 Skin hydration. 1. Occludent prevents TEWL. 2. Humectants fill intercellular spaces. 3. Lipids, liposomes and other active ingredients are taken up by lamellar bodies (extracellular lipids in the stratum granulosum).

CUTANEOUS ABSORPTION

Products penetrate the skin by passive diffusion through the epidermis at the target site. Many cosmetic products are designed to sit on the skin and not enter tissue fluid, though they may be beneficial to the dermal layers. Pharmaceutical products such as hydrocortisone may be applied to target blood vessels and interstitial fluid during a histamine reaction; it would be of no use in the epidermis. Likewise, cosmetic products that are lipophilic are of no use in interstitial fluid, so it is ideal that they do not descend to the dermis. The skin is practically impermeable when dry because substances need water or oil to move by passive diffusion. The rate of movement of a substance through the skin depends on:

- the size of the molecule
- the shape of the molecule
- the ability of the molecule to interact with the chemistry of the skin
- the ability of the molecule to pass through the cell membrane
- the ability of the molecule to attach or bind with receptor proteins on the cell membrane.

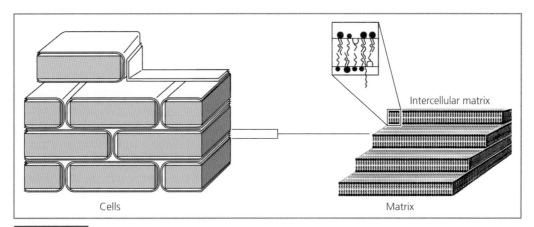

Figure 21.14 The 'brick-and-mortar' structure in the stratum corneum

ABSORPTION PATHWAYS

Absorption of the product by the epidermis can occur in a number of ways:

- *Via the hair follicle, sebaceous gland, eccrine or apocrine sweat gland.* The cells of the hair follicle are targeted with lipophilic ingredients for anti-acne ingredients and hair growth products. The sweat gland absorbs hydrophilic ingredients best because there are no corneocytes to navigate.

- *Between the cells.* There are approximately 30 nanometres (nm) of space between intercellular lipids. This is the preferred route for most products if they contain lipophilic ingredients. The skin's lipids are much more permeable than the cell membrane for oil-based ingredients. Hydrophilic ingredients must pass through 'gaps' in the lipids. Retinol, if it can reach the stratum germinativum, is an example of a lipophilic ingredient shown to penetrate the cell membrane and act with some anti-ageing effect by affecting cell differentiation and proliferation.

- *Through the cells (transcellular).* The dead keratinocytes, although hard and relatively dehydrated, may be softened with the use of chemical surfactants and steam. The cells of the stratum corneum have about 50 per cent less water than the underlying cells, so diffusion is much less effective through these cells. Even so, hydrophilic substances can travel this path, because there is more water within the cells. Exfoliation prior to product application allows for better exposure to hydrated cells.

CUTANEOUS ADSORPTION

Some substances are known to not penetrate the skin but instead provide benefit to the skin by attaching to the stratum corneum. Large molecules with humectant properties, such as collagen and other NMFs, bond to the corneocytes, intercellular lipids, hair or nail. The strength of the ingredient's bond with the skin is qualified by how 'substantive' it can be. If a product cannot be washed away easily it is said to be *substantive*.

THE USE OF VEHICLES AND CARRIERS IN PERCUTANEOUS PENETRATION

The chemical must reach its target site. To use retinol as an example: the stratum germinatium is its target site, and so if it is to penetrate further it will be systemically absorbed by the blood capillaries. Percutaneous absorption can be enhanced in a number of ways. These include:

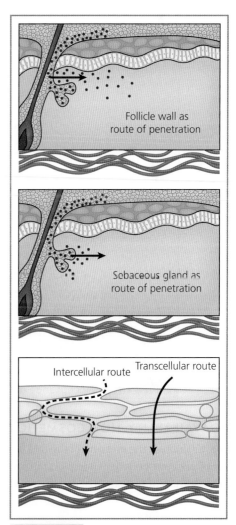

Follicle wall as route of penetration

Sebaceous gland as route of penetration

Intercellular route   Transcellular route

Figure 21.15 Routes of the absorption of chemical ingredients

- *Chemicals.* Certain chemicals or enzymes are used that modify epidermal lipids or increase hydration, thus paving the way for product penetration. Examples include alpha-hydroxy acids, propylene glycol and urea.
- *Electricity.* Galvanic iontophoresis uses electrical forces to 'push' the product into the skin. This is covered in more detail in Chapter 22 Specialist facial treatments.
- *Micro- and nanoparticles and the application of heat.* The vasodilation of the capillaries produced by heating has a hydrating effect on the surface of the skin, which increases the rate of passive diffusion. Beauty therapists will use electrical equipment, such as the steamer or massage, to increase circulation to the skin. Micro-particles in an ingredient are 0.1–100 micrometres in size. Nanoparticles are between 1 and 100 nanometres in size. A micrometre is one millionth of a metre, and a nanometre is one billionth of a metre. There is debate as to whether nanoparticles are dangerous to health as the product can penetrate much deeper into the skin, potentially into the bloodstream, although naturally occurring materials will probably be broken down by normal metabolism. It is believed that nanoparticles larger than 40 nm in diameter will not pass through the stratum corneum. Lipids are often delivered as nanoparticles for dry skin and as liposomes for oily/acneic skin. The use of small particles can increase absorption by vastly increasing the effective surface area of the product.
- *Liposomes.* These are used to encapsulate ingredients to be accepted into the cells. The active ingredient is surrounded by a phospholipid bilayer, much like the cell membrane. The cell membrane of the skin cells will readily accept/merge with the liposome's bilayer, building it into its own membrane, allowing the ingredients to enter the cell (see **Figure 21.16**). The phospholipids are often obtained from egg yolk, and lecithin from soybean. Active ingredients delivered by liposomes are often the more unstable chemicals, including vitamin E, vitamin C, aloe vera and jojoba. Liposomes should be under 200 nm for effective penetration of the skin.
- *Through penetration enhancers.* These include silicones and carboxylic acids.

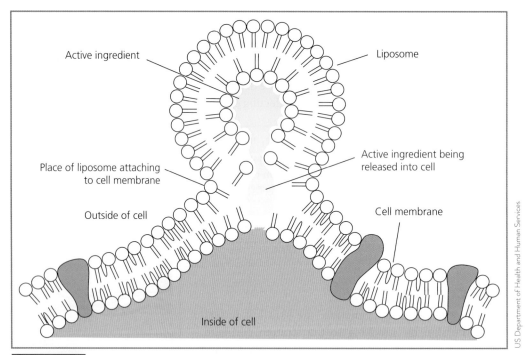

US Department of Health and Human Services

Figure 21.16 Acceptance of liposomes into a cell

BIOAVAILABLITY: HOW COSMETIC CHEMISTS MEASURE THE RATE OF PENETRATION

Bioavailability is the measure of the rate it takes an active ingredient to reach and become available at the target site. The rate of penetration is dependent on the percentage of ingredient applied and may be assessed by scientific experiments. The key areas of the skin that can be tested are outlined below.

The skin's surface

The skin's surface may be easily tested for biological activity with scrapes (with a dull blade), acetate tape, a trichogram (which observes the hair), biopsy, breeding cultures and allergy testing.

The receptor sites

Molecules found on the outside of a cell membrane of the target area of the skin. If the target area was to perform a physiological process, then it can be tested (herein lies the 'therapeutic cosmeceutical' claims debate).

The body's removal of the product

The ingredient may be metabolised by the skin cells or cells of the immune system or removed via the circulatory or lymphatic system. For example, to demonstrate the diffusion or flow-through of the ingredient through the skin, a cosmetic chemist may place a piece of skin over a plasma-like fluid, then apply a product to that skin. The chemist will measure the active ingredient in the product and later in the 'plasma' fluid as a function of time.

BIOLOGICAL ACTIVITY

Biological activity is the action the product has on the skin. It should be tested to see whether it has the desired effects. Here are some examples of desired effects and some ways scientists test before and after application of the product:

- Skin hydration may be tested with an electrical appliance directly to the skin. It will assess the content of NMF within the stratum corneum by measuring the skin's electrical resistance.
- Skin elasticity may be tested with an optical device that will pull the skin under vacuum momentarily and release it, recording the time taken to 'bounce back'.
- Frictional resistance indicates oiliness or dryness of the skin. When an oily product has penetrated, the skin will be less oily. There are machines that measure the frictional resistance to estimate the rate of product penetration.
- TEWL may be measured with an evaporimeter, which measures the atmospheric humidity immediately from the skin. This is helpful in determining the effectiveness of a product in sealing in moisture.

Cosmetic chemistry in beauty products

Knowledge
Evidence

With knowledge of the chemicals and their interactions, beauty therapists have a better understanding of how beauty products work on the skin.

COSMETIC EMULSIONS

Emulsions are a blend of oil and water where, ordinarily, oil and water do not combine. Emulsions tend to have a milky appearance. Incidentally, cow's milk is an example of a natural emulsion. It is well known that the cream will always rise to the top, due to its high lipid content. If you shake it fast enough, the oil and water will emulsify, but only for a short period.

The oil and water in emulsions are held together with the addition of an emulsifier. This will make the emulsion stable, ensuring the oil droplets don't clump together and sit at the top of the product. (Surfactants and hydrocolloids are discussed later in this chapter.)

Emulsions can be oil in water (o/w) or water in oil (w/o). The oil in water emulsion will feel less oily, as in 'lotions' and 'creams' and the water in oil emulsion will feel more oily, as in 'emollients', 'sunscreens' and 'ampoules'. Understandably, the oil in water preparation is more popular in skincare cosmetics.

MULTIPHASE EMULSIONS

A single-phase emulsion is a simple o/w or a w/o emulsion. Products can also have multiphases, such as o/w/o. The benefits of using a multiphase emulsion include:

- the slow release of the active ingredient
- the protection of unstable ingredients, such as the antioxidants vitamin C and E, from heat and oxidation
- the delivery of one or more chemicals that do not naturally mix together
- its tendency to have a finer texture.
  Microemulsions are usually o/w and are less than 200 nm in diameter. Benefits include:
- smaller droplets providing enhanced absorption through the skin barrier
- increased stability and shelf life in heat
- appearance and texture being finer for a smoother application
- preventing irritation by the use of less surfactant.
  The most common uses for microemulsions are for soothing ingredients, moisturisers and sunscreens. Fragrances and essential oils are readily mixed into the cosmetic preparation.

Hydrocolloid emulsions are emulsions that do not use surfactants. They are long chains of carbohydrates that bind to water and interlace between the oil droplets, preventing the oil droplets from bonding together. At their best, they suspend the lipids in a 'meshwork', which gives the emulsion a gel-like consistency. Pectin, which is used to make jam, is a good example of a hydrocolloid.

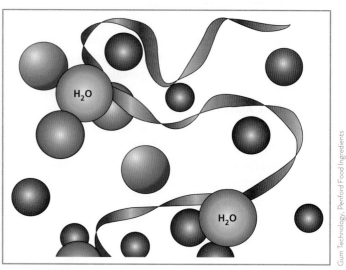

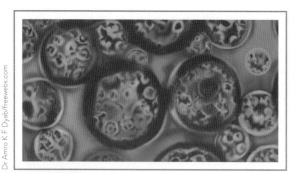

Figure 21.17 Multiphase emulsion: water in oil in water (w/o/w emulsion)

Figure 21.18 Hydrocolloid emulsion: interlacing starch (in orange) with water droplets attached, suspending droplets of oil-based droplets (blue, purple)

SURFACTANTS

When a surfactant interacts with a lipid, it modifies the surface characteristics so the lipid disperses more readily in water. Surfactants reduce *surface tension* between two liquids, which makes them very good emulsifiers; they can be used for a range of cosmetic purposes, such as in the manufacture of liposomes, cleansers and lotions. Cosmetic chemists refer to different types of surfactants as emulsifiers (thickeners), detergents, dispersants and wetting agents.

All surfactants have a hydrophilic head and a hydrophobic tail. The hydrophilic head binds with the water in the emulsion and the hydrophobic (lipophilic) tail binds with the oil. The net effect is to modify the oil so that it presents a hydrophilic surface to the solution.

Examples of commonly used emulsifiers in cosmetics are soy lecithin, egg yolk, triethanolamine (TEA) sodium lauryl sulphate, sodium laureth sulphate, polysorbate, cetyl alcohol, stearyl alcohol, sodium lauroyl glutamate, sorbitan monostearate, sorbitan trioleate, triglyceryl monooleate and PEG-7 glyceryl monococoate. These surfactants are often used in a blend.

Detergents are surfactants that cleanse the skin by clinging to the lipids and debris to be washed away. Surfactants are commonly used as detergents in cleansers, once dubbed the 'soap-less soap'. This is of benefit because detergents can be just as useful when they are pH balanced. A soapy foaming cleanser will leave an insoluble salt film on the skin's surface. Over-drying the skin can encourage hyperactive sebaceous activity in an oily skin and cause a dry skin to become flaky.

Some inexpensive facial cleansers have used formulations that include the detergents sodium lauryl sulphate and sodium laureth sulphate in high quantities.

Soaps are made when a fatty acid reacts with an alkali. The alkali is often sodium hydroxide (lye). A very strong alkali is combined with a fatty acid such as animal fat or olive oil. The fatty acid is said to undergo 'saponification'. Saponification occurs when a triglyceride in the presence of a strong alkali (such as lye) is broken into free fatty acids and a glycerol. The problem of the use of soap on the skin is the 'scum' it leaves. The scum is the result of the salt formed (e.g. sodium stearate) and its reaction with the minerals in water (e.g. calcium).

- *Ionic and non-ionic surfactants* are the categories of surfactants and are based on the balance of their ionic charge. They react with epidermal lipids, and irritate the skin at various levels.
- *Anionic surfactants* have a polar head that is negatively charged. The functional group, the sulphates, includes sodium lauryl sulphate (SLS), sodium laureth sulphate (also known as sodium laurel ether sulphate, SLES) and sodium myreth sulphate. The functional group, the carboxylates, includes sodium stearate (soap). Their residue leaves a rough, dry feeling to the skin. Anionic surfactants have medium irritation levels.

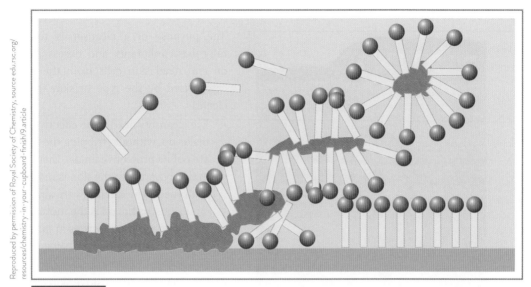

Figure 21.19 The detergent's non-polar tail (lipophilic) binds to the grease at the skin's surface, while its polar head (hydrophilic) causes the grease to mix with water and wash away

Reproduced by permission of Royal Society of Chemistry, source edu.rsc.org/resources/chemistry-in-your-cupboard-finish/9.article

- *Cationic surfactants* have a positively charged ionic head, and are often from the quarternary ammonium group (quats). Quats are good antiseptics, such as benzalkonium chloride, which is often used in hand sanitisers. Cationic surfactants are very irritating to the skin, but may be used to neutralise an anionic product. Hair conditioner contains a cationic surfactant that neutralises anionic shampoo surfactants, just as fabric softeners neutralise laundry detergent!

- *Amphoteric surfactants* are an acid and base at the same time. Water falls into this category. They can be used to emulsify, but do not perform a specific action in the context of beauty and skincare, except for use as a foaming stabiliser.

- *Non-ionic surfactants* are often used for mild cleansers because they do not ionise. Examples in cosmetics include TEA, lauryl glucoside, cocamide MEA, glyceryl laurate and polyethylene glycol esters (PEGs).

24 Note that although non-ionic surfactants do not ionise, these chemicals can still be dangerous to health. Look up three of the listed non-ionic surfactants and write a half-page summary of the current research.

To see the surfactant chemicals listed above at a glance and to identify their significance in a cosmetic product is daunting and it is very difficult to say with any authority what effect the surfactant will have on the skin. The product will likely contain a number of surfactants combined at different stages of production. We can only determine what that surfactant does in a given chemical reaction and take guidance from the product manufacturer on how best to use the product. Can it be used with other products in that line? Can it be mixed with products from another manufacturer? In addition, we can learn about the safety of cosmetic ingredients – low-risk surfactants are expensive ingredients, so cheaper and more toxic ingredients are frequently promoted by interested lobby groups.

A cosmetic that is said to be pH balanced should be the same pH as the skin it aims to treat, usually 4.5–5.5 depending on the skin type and condition. Some more acidic products, such as AHAs, are applied briefly to treat the skin.

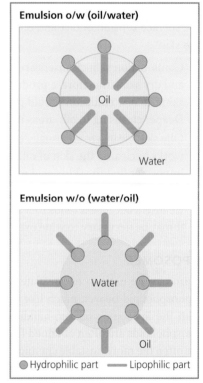

Emulsion o/w (oil/water)

Oil

Water

Emulsion w/o (water/oil)

Water

Oil

● Hydrophilic part ━━ Lipophilic part

www.cargilltexturizing.com

Figure 21.20 Oil in water ('water-based') and water in oil ('oil-based') emulsions both use a surfactant that has a hydrophilic head and a lipophilic tail

RANGE OF EMULSIONS

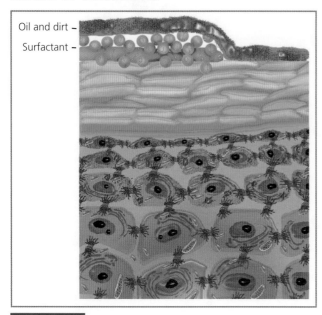

Oil and dirt –
Surfactant –

Figure 21.21 Surfactants. Detergents are the main type of surfactant used in skin cleansers. They reduce the surface tension of dirt and oils and lift them from the skin

The purpose of a cleanser is to remove make-up, microbes, pollutants and excess skin debris such as oil and dead skin cells from the skin. Cleansers are categorised by the type of skin they are intended to clean.

- Eye cleansers need to be effective in removing make-up, without stripping the fine skin of the area of its protective lipids. There are few oil glands in the eye area, so the skin is more vulnerable to dryness. The cleansers are almost always w/o emulsions, as the oil is the more effective solvent for the mineral oils and waxes in make-up. In addition, it is obviously important that such preparations be non-irritant to the eye.

- Dry, mature and sensitive skin cleansers tend to be cream w/o emulsions. They are milky in consistency with high lipid content to prevent the skin from dryness. Some of these cleansers may be used without water because the oil is the active solvent rather than the surfactant. This is particularly beneficial for the rare sensitive skins that can react to the minerals in tap water.

- Combination skin cleansers are lotion o/w emulsions. The skin has the needs of any combination of oily, normal and dry skin. Therefore, a product line selling a 'normal skin cleanser' can also be appropriate for a 'combination skin'. An o/w emulsion should provide a balanced cleansing lotion that maintains a normal or combination skin. Deeper cleansing to oily areas may be done with an exfoliant or absorbent mask.

- Oily skin cleansers are usually detergent-based cleansers. They do not use oil, and the surfactant (detergent) works to cleanse the skin of oil, dead skin cells and bacteria. Detergent cleansers have a neutral pH (pH 7), so it is important that the skin's acid mantle is restored with a toner.

- Foaming (soap) cleansers contain strong alkali surfactants that form a lather. Soap is not common in professional cleansers because it tends to leave a film on the skin, which disrupts the skin's acid mantle. Where soaps are used, it may be necessary to wipe over the skin with an alcohol-based toner to remove it.

LIPOSOMES

Lipsomes are used to encapsulate ingredients to be accepted into the cells. The active ingredient is surrounded by a phospholipid bilayer, much like the cell membrane. The cell membrane of the skin cells will readily accept/merge with the liposome's bilayer, building it into its own membrane and allowing the ingredients to enter the cell. The phospholipids are often obtained from egg yolk, and lecithin from soybean. Active ingredients delivered by liposomes are often the more unstable chemicals, including vitamin E, vitamin C, aloe vera and jojoba. Liposomes should be under 200 nm for effective penetration of the skin.

Cosmetic ingredients and regulations

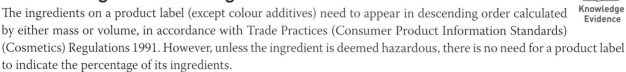
Knowledge Evidence

The ingredients on a product label (except colour additives) need to appear in descending order calculated by either mass or volume, in accordance with Trade Practices (Consumer Product Information Standards) (Cosmetics) Regulations 1991. However, unless the ingredient is deemed hazardous, there is no need for a product label to indicate the percentage of its ingredients.

For more on these standards, visit the Australian Competition and Consumer Commission's website at **http://www.productsafety.gov.au.**

NATURAL MOISTURISING FACTOR (NMF)

The chemicals in the skin that help to keep the surface hydrated and intact are collectively known as the natural moisturising factors (NMFs). The three main NMFs in the skin, amino acids, lactic acid and urea, build up around the cells in the stratum corneum as a result of the keratinisation process. Some NMFs, such as urea, hyaluronic acid, sodium PCA and glycerol, naturally bind to water at the surface of the skin, supporting the stratum corneum. Lactic acid increases the production of ceramides to further enhance barrier function. Many of these chemicals may be found in cosmetic ingredients.

MOISTURISERS

Moisturisers are used to increase and retain water in the skin. NMFs in the skin are emulated by cosmetic manufacturers to hydrate the skin. NMFs were discussed earlier in this chapter. There are two major types of ingredients that help to hydrate the skin:

- humectants, which are used to help build and retain moisture in the skin
- occludents, which help prevent trans-epidermal water loss (TEWL).

Humectants

Humectants are molecules that can bind to and hold hundreds of water molecules. This is how the skin retains moisture naturally, and cosmetic chemists add skin-like humectants to their moisturisers. Humectants act by binding water molecules to atoms of (usually) oxygen that are part of the humectant molecule. Types include:

- glycerol – polyhydric alcohol (it has three -OH groups). This is probably the most commonly known humectant
- sorbitol – a much milder and less aggressive humectant than glycerol
- polyethylene glycols (PEGs) – PEG400 is often used to keep a product moist in a container
- urea – often sold as 'urea cream' for people with dermatitis
- proteins – denatured proteins, polypeptides and amino acids. While they sometimes call these skin foods, they in no way provide any nutrients to healthy intact skin. They have an important benefit in that they are substantive to the skin
- glycosaminoglycans (GAGs) – usually as the sodium salt of hyaluronic acid – NaHLA.

Occludents

Occludents act as hydrophobic compounds that seal and trap free water molecules underneath its layer. Types include hydrocarbons, like petroleum jelly, and lipids, such as lanolin and vegetable oils.

Sodium lauryl sulphate (SLS) and sodium laureth sulphate (SLES)

These are inexpensive surfactant and detergent ingredients that lather and emulsify lipids. However, in relatively high concentrations SLS and SLES are known to irritate and dehydrate the skin.

EMOLLIENTS

'Emollient' is the term given to ingredients that soften and smooth the surface of the skin, giving the feel and appearance of hydration (see Figure 21.22). This sounds a lot like the definition of a 'moisturiser'.

The difference is that the term 'moisturiser' is used to describe the cosmetic product, and emollient is used to describe the specific ingredient.

Some emollient ingredients include lanolin, shea butter, stearic acid, isopropyl myristate and cetearyl alcohol.

As much as humectants are very efficient at 'plumping' the skin with moisture, the occludents are important in moisturisers to lock that water in. Dry skin moisturisers have a higher proportion of occludents due to a lack of intercellular substances. Chemicals used as occludents include hydrocarbons, lipids and mineral oils. Some occludents used in cosmetics today are listed below.

- Paraffin is often used in medical preparations for eczema and other skin irritations and it is an

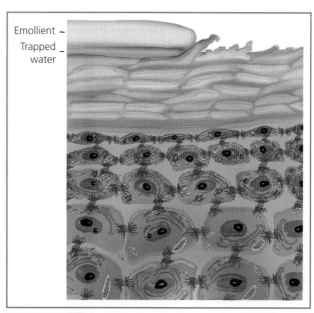

Figure 21.22 Emollients trap moisture in the skin by the process of occlusion

inexpensive occludent in moisturisers. It gives dry skin a smooth, protected feel. Paraffin is, contrary to some popular belief, non-comedogenic.

- Lanolin is the oil found on sheep's wool. The chemical binds with the skin's lipids well. It has largely been replaced by paraffin and petrolatum in cosmetics because of its controversial reputation as a skin irritant.
- Shea butter is a popular occludent as it is a naturally occurring oil. It contains a high level of unsaponifiable lipids, which means that after processing, the emollient qualities will still remain.

The skin around the eyes and lips is much thinner than on the rest of the face, so absorption is very quick. Various product lines offer specialist creams, masks and serums. Likewise, the skin of the palms of the hand and soles of the feet is less permeable due to the thicker epidermis; in particular, the stratum lucidum. This layer contains thick, clear cells that are more closely packed, allowing less product penetration. In addition to this, the feet can have a built-up of dead skin from calluses. Foot creams are formulated to be rich in emollients to enhance absorption.

ANTIOXIDANTS

Antioxidants prevent the damage caused by free radicals. Free radicals are chemicals resulting from chemical reactions that have been left with one electron in their outer shell, making them very reactive molecules. They roam around the body, seeking to pair that electron, causing damage to healthy cells. This is often exaggerated as the production of free radicals is a normal part of metabolism and plays an important part in killing foreign cells. This can cause disease, but it concerns the beauty therapist when the results are skin ageing.

On a chemical level, a molecule in the skin 'oxidises'; that is, it reacts with oxygen, causing it to become an unstable 'free radical'. It then has a knock-on effect: the unstable free radical reacts with more healthy tissue, creating more free radicals, although naturally occurring antioxidants normally control this process.

The body's natural defence against free radical damage is antioxidants. In particular, vitamins A, C and E work in the skin to disable the free radicals. They donate an electron to pair with the free radical's sole electron, thus ending the destructive cycle. Antioxidants are unique in that they can remain relatively stable when they lose an electron.

Vitamin C (l-ascorbic acid)

Vitamin C helps to stimulate collagen production, supports capillary walls and is effective in reducing the pigment in the epidermis by inhibiting melanocytes and deactivating free radicals. The trouble with using vitamin C in cosmetics is that it is lipophobic (repels oil, hydrophilic) and very reactive to UV light and air. It is most often encapsulated in a liposome.

Vitamin E (alpha tocopherol)

Vitamin E is the major antioxidant in the body. In cosmetics, it is relatively efficient in delivering the molecule as it is not resisted by the epidermal lipids and reaches the dermis. Vitamin E in the skin can become inactive, but can be supplemented with vitamin C (l-ascorbic acid), retinoids and some other antioxidants.

Retinoids (vitamin A derivatives)

Retinoids are used in beauty primarily for acne and ageing skin concerns. They have been shown to regulate epidermal cell differentiation, cell death and cell proliferation. Retinoids also protect collagen against degradation by stopping the enzymes that break down proteins (metalloproteinases). There are four types of retinoids available that can be applied topically (on the skin), with varying effects and potency. The potency varies according to how many chemical changes the retinoid has to go through to be 'biologically available'.

1 Retinoic acid (tretinoin) is the most potent and prescription-only, such as Retin-A and Roaccutane. It is recommended to use it at intervals as it weakens the skin over time.
2 Retinaldehyde is the second-most potent, and takes only one chemical change to be biologically available. The effect is more skin thickening and elasticity. It is expensive to source, so less common. Use with care in conjunction with BHAs for acne, as it does not slough off the skin like retinol does.
3 Retinol is biologically available as retinoic acid after two chemical conversions (to retinaldehyde, then to retinoic acid).
4 Esters of retinol: retinyl palmitate, retinol propionate. These are less potent, and involve a three-step chemical conversion to become retinoic acid.

Caution needed with antioxidants

On the whole, application of antioxidants to the skin has a limited effect as any that are absorbed are likely to be more widely distributed by normal blood flow. Some of these agents are lipophilic and may be retained by fat deposits in the skin, but they are not very effective there. Basically, anything that penetrates the skin to living cells is going to be exposed to blood and lymph and so will be carried away in these fluids.

Taking antioxidants by mouth has not been investigated for its effect on skin, but investigations of their effect on cardiovascular disease and mortality has not shown any significant beneficial effect and in some cases has been harmful. So, when antioxidants are applied they have a more immediate local action, but will probably disperse with time. Antioxidants such as vitamin A are the most toxic to the skin when used in excess.

25 What is an emollient?

_____

26 What is an occludent? Give one example.

_____

27 What is a humectant? Give one example.

_____

28 How can you tell whether an emulsion is water-based?

_____

29 What holds an oil and water together?

_____

30 Which of the following antioxidants is not an oil?

    A Vitamin A (retinol)

    B Vitamin E (tocopherol)

    C Vitamin C (l-ascorbic acid)

_____

31 Are waxes simple or complex lipids? Give an example.

_____

32 What are some types of AHAs that are available?

_____

33 What enzymes are used to exfoliate the skin?

_____

34 When making a scrub exfoliant, could you use the same ingredients for the face as you would for the body? Why or why not?

_____

Chemical formulations of facial products

Knowledge
Evidence

MASKS

When the client is relaxed, there is more circulation in the skin so it is a good time to deliver active ingredients ('actives'). Masks are preparations that are designed to be applied for a period of around 20 minutes and then removed. They can be used alone or to deliver other actives as creams or serums. There is a vast range of masks available, but they will all have a distinctive base:

- Wax-based masks are made from ingredients such as beeswax or paraffin wax, and applied warm. The use of gauze is usually required to aid application and removal. The purpose is to prevent TEWL and to increase circulation to encourage product penetration.
- Rubber-based masks are often made with a gum, such as guar gum or cashew gum, which gives them a rubbery, elastic feel. Sometimes they can be made with latex. They are water-based, with ingredients that are moisturising,

decongesting or astringent, among many other properties. The mask tightens as it cools, which is desirable in an anti-ageing or pore-tightening preparation.

- Hydrocolloid-based masks contain active ingredients with moisture delivered in a hydrocolloid emulsion (as discussed earlier). As the skin's warmth increases, the pectin in the hydrocolloid collapses, the moisture penetrates the skin with the active ingredients and the skin is left feeling tight, with the remaining fibres at the surface.
- Earth-based masks are usually for oily skin conditions. Fine clays such as Fuller's earth, magnesium sulphate or kaolin work on the surface, absorbing oil and sticking to the dead skin cells, which are then washed away. Added actives might be astringent or calming ingredients.
- Protein masks are often used to replace the NMF amino acids in a mature skin. Eggs are rich in amino acids. Plant proteins and enzymes are other ingredients that might be used.
- Fruit and vegetable preparations may be used in a mask and the range of ingredients is vast. For example, lemon juice is skin lightening and paw paw contains a keratolytic enzyme.

EXFOLIANTS

The intention of an exfoliant is to slough off excess layers of the stratum corneum to reveal healthier cells, to remove excess oil and debris and to aid in further product penetration. The stratum corneum has around 25 layers of cells, and salon exfoliants that are designed to be used one to two times per week remove around five layers of skin cells. Exfoliants can be mechanical or chemical in their action. The microscopic process of exfoliation involves breaking apart desmosomes, which are the proteins attached to the cell membranes of corneocytes. They are anchored to the cytoskeleton from within the cell and they clasp each other with peptide bonds that protrude outwards. The cells naturally 'desquamate' due to a natural enzyme, called *chymotrypsin*, that breaks the peptide bond. It is the job of the beauty therapist to break the peptide bond.

Mechanical exfoliants

Products that break the desmosomes with physical force can be classified as mechanical exfoliants:

- Scrubs use grains, made with crushed vegetable material, to remove skin and stimulate circulation. Natural ingredients such as ground almond or rice flour are usually added to an emulsion. Products can often be made with synthetically made plastic or dissolving spheres for fragile skins.
- Friction-off exfoliants are products made with clays such as kaolin or Fuller's earth, which adheres to the dead skin cells. As it dries, it may be rinsed off with water, but usually with dry friction of the fingertips.

CHEMICAL FACIAL PEELS

Rather than mechanically scrubbing the skin, the active ingredients in chemical peels break the intercellular bonds that hold the skin cells together. The term 'chemical peel' is synonymous with peeling the skin with medium- to high-strength concentrations, which is within the realm of medical and dermal therapists. See Figure 21.23 for a list of ingredients.

Facial chemical peels can be used at relatively high concentrations, though there are certainly limits to the strength that is appropriate in the salon. Dermal therapists are better qualified to administer the facial peels that require post-treatment products and advice.

35 Why do you use a mask when the client is relaxed? How long would you normally use a mask for?

----

36 What is the difference between an exfoliant and a chemical facial peel?

Figure 21.23	Chemical facial peel ingredients
alpha-hydroxy acids (AHAs)	Up to 30 per cent for beauty salon use. Derived usually from natural sources, such as sugar cane (glycolic acid) or a gentler form from milk (lactic acid), which is the preferred AHA for aged skin. AHAs dissolve dead skin cells at the surface of the skin. They help product absorption of active ingredients, stimulate cell proliferation and also stimulate the production of hyaluronic acid, one of the skin's glycosaminoglycan humectants.
beta-hydroxy acid (BHA)	4–12 per cent for salon use. Unlike AHAs, the BHA is a relatively larger molecule that works intensely on the surface of the skin to clear pores of dead skin cells, oil, dirt and bacteria. It is often used to treat problem/acneic skin and the associated hyperkeratinisation.
TCAs (trichloracetic acids)	Up to 20 per cent for beauty salon use. They are good for skin rejuvenation, wrinkles, scarring and pigmentation and are preferred for darker skins over phenol. Phenol peeling is within the realm of dermal and medical therapies.
enzymes	Primarily papain, derived from papaya, and bromelain, from pineapple; both digest dead skin cells of the stratum corneum. These are often added to friction off exfoliants for a 'double-action', which is especially useful for skins that have fragile capillaries.
surfactant ingredients	Added to exfoliants to bind and lift the corneocytes. TEA and glyceryl stearate SE (self-emulsifying) are popular desquamating surfactants.

37 When applying and removing the mask:

a If using a commercial mask product, what should you always do first?

_____

b If applying multiple masks, which one would you apply first?

_____

c Describe the method you would use for removing a mask with towels.

_____

_____

_____

38 Give an example of a finishing product.

_____

LO21.7 REVIEWING THE FACIAL TREATMENT

Performance Evidence

Selecting products to complement the skin

Seek the manufacturer's guide on skin types, skin conditions and skin disorders. Product lines are commonly available for skin types and conditions such as mature, oily normal, combination, dry, dehydrated and sensitive. Some brands may pay particular attention to allergies, fragile capillaries, rosacea, acne and pigmentation. When a product or treatment contains actives for one skin condition, it is likely to be contraindicated to another. For example, a skin that is prone to eczema and allergies will absorb the product much more quickly than a normal skin when the epidermis is damaged.

Knowledge Evidence

HYDROCARBONS

As hydrocarbons do not combine with the skin's natural chemicals, it is assumed that these chemicals may reside in the skin, causing unknown side effects. This is practically impossible to prove, so there is no regulation on their usage unless long-term use has given rise to concerns.

FORMALDEHYDE

There have been concerns expressed about the use of high levels of formaldehyde in cosmetics, such as in nail hardeners, eyelash glues, hair tints and some moisturisers.

Formaldehyde causes skin sensitivity and irritation, and there are concerns that it may be carcinogenic. However, formaldehyde is a product of normal metabolism and would normally only be of concern in high concentrations, as a solution or vapour. Some older-model chemical sterilisers contain dangerous levels of formaldehyde.

ALLERGIC REACTIONS

Knowledge Evidence

Check that the client does not have allergies to the ingredients. Manufacturers are required by law to list known allergens on the product label. Certain cosmetic ingredients are known to provoke allergic reactions in some people, causing irritation, excessive erythema, inflammation and swelling.

Known cosmetic allergens include:

* lanolin (adulterated), sheep's sebum, obtained from sheep's wool and added to many cosmetics as an emollient. Lanolin BP is not adulterated and is not a known allergen
* strong alcohols, used as a detergent or astringent
* gums, used as an adhesive and binding agent
* pearlising agents, such as aluminium and mica; these give a shimmer effect to some products
* mineral oils, including paraffin, castor oil, petrolatum
* eosin dye, used in make-up and perfumes
* cobalt (blue), carmine (red) – lake colours used in make-up
* parabens, used as an antiseptic or preservative
* perfumes and fragrances – natural essential oils or synthetic fragrances.

1 Select a cosmetic product and write a half-page report on its product safety labelling. Include:

* details of the ingredients list. Do you recognise any dangerous ingredients at the top of the list?

* indications on the label of any allergens or chemicals that may cause an adverse reaction

* how well ingredients are displayed so that the consumer is fully aware of the product's contents.

Client feedback and adverse effects

2 Would you only look for feedback from a client at the end a facial treatment?

----

3 There are three main things that you are looking for when discussing the treatment and gaining feedback after a treatment. What are they?

----

4 Go online and find three images for each of the following conditions, so that you can familiarise yourself with their appearance:

- allergic reaction

- erythema

- skin response to irritation or trauma (e.g. being pinched)

- skin blemishes

- skin inflammation.

Make sure each picture shows the condition on a different Fitzpatrick skin type.

LO21.8 RECOMMENDING POST-TREATMENT SKINCARE REGIMEN

Performance
Evidence

Aftercare advice

Knowledge
Evidence

1 Post-treatment advice for a facial is given in addition to the product prescription. It covers lifestyle, dietary requirements and other positive changes that can improve the skin.

a Suggest two items of post-treatment advice for mild acne vulgaris.

---

b Suggest four items of product advice that you would incorporate into the product prescription for mild acne vulgaris.

---

2 You should recommend future treatments suited to the client's needs for at least the next four visits. The future treatment plan is an assurance that you intend to focus on their skincare needs over the long term, and this helps you retain clients. In the table below, match the following client skin concerns with the future treatment plan, choosing from:

- oily, dehydrated

- mature, diffuse red

- mild acne, dehydrated and sensitive.

TREATMENT OBJECTIVES (I.E. CLIENT SKIN CONCERN)	FUTURE TREATMENT PLAN		
	FACIAL TYPE	DURATION	FREQUENCY
	Hydrating Vitamin C energising Anti-ageing	45 minutes 1 hour 1 hour	After one week × 2 After one month × 1 After one month × 1
	Hydrating and soothing Hydrating, soothing and extractions	45 minutes 1 hour	After one week × 2 After one week × 2
	Deep cleanse plus enzyme peel Mineralising and hydrating facial	45 minutes 1 hour	After one week × 2 After one month × 2

Nutrition and skin health

Knowledge Evidence

3 What types of foods do you know can improve skin health?

_____

_____

4 What can food allergens do to the skin?

_____

5 Can a food allergen contraindicate a client to a product? Explain your answer.

_____

_____

Lifestyle factors on skin

6 Each of the following lifestyle factors can have a profound effect on the skin's health and appearance. Choose one and make an A4 poster showing its effect on the skin, and what a lifestyle change can do. Present your poster to the class.

- alcohol consumption
- climate
- exercise routine
- hobbies
- nutrition
- sleeping patterns
- tobacco consumption
- type of employment

LO21.9 CLEANING THE TREATMENT AREA

Knowledge Evidence

1 List four items that you would need to ensure you clean at the completion of a facial treatment that are specific to that treatment.

_____

_____

_____

_____

2 Why would you need disinfectant solution prepared and within reach throughout and at the completion of a facial treatment? What might be an alternative?

_____

3   How might a salon aim to be more sustainable in its use of product, water and power with facial treatments?

_____

_____

_____

_____

_____

_____

4   Why do we think of single-use items as single-use? Why can we not reuse them?

_____

_____

_____

CHAPTER 22: SPECIALISED FACIAL TREATMENTS

After completing this chapter, you should be able to:

LO22.1 establish client priorities by a thorough consultation and skin analysis

LO22.2 design and recommend specialist facials

LO22.3 prepare for the specialised facial treatment according to salon protocol and relevant health and safety legislative requirements, and by following sustainable work practices

LO22.4 cleanse the skin using an ultrasonic or desincrustation machine

LO22.5 soften skin with exfoliation and steam to remove minor skin blemishes, extract blockages, and complete the treatment with a direct high-frequency machine

- apply basic peels, selecting, applying and removing the formula according to the manufacturer's instructions and applying soothing post-treatment products and advice
- apply microcurrent for anti-ageing benefits

LO22.6 complete the specialised facial treatment with a mask and post-treatment care

LO22.7 administer post-treatment aftercare advice, homecare and future treatment recommendations, and record outcomes

LO22.8 clean the treatment area in preparation for the next service.

INTRODUCTION

Specialised facials are performed with the skills and knowledge learnt in Chapter 21, but with additional skills and abilities in consultation for the identification and treatment of minor skin disorders within the scope of beauty therapy. You will use professional techniques and equipment, including massage movements, electrical machinery and specialised products. Mechanical and electrical facial treatments – electrotherapy – provide intensified results compared with what can be achieved manually.

The learning activities in this chapter allow you to demonstrate your knowledge and understanding of the unit of competency *SHBBFAS006 Provide specialised facial treatments*, using the performance and knowledge evidences.

LO22.1 ESTABLISHING CLIENT PRIORITIES

Performance Evidence

Contraindications

1 Refer to Figure 22.7 in your textbook and Appendix A Contraindications and restrictions. You may need to also do some research online. Complete the following table for contraindications that restrict or prevent a specialist facial treatment. Use the image or description to provide the condition's name, its relevance to the specialist facial treatment and the action you would take if a client had the condition.

Performance Evidence

Knowledge Evidence

SYMPTOM	CONTRAINDICATION	RELEVANCE TO FACIAL TREATMENTS	ACTION
A small device implanted into a person with arrhythmia. The cords send low-energy electrical pulses to prompt the heart to beat at a normal rate.			

SYMPTOM	CONTRAINDICATION	RELEVANCE TO FACIAL TREATMENTS	ACTION
Disorder of the central nervous system in which every nerve cell in the brain is disturbed, which causes seizures involving either abnormal behaviour or a complete loss of consciousness. Seizures may be: • partial seizure: strange or random bodily behaviour • generalised seizures: the person is completely unconscious, either in absences or with bodily jerks and twitches.			
Alamy Stock Photo/Science Photo Library			
Shutterstock.com/madeinitaly4k			

Formulation, function and action of cosmetic formulations and ingredients in treatment products

Knowledge Evidence

2   What is contact irritant dermatitis?

_____

3   What types of products contain irritants that people with dermatitis should avoid? (Refer to Figure 21.29 in Chapter 21 of your textbook.)

_____

_____

_____

4 What does a surfactant do?

5 What skincare products can surfactants be used in?

Relevant medical history and medications

6 List some types of medications that are relevant to the skin (i.e. cause skin thinning or inflammation).

7 When a client with rosacea has oily skin, can you perform ultrasonic cleanse on them? Explain your answer.

8 Is diffuse red skin most often oily or dry?

9 Does diffuse red skin most often have good or bad circulation? Why?

10 What are the three ways you can identify oily skin in a skin analysis?

11 What types of medications can give a client dry skin?

Client requirements

12 What would you expect to see from a client's skin if they are classified as diffuse red?

13 Which machines would you use for diffuse red skin in a specialist facial treatment?

LO22.2 DESIGNING AND RECOMMENDING SPECIALISED FACIALS

Performance
Evidence

Effects and benefits of the treatment and products used

1 When providing the electrical treatment, you need to explain the expected sensation of the machine, the professional skin care products used, and the effects and benefits of the treatment. Complete the table below by describing the procedure for each piece of equipment in a way the client can easily understand.

Knowledge
Evidence

EQUIPMENT	EXPECTED SENSATION	PRODUCT USED	EFFECT	BENEFIT
Desincrustation				
Iontophoresis				
Direct high-frequency				
Microcurrent				
Ultrasonic cleanse				
Sonophoresis				

2 Research the below skin types and conditions and refer to Figure 22.2 in your textbook. Complete the table below with notes on what a client's main concern would be, the machines and professional products used in treatments, product prescription, and lifestyle and aftercare advice.

SKIN TYPE/ CONDITION	MAIN CONCERN (SYMPTOMS)	MACHINES USED	PROFESSIONAL PRODUCTS	PRODUCT PRESCRIPTION	LIFESTYLE/ AFTERCARE ADVICE
Prematurely aged					
Diffuse red					
Seborrhoea					
Acne vulgaris					

SUITABILITY OF ELECTRICAL TREATMENTS TO CLIENT AND RECORDING OUTCOMES

Knowledge Evidence

3 If a client tells you they are feeling numbness during a galvanic treatment, should you proceed? Explain your answer.

_____

_____

4 How would you record the outcome (the client's numbness) in a way that communicates it clearly for the next therapist?

_____

_____

5 Can you perform desincrustation on a thin skin?

_____

Effectively treating clients

Performance
Evidence

6 During the planning consultation, the client tells you that there are some products that cause their skin to get very red and hot, and they would prefer those ones weren't used. Where on the client record card for specialised facials would you place this information?

LO22.3 PREPARING FOR SPECIALISED FACIAL TREATMENTS

Performance
Evidence

Identifying electrical risk situations and taking remedial action

Knowledge
Evidence

Refer to the section on 'Basic electricity' in the Workbook for Online Chapter 8, and answer the following questions.

1 It is important to understand where static electricity can be dangerous in the salon. What is electrostatic discharge?

2 What can you do to prevent electrostatic discharge when preparing for electrotherapy treatments?

3 What is an alternating current? Explain how the machine makes an AC.

4 Which electrical treatment uses alternating current and static electricity?

5 What is the frequency range for the treatment/machine identified in Question 4?

6 Briefly explain how a capacitor can be similar to a battery.

7 What is a direct current? How do you ensure the direct current flows?

8 Is microcurrent AC or DC?

9 Do the electrons flow along the wire from an area of negative charge to positive, or positive to negative?

10 Which current is coming from the power station? AC or DC?

11 What is installed into a machine to prevent a short circuit?

_____

12 What is installed into the wiring system in the wall to prevent a short circuit?

_____

13 Which electrical treatment turns electrical energy into chemical energy?

_____

14 Does the therapist hold the anode or the cathode in desincrustation?

_____

Presenting yourself, according to organisational policy

Knowledge Evidence

15 Why is it important for the beauty therapist to:

   a not be wearing any piercings or jewellery during specialist facial treatments?

_____

   b wear rubber-soled shoes during specialist facial treatments?

_____

_____

   c launder their uniform daily?

_____

_____

Complying with health and hygiene regulations and requirements

Performance Evidence

LEGISLATION AND ORGANISATIONAL POLICIES

16 What do you need to record on the client record card when the client wishes to continue with treatment despite having a piercing?

Knowledge Evidence

_____

_____

17 How often should you have machines checked by a registered and licensed electrician?

_____

18 If there is a fault in a machine, what is the procedure and to whom do you report the incident and how?

_____

_____

_____

_____

19 Where do you place the machine on the treatment trolley?

_____

20 How do you dispose of the cottonwool used for desincrustation?

LO22.4 CLEANSING SKIN USING ULTRASONIC OR GALVANIC

Performance
Evidence

Knowledge
Evidence

Safety around energy

1 Review the 'Step-by-step: Machine check' in Appendix C Treatment area and equipment. Unjumble the list of steps you need to follow when performing a machine safety check. Number the steps so they are in their correct order.

Check plugs, switches and cords for stability, kinks and fraying.	
Clean attachments that have been in contact with your skin so they are ready for use on the client.	
Test all attachments, cords and electrodes, that connections are secure and current flows evenly, as appropriate and according to the manufacturer's instructions.	
Position machine so that it is stable on the top tier of the trolley, and above and away from water.	
Check that indicator dials and lights are working.	
Wash and dry your hands thoroughly.	

2 When you are performing a machine check on your galvanic machine, how do you know which output is positive and which is negative? Is it the same for every machine?

3 When performing desincrustation, how do you ensure there is sufficient saline on the skin?

4 What is the danger of using direct current on dry skin?

5 What is different about performing a machine safety test on the ultrasonic machine?

6 When testing the high-frequency machine, what do you need to see, hear, smell and feel?

- See

- Hear

- Smell

- Feel

7   If you have tested the machine and there is a fault, what should you do?

8   A fault in a machine has been repaired, and then tested by you and everything was okay. However, if a client is then harmed by this machine, who is liable?

Ultrasonic machine

Ultrasonic literally means 'beyond sound'. The human ear hears sound frequencies up to 20 000 hertz (Hz). Ultrasound includes all frequencies above what humans consider 'audible', so the ultrasonic waves in the probe are not audible to us. Sound energy is vibrational. Consider the vocal cords and how they vibrate (you can feel it as you speak), or the vibrations made by a drum.

Knowledge Evidence

ENERGY FORMS

Sound energy is emitted into the probe via a transducer. The transducer component in the cosmetic ultrasonic machine converts electrical energy into sound energy. For more about energy forms, refer to the section 'Energy' in the Workbook for Online chapter 8. Sound waves make vibrations in particles.

OPERATIONAL CHARACTERISTICS OF THE MACHINE

One simple type of acoustic transducer contains a piezoelectric ceramic, crystal, plastic or other material such as lithium niobate, lead titanate, barium titanate or lead metaniobate.

Figure 22.1 shows the transducer made with piezoelectric material in the hand-held probe. Sound waves are measured in frequencies (Hz). Unlike the pattern of the high-frequency waves that travel in a transverse pattern, sound waves are longitudinal pressure waves. This means that the probe must be held flat on the skin because the pressure waves are travelling longitudinally from the transducer (which, unless otherwise stated by the manufacturer, would be perpendicular to the surface). Sound travels fastest in a solid, and slowest in gas (air). For a video representation of sound energy transmission, go to YouTube at: **https://www.youtube.com/watch?v=GkNJvZINSEY**.

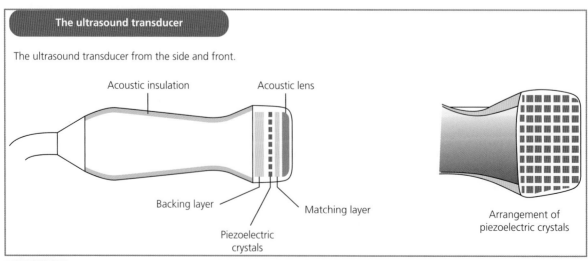

The ultrasound transducer

The ultrasound transducer from the side and front.

Acoustic insulation

Acoustic lens

Backing layer

Matching layer

Piezoelectric crystals

Arrangement of piezoelectric crystals

Figure 22.1   Ultrasound transducer in hand-held probe

SKIN BARRIER PROPERTIES

The skin has what is known as a relatively high *acoustic impedance*. It means the skin has a relatively high resistance to the vibrations of the ultrasonic wave. With nothing on the skin, the longitudinal ultrasound waves reflect off the skin from the probe (via the transducer) into the air. Because the wave is a longitudinal wave, the energy doesn't scatter much, it almost completely reflects, so the client will likely feel nothing (rather than a weaker sensation) when there is insufficient product on the skin. When we perform ultrasonic cleanse, we use water (or water and an emulsifying cleanser that helps to hold the water on the skin). For sonophoresis, it is common to use conductive gel over the serum or a similar product. Conductive gel has an acoustic impedance similar to the dermis, so it can carry the vibrations through the epidermis without reflection into the air. Conductive gels are hydrocolloids; they are water-based, so if the product starts to dry out, add water rather than more product. Conductive gels often contain:

- water
- glycerine
- propylene glycol.

The characteristics of an ultrasound wave are pictured in **Figure 22.2**. When the pressure phase is high, the particles move together, cells shrink and compress (compression phase, shown in red), and when the pressure phase is low, the particles move apart, cells expand and implode (rarefaction phase, shown in blue).

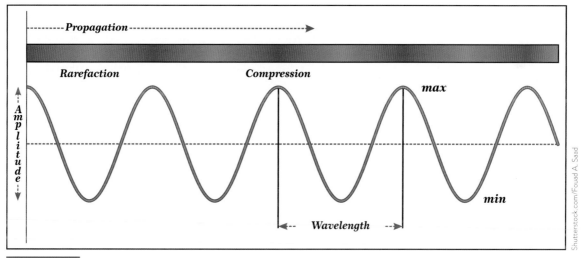

Figure 22.2   Characteristics of an ultrasound wave

In the context of the cells of the epidermis, the membranes are stretched in the rarefaction phase, which allows substances to pass easily in and out of the cell. When in the compression phase the cell membranes tighten, which in turn constricts the capillaries. This is good for inflammation (in sensitive skin) and dilated capillaries (in diffuse red skin) as it prevents the flow of inflammatory substances into the skin and strengthens capillary walls.

The image below shows the longitudinal sound wave pattern.

9   Indicate in the image of the longitudinal wave pattern where the compression phase is with a 'C' and the rarefaction phase with an 'R' under each arrow.

10 Draw a larger or smaller circle above the image to indicate what the cells are doing at each phase – where cells are compressed (draw a small circle) or imploded/expanding (draw a large circle).

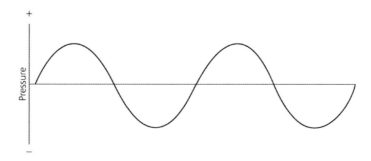

THERMAL AND NON-THERMAL EFFECTS AND BENEFITS

Ultrasonic waves have a thermal effect (warming of tissues) if left on for too long or with a high intensity. This thermal effect is used by ultrasonic machines in other industries to achieve certain benefits but these are not the focus of beauty therapy. See **Figure 22.3** below for a description of the thermal and non-thermal effects and benefits of ultrasonic relevant to the cosmetic machine.

The reduction of mechanical and vibrational energy in skin tissues is known as 'attenuation'. When sound energy is attenuated in the skin, heat accumulation in tissues is reduced. However, if there is too much sound energy input thermal effects increase, causing adverse effects such as tissue damage from the complete implosion and destruction of the cells.

Factors that influence the heat accumulation include:

- mode: continuous and pulsed
- product
- treatment duration
- intensity
- frequency.

THERMAL EFFECTS	BENEFITS	NON-THERMAL EFFECTS	BENEFITS
Increases body temperature by 1 °C. Increases cell metabolism and healing. Improves collagen extensibility and fibroblast activity. Stimulates circulation.	Healing Prevents ageing Rejuvenating	*Sonophoresis:* • Modifies inflammatory responses. • The sound waves cause the cells to expand, making the membrane more 'porous' (sonophoresis). • Produces microbubbles in tissue fluid that rapidly oscillate through the fluid in and around the cells, improving cell membrane permeability. *Ultrasonic cleanse:* • Mobilises and dislodges particles (dead skin cells, impurities and oil) on skin's surface and suspends it in the medium (water and cleansing solution).	Calms and soothes skin. Aids penetration of serums. Improves movement of substances through semi-permeable membranes in epidermis. Increased hydration. Deep cleanse. Removes bacteria (acne vulgaris).

Figure 22.3 Benefits of ultrasonic machine according to thermal and non-thermal effects

Mode

There are two machine settings used: continuous and pulsed mode.

1 *Continuous wave – used for ultrasonic cleanse:* the continuous pulse has thermal effects. It has a 100 per cent duty cycle (on-time) and stays continuous throughout the treatment. The ultrasonic cleanse should not exceed the recommended treatment time (approximately 3 minutes for the full face).

2 *Pulsed mode – used for sonophoresis:* pulsed mode means the wave intensity is interrupted periodically. The machines have approximately 20–50 per cent duty cycle (on-time), which gives the probe time to cool down and no thermal effects. This means the treatment time can be longer.

Product

Non-polar (oil) and small polar molecules (water) can readily pass through the cell membrane, so the serum can diffuse across it by passive transport (down the concentration gradient). Diffusion is aided when the cells are expanded (rarefaction phase) during the sonophoresis treatment, making the cells more porous ('sono-*phor*-esis'). Sonophoresis for serum infusion is typically followed by iontophoresis to further drive the product into the skin.

Intensity

The intensity of the sound beam is the strength, and it can greatly increase the thermal response in the skin. The intensity of an ultrasonic machine is relatively insignificant, though it is slightly higher for the infusion mode because the target area is deeper in the skin. The amperage is measured in watts per cm² for ultrasonic because it fluctuates from positive to negative. The precise intensity is difficult to measure because it is alternating current and pressure, but it is estimated at approximately 0.25 w/cm² to 1.5 w/cm². The intensity parameters are usually pre-set, though some machines can allow you to adjust the parameters.

Frequency

Frequencies used in beauty therapy are between 20 KHz and 1 MHz. The higher the frequency, the more superficial the treatment. The danger of having a higher frequency at a superficial level (i.e. during the continuous pulse for the ultrasonic cleanse) is that it can increase the thermal effects. It has been found that 20 KHz machines are ideal for product penetration and reducing thermal effects. It is important to follow the manufacturer's instructions with regard to product use and treatment duration.

Refer to the sections on 'Energy' in the Workbook for Online Chapter 8, and 'Energy forms' above to answer the following questions.

11 What form of energy comes out of the probe and onto the skin?

_____

12 What component in the machine converts the energy in the machine?

_____

13 What does acoustic piezoelectric material do to energy that is passed through it?

_____

14 Is the sound wave a transverse or longitudinal wave?

_____

15 What does ultrasonic or ultrasound mean?

_____

_____

16 Explain how the ultrasonic waves cleanse the skin at a cellular level.

_____

17 How does sonophoresis infuse products into the cells?

_____

18 Why is it important to follow manufacturers' instructions? What makes the difference if the frequency varies from machine to machine?

_____

_____

19 What does tissue damage (or skin trauma) look like?

_____

20 Which machine would you follow the sonophoresis with to enhance product penetration?

_____

21 Would you ideally use an oil-based or a water-based serum? Give justification for your answer.

_____

22 What are the factors that influence heat accumulation, which result in a thermal effect?

23 What is the adverse effect caused when there is too much warming of body tissues?

24 The skin has what is known as a relatively high *acoustic impedance*. What does this mean, and what is the effect when using the ultrasonic machine in specialised facial treatments?

25 What are three types of electrodes that might be found on a galvanic machine?

LO22.5 REMOVING MINOR SKIN BLEMISHES AND INFUSING SERUMS

1 Once the skin has been deep cleansed, what should be done to the skin next?

2 What can be done if the skin's surface is dry?

3 Watch the video at **https://www.youtube.com/watch?v=yBd2ln_3C2U** to answer the following questions about exfoliation.

   a What is exfoliation?

   b What is one reason we should exfoliate our skin?

   c When is the general time when it would be good for people to be regularly exfoliating?

4 What equipment will you need to have ready before you perform any extractions of minor skin blemishes? Check Chapter 22 to help you with this.

_____

_____

_____

_____

_____

5 Should you assume that if a client books in for any facial treatment that they would like extractions performed? If not, what should you do?

_____

Infuse serums using iontophoresis or sonophoresis

Knowledge
Evidence

6 Serums used with iontophoresis contain active ingredients that are electrolytes. What is an electrolyte?

_____

7 The idea that opposite forces attract means that cations (+) will always move towards the cathode (–). If we know that like forces repel, and you are working with a cathode (–), what sort of serum would you use? How does that help in a specialised facial treatment?

_____

_____

8 What is one effect of using the sonophoresis?

_____

_____

9 The modality of the sonophoresis is pulsed. What does this mean and why does it matter?

_____

_____

Applying basic peels

The basic peel, or 'cosmetic' and 'cosmeceutical' low- to medium-strength peels, are within the scope of practice of beauty therapy.

SELECTING, APPLYING AND REMOVING FORMULA

1 Choose the best ingredients in a peel for the following skin types:

a oily skin with heavy scarring

_____

b mature skin

_____

c photo-aged skin

_____

2   Why should you never have the steam turned on during application?

_____

3   Before you start a peel treatment with a new product, you should always read the safety data sheet and contact the manufacturer. What are you asking the manufacturer when you contact them?

_____

4   What temperature water should you use when removing product from the client's skin?

_____

5   What is the neutralisation equation and what is its relevance to a peel? When would you use it?

_____

_____

_____

POST-TREATMENT PRODUCTS AND ADVICE

6   Why is it important to use soothing and protecting products after a peel?

_____

_____

7   What three product types should you definitely use after a peel?

_____

8   List four pieces of advice that you could provide the client following a peel treatment.

_____

_____

_____

_____

Microcurrent

9   List five general effects of microcurrent on the face.

_____

_____

_____

_____

_____

10 What fraction of an amp is a microamp?

_____

11 Which contraindication relates to microcurrent's ability to stimulate mitosis in the stratum germinativum?

_____

LO22.6 COMPLETING THE TREATMENT

1 List the three ways you can apply a mask following skin treatment and serum infusion.

2 How would you remove a mask that you can peel off?

3 Complete the table below by describing how each aftercare product helps to soothe and induce vasoconstriction following a specialised facial treatment.

AFTERCARE PRODUCT	DESCRIBE HOW IT HELPS AFTER A SPECIALISED FACIAL TREATMENT	DESCRIBE HOW IT HELPS INDUCE VASOCONSTRICTION
Serum		
Oil-based product		
An additional mask		

LO22.7 PROVIDING POST-TREATMENT ADVICE

Nutrition and the skin

Knowledge Evidence

1 You will find that almost every vitamin and mineral essential to human health is needed for the skin. Research the vitamins and minerals that are needed in the diet to maintain healthy skin for some common skin conditions and list them in the table below.

SKIN CONDITION	VITAMIN OR MINERAL	FOODS TO EAT	EFFECT	BENEFIT
Prematurely aged				
Diffuse red				
Seborrhoea and acne vulgaris				

2 How do you diagnose dehydration in a skin analysis?

3 Does oily skin get dehydrated? Explain your answer.

4   If a person has had a bypass surgery, when would it be appropriate to use electrical machines?

5   How does fibre help the digestive system to work more efficiently?

The role and limitations of the beauty therapist

It is important that the beauty professional knows where responsibilities lie. That is, what is expected of the beauty therapist, and what conditions must be referred to a medical professional. The beauty therapist is able to correctly identify that a person has nutritional needs, but is not qualified to diagnose nor treat them. If the task of deciding who is best to diagnose the condition is onerous, the client should be referred to a GP.

Knowledge Evidence

The performance and outcomes of all beauty treatments are affected by the client's nutrition and physical health. As a part of the client consultation, the beauty therapist asks the client questions relating to nutrition and lifestyle.

6   Suggest five questions other than those mentioned that you could ask your client relating to physical activity or nutrition that could help you determine their health.

7   You may find that lifestyle and skin nutrition becomes a large part of the 'small talk' you make with clients. It is best practice to note down lifestyle factors and recommendations on the client record card that can assist with current and future treatment plans.

   a   Why might your skin look better in the morning when you wake up than in the evening when you return home from work?

   b   Compile a list of lifestyle factors that can contribute to poor skin health.

Client feedback

8   It is important that as a student you receive treatments as well as practise them so that you know how the machines feel. Explain how you would obtain client feedback when you first apply the electrode to the skin and start to apply current for iontophoresis.

Adverse reactions

Knowledge
Evidence

9 Use Figure 22.12 in your textbook, The benefits and adverse effects of specialised facials, and write in the table below what actions you would take.

MACHINE	POSSIBLE ADVERSE EFFECT	ACTION
Ultrasonic	Flaking (cleanse)	
Desincrustation	Sensitisation	
Iontophoresis	Excessive erythema	

Lifestyle factors that can affect the skin

Knowledge
Evidence

10 Complete the following table of lifestyle factors and client concerns by providing lifestyle and dietary advice, and recommending future treatments for the client types listed. Consult with your teacher, product and equipment manufacturers or visit local salons to gain an idea of expected future treatments.

LIFESTYLE FACTOR	CLIENT CONCERN	RECOMMENDATION	FUTURE TREATMENTS
Alcohol consumption	Male with rosacea on nose (rhinophyma)		
Climate	Sudden acne breakout with dehydration when moved from overseas tropical climate to dry cold climate		
Exercise routine	Sluggish skin, poor circulation		
Hobbies	Just retired but can't relax		

LIFESTYLE FACTOR	CLIENT CONCERN	RECOMMENDATION	FUTURE TREATMENTS
Nutrition	20-year-old, just moved out of home, now has acne vulgaris		
Sleeping patterns	Has a baby, stressed skin		
Tobacco consumption	Bluish skin, sluggish, poor circulation		
Type of employment	Follicles constantly getting blocked with dust at work (construction)		

Aftercare advice

Knowledge Evidence

11 Complete the following table by outlining the aftercare advice (not product advice) and future treatment plan to meet the client objective with a specific machine.

MACHINE	CLIENT OBJECTIVE	AFTERCARE ADVICE	FUTURE TREATMENT PLAN
Microcurrent	Anti-ageing		
Direct high-frequency	Pustules and papules spot treatment		
Desincrustation	Seborrhoea		

MACHINE	CLIENT OBJECTIVE	AFTERCARE ADVICE	FUTURE TREATMENT PLAN
Iontophoresis	Diffuse redness		
Ultrasonic cleanse	Acne vulgaris		
Sonophoresis	Sensitive skin		

12 At the end of the specialised facial treatment the client tells you they are very happy with how their skin feels, and they would like to book in for future treatments as you have recommended. Where on the client record card for specialist facial treatments would you record this information?

_____

_____

Design and recommend future treatments

13 There are general treatment recommendations for all clients with each machine. Future treatment planning depends largely on the client's requirements and suitability to treatments, which includes budget, willingness to attend and time commitments. Recommend a future treatment plan for the following clients. Remember to state the type of facial (e.g. deep cleanse, anti-ageing), the frequency of treatments (weekly, fortnightly) and duration (e.g. 30-minute facial, 75-minute facial)

Knowledge
Evidence

a Sonophoresis and iontophoresis for a mature-aged client, willing to spend money on the treatments and commit time, but on a pension (meagre income).

_____

_____

b Sonophoresis for a client with inflamed acne, and a very low budget and not willing to have a full facial.

_____

c Client with the budget and time to spend on skincare, wanting anti-ageing treatments and relaxation.

_____

_____

LO22.8 CLEANING THE TREATMENT AREA

1 List five single-use items that might be used in a specialised facial treatment.

_____

_____

_____

_____

_____

2 Preparing the treatment area for specialised facials is also about monitoring the electrical equipment thoroughly. How often should you test machines?

_____

3 What would your actions be if there were an issue with a machine?

_____

_____

4 What is one way you could be more sustainable in the salon with products, water and power?

   a product

_____

_____

   b water

_____

_____

   c power

_____

_____

PART 3

THE SCIENCE OF BEAUTY

SECTION 8
SKIN AND BODY SCIENCE   370

23	Promote healthy nutritional options in a beauty therapy context	371
24	Body structures and systems	387
25	Skin science	419
26	Incorporate knowledge of anatomy and physiology into beauty therapy	454

SECTION 8

SKIN AND BODY SCIENCE

23 Promote healthy nutritional options in a beauty therapy context

24 Body structures and systems

25 Skin science

26 Incorporate knowledge of anatomy and physiology into beauty therapy

CHAPTER 23: PROMOTE HEALTHY NUTRITIONAL OPTIONS IN A BEAUTY THERAPY CONTEXT

After reading this chapter, you should be able to:

LO23.1   identify the principles of nutrition and the role of food and nutrition and its effects in beauty therapy treatments, and apply your knowledge of nutritional needs across the lifespan

LO23.2   apply knowledge of the body's systems to beauty therapy treatments

LO23.3   provide advice on dietary guidelines.

INTRODUCTION

Promote healthy nutritional options to your clients to ensure they can obtain the best beauty treatment outcomes. In this chapter you will learn how to provide appropriate information to your client about how nutrients make their way to the skin, what can happen to the skin when there is not enough nutrition, what you can do to encourage your client to improve their general wellbeing, and when and who to refer the client to for specialised nutritional advice.

The learning activities in this chapter allow you to demonstrate your knowledge and understanding of the unit of competency *SHBXCCS006 Promote healthy nutritional options in a beauty therapy context*, using the performance and knowledge evidences.

LO23.1 PRINCIPLES OF NUTRITION

Knowledge Evidence

Nutrition principles

Knowledge Evidence

1   Define 'nutrients'.

_____

2   Finish the sentence: The distinct groups of nutrients include: proteins, c_____, lipids (including cholesterol), vitamins and m_____.

Nutrition in regards to treatment procedures

Knowledge Evidence

3   When a client has milia, how can you associate it with nutrition? Could it be:

   A   food intolerance

   B   allergy to product

   C   food allergy

   D   none of the above

_____

4 Which of the following nutrients are also product ingredients that can be beneficial to *dehydrated* skin?

   A Dietary fibre

   B Peptides

   C Prebiotics

   D All of the above

5 Which of the following nutrients is of great benefit in body treatment products but should be consumed in small amounts via the digestive system?

   A Vitamin C

   B Salt

   C Vitamin B (any)

   D None of the above

Nutritional composition of a range of commonly available foods

Knowledge Evidence

6 To understand the best food sources for certain nutrients, read the information at the following link: **https://www.betterhealth.vic.gov.au/health/healthyliving/Vitamins-and-minerals**. Draw a line to match the following nutrients to the ideal food sources.

NUTRIENT
Vitamin: ascorbic acid
Protein
Mineral: calcium
Carbohydrate: complex
Lipids: cholesterol

IDEAL FOOD SOURCE
Dairy foods
Avocado
Rice
Oranges
Legumes/nuts

PROTEINS

7 Which of the following is *not* a structural protein?

   A Collagen

   B Elastin

   C Keratin

   D Enzyme

8 What is a hormone?

9 What does the protein arginine do in the skin?

CARBOHYDRATES

10 What is the key function of carbohydrates in the body?

11 Carbohydrates can be simple (sugars) and complex (starches). Which of the following is better for the body to eat as dietary food?

    A Monosaccharides

    B Disaccharides

    C Polysaccharides

    D All of the above

12 Which complex carbohydrate type makes good dietary fibre, and why?

13 What is the glycaemic index (GI)?

14 Acne vulgaris and ageing can both be affected by high GI. Briefly explain how high GI makes each condition worse.

FATS (LIPIDS)

15 What happens to fatty acids when they are not needed for immediate use?

16 What is the name of the enzyme that breaks down triglycerides into free fatty acids?

17 List seven benefits of having fats (lipids) in your diet.

18 What is the difference between the chemical configuration of saturated and unsaturated fat?

19 What does the 'kink' due to the double-bond or triple-bond configuration mean for unsaturated fats and oils?

_____

_____

_____

20 Give an example of a saturated fat and an unsaturated fat used in beauty therapy treatments.

_____

_____

21 Why should people avoid foods that contain trans-fats?

_____

VITAMINS: ROLE AND FUNCTION IN THE HUMAN BODY OF THE 13 ESSENTIAL VITAMINS AND EFFECT ON PERSONAL APPEARANCE

Research the role of vitamins on personal appearance in the section 'Vitamins' in Chapter 23 of your textbook.

22 Complete the table below to identify what the vitamins do in the skin or to enhance skin appearance. The following vitamins assist directly with skin health and appearance.

ESSENTIAL VITAMIN	ROLE/FUNCTION IN THE BODY	WHAT DOES IT DO IN THE SKIN OR TO ENHANCE SKIN APPEARANCE?
Vitamin A (retinoic acid, retinol and retinyl palmitate)	Retinoic acid assists the body with vision and reproduction.	
B2 (riboflavin)	Riboflavin is primarily involved in energy production and helps vision and skin health.	
B3 (niacinamide)	Niacin is essential for the body to convert carbohydrates, fat and alcohol into energy. It helps maintain skin health and supports the nervous and digestive systems.	
B6 (pyridoxine)	Pyridoxine is needed for protein and carbohydrate metabolism, the formation of red blood cells and certain brain chemicals. It influences brain processes and development, immune function and steroid hormone activity.	

ESSENTIAL VITAMIN	ROLE/FUNCTION IN THE BODY	WHAT DOES IT DO IN THE SKIN OR TO ENHANCE SKIN APPEARANCE?
B7 (biotin)	Biotin (B7) is needed for energy metabolism, fat synthesis, amino acid metabolism and glycogen synthesis. High biotin intake can contribute to raised blood cholesterol levels.	
Vitamin C (ascorbic acid)	Vitamin C is an important agent in the manufacture of collagen. It helps to strengthen capillary walls and is anti-inflammatory.	
Vitamin D	Vitamin D is key to healthy skin and taken as internal support rather than topical application. It will promote healthy skin, hair, thyroid and immune function. It also assists absorption of calcium and phosphorus, improves muscle strength and is essential for bone health.	
Vitamin E (tocopherol)	Vitamin E can protect against collagen cross-linking and lipid peroxidation.	

23 To complete the table below, identify a role and function *in the body* for every essential vitamin. These vitamins indirectly benefit the skin. Information on B vitamins can be found at the following link: **https://www.betterhealth.vic.gov.au/health/healthyliving/vitamin-b** and a link to information on vitamin K (potassium) is **https://www.betterhealth. vic.gov.au/health/healthyliving/Vitamins-and-minerals**.

ESSENTIAL VITAMIN	ROLE/FUNCTION IN THE BODY
B1 (thiamin)	
B5 (pantothenic acid)	
B9 (folic acid or 'folate')	
B12 (cyanocobalamin)	
Vitamin K (potassium)	

MINERALS

24 Complete the following table by identifying some of the minerals we consider 'nutrients' that maintain body function and their food source (as listed in the section 'Minerals' in Chapter 23 of your textbook).

MINERAL	FUNCTION IN THE BODY	FOOD SOURCES
	Often added to table salt, so that the human population is not deficient. It is very important to the body, as it forms thyroid hormones which control the rate of metabolism and physical development.	
	Required for haemoglobin, which transports oxygen throughout the body. In clients with anaemia (deficiency), pale and dry skin, poor hair texture and pruritis (itchy skin) are common, as is koilonychia, or spoon-shaped nails.	
	Required by enzymes that maintain proteins, such as collagen, elastin and keratin. It plays an important role in immunity and cell growth. It is anti-inflammatory and decreases oxidative stress; deficiency can lead to dermatitis, nail defects and poor wound healing.	
	Helps build blood cells along with iron. It is involved in connective tissue formation and a deficiency can affect collagen formation in the skin.	

WATER

25 In one sentence, describe what water does for all of the tissues of the body.

_____

_____

26 Of the following specific benefits of water, which has to do with how much blood is pumped through the heart?

A Transports nutrients and oxygen into cells from the bloodstream

B Transports waste and toxins out of cells to the bloodstream

C Maintains blood pressure

D Helps to control oedema

27 Of the following specific benefits of water, which has to do with taking pressure off the liver?

A Helps to metabolise fat

B Counterbalances excess salt in the diet

C Helps relieve constipation

d Hydrates the skin

DIETARY FIBRE

28 True or false: Fibre is important in the diet because it helps *to turn the waste into liquid* that travels through the intestines, allowing it to move through effectively.

---

29 True or false: Dietary fibre is good at maintaining the correct level of intestinal flora because of its resistance to digestive enzymes.

---

30 True or false: Insoluble fibre is not digested by the body.

---

31 What is a probiotic?

---

32 What is a prebiotic?

---

33 Soluble fibre is *not*:

A a prebiotic

B only partially absorbed by the small intestine

C fibre that helps the body to reabsorb water, which helps the faeces to move

D polysaccharides oligofructose and inulin, for example

CHOLESTEROL

34 What does cholesterol provide for cell membranes?

---

35 Which hormones does cholesterol play a part in building?

---

36 Which fat-soluble vitamins does cholesterol help to transport?

---

37 What is the name of the acid cholesterol helps to create in the liver?

---

38 Why are low density lipoproteins (LDLs) considered 'bad'?

A They allow the HDLs to be transported in the bloodstream.

B They are able to remove HDLs from the inner lining of arteries.

C They can cause a blockage of the arteries because they build up on the inner lining.

D They help to prevent coronary heart disease.

Federal, state or territory legislation and local health and hygiene regulations relevant to food and nutrition

Knowledge Evidence

39 Which Privacy Principle in the *Privacy Act 1988* is relevant when you are reviewing information on a client record card, to ensure the personal information collected is accurate, up to date and complete?

40 The *Work Health and Safety Act 2011* covers work health and safety with regards to duty of care. What are the Standards that cover food hygiene in the salon; for example, when food is served to the clients?

Identify and apply nutrient needs and health problems across the lifespan to beauty therapy treatments

Performance Evidence

In this section, you can learn to apply the principles of the healthy living pyramid to skincare for clients of various ages and with varying health problems.

Healthy food and nutrition guidelines for Australians

Knowledge Evidence

1 View the visual representations of the Australian Dietary Guidelines at both of the following websites:

1 NHMRC's Australian Guide to Healthy Eating at **https://www.eatforhealth.gov.au**

2 Nutrition Australia's Healthy Eating Pyramid at **https://nutritionaustralia.org**

List two differences between the NHMRC's Australian Guide to Healthy Eating and Nutrition Australia's Healthy Eating Pyramid.

2 They are both good visual representations of what our daily intake should be. Which one do you consider best visually represents healthy food choices? Explain your answer.

3 Review the NHMRC's Australian Guide to Healthy Eating.

a According to Guideline 1, what do older people need to do?

b What does Guideline 3 say about coconut and palm oil?

c According to Guideline 3, which of the following products contain added sugar?

A Fruit drinks

B Vitamin waters

C Energy and sports drinks

D All of the above

d What is Guideline 5?

Guidelines for specific nutrients and recommended dietary intakes

Knowledge Evidence

4 At **https://www.eatforhealth.gov.au**, research serve sizes for different age groups, under 'Food essentials and the five food groups'. Complete the table below to identify individual nutritional requirements for different age groups and genders.

NUTRIENT	FOOD GROUP	POPULATION GROUP	HOW MANY SERVES OF THIS FOOD GROUP IS REQUIRED PER DAY FOR THE AGE GROUP(S)?
Calcium	Dairy and alternatives	All ages	
Iron	Lean meat	All ages, increase serves for pregnant women, menstruating women or endurance athletes	
Zinc	Lean meat and poultry, fish, eggs, tofu, nuts and seeds and legumes/beans	All ages (increase serves with age)	
Dietary fibre	Vegetables and legumes/beans	Adults	
Vitamin E	Grain (cereal) foods, mostly wholegrain and/or high fibre cereal varieties	Adults	

5 Recommended dietary intake (RDI) is a measure of the amount of daily nutrient intake required for healthy Australians. Where do you often see the RDI percentage?

6 True or false: The RDI tells you whether the food provides the required dose of that nutrient.

Role of nutrients in managing ideal weight

Knowledge Evidence

7 Estimated energy requirements (EER) is the amount of energy intake to maintain energy balance. What are the four factors that can vary a person's EER?

8 What happens when you eat less than your EER? Complete the following words.
   n_____ e_____ b_____

9 Which of the following options are the potential effects on the body of a depletion in nutrients and energy? Select all that apply.

   A Dehydration that consequentially causes dry, flaky skin and diarrhoea

   B Lacklustre hair and sallow, gaunt skin

   C Tightened, firmer skin

   D Slower basal metabolic rate (BMR)

10 How should you advise the client to lose weight healthily, and in a way that does not have a negative impact on the skin?

_____

_____

Nutritional needs and health problems across the lifespan

11 Refer to Figure 23.6, 'Nutrient requirements as we age' in your textbook and complete the table below to identify nutritional needs across the lifespan, including some nutrient needs for age-related health problems.

Knowledge
Evidence

AGE GROUP	KEY BODY SYSTEM FOCUS	ADDITIONAL NUTRIENT FOCUS
0–3 years Infancy	This age group is developing and growing at a very fast rate. More energy-rich foods are required, especially after the child begins walking.	
4–9 years Childhood	This age group requires more energy as it is a very active phase. Growth and development requires more nutrients to build bones and other structures.	
10–20 years Adolescence	The key focus in this age group is growth. About 40 per cent of skeletal growth occurs in adolescence. Women need zinc for a healthy reproductive system.	
21 onwards Adulthood	A fully developed adult should have an efficiently functioning body with nothing to worry about. In reality, factors other than diet affect the way our body cells function. Healthy eating choices coupled with exercise should help the body resist illness.	
Pregnancy and breastfeeding	The growing foetus will invariably take what is needed, leaving the mother depleted of essential nutrients. 'Eating for two' is no longer the best way to ensure a healthy baby; it will likely lead to weight gain and other complications.	
41–69 years Menopause (women only)	Menopause is normally accompanied by symptoms such as stress and hot flushes due to the intense hormone fluctuations in the body. At menopause, the hormones produced in the ovaries stop working and the adrenal glands take the full role of producing the 'female hormones' oestrogen and progesterone.	
Late adulthood (65 years +)	Elderly individuals require less nutrients because there is no growth. In particular, the energy and protein requirements are considerably decreased.	

Common diet-related health problems

Knowledge
Evidence

OBESITY AND WEIGHT MANAGEMENT

12 List three of the main issues for clients with obesity that concern the beauty therapist.

_____

_____

_____

13 If a client with obesity attempted to eat dramatically less than their EER in an attempt to lose weight, what could be the potentially dangerous result?

_____

EATING DISORDERS

14 List four types of eating disorders.

_____

15 What type of personal support is recommended for a client with an eating disorder?

_____

_____

16 Which is the best time to approach a person about an eating disorder:

A _If_ the client feels comfortable in talking about it with you, respecting privacy and confidentiality.

B _When_ the client is ready to talk about it to you, and be sure to note it on the client record card.

_____

FOOD INTOLERANCE

17 What are four possible symptoms on the skin of food intolerance?

_____

_____

_____

18 From the list of food intolerances in the section 'Food intolerances' in Chapter 23 of your textbook, select one type of food intolerance and identify its:

a underlying cause

b signs and symptoms, especially any that pertain to the skin

_____

_____

_____

LO23.2 APPLYING KNOWLEDGE OF BODY SYSTEMS TO BEAUTY THERAPY TREATMENTS

Performance Evidence

Knowledge Evidence

Basic knowledge of body systems

Review knowledge of body structures and systems in Chapter 24 of your textbook, and answer the following questions.

1 Draw a line to match the organ system with its function.

ORGAN SYSTEM	FUNCTION
Muscular system	Forms a strong framework that supports the softer tissues and maintains the shape of the body.
Skeletal system	Provides contraction and movements to body parts. It holds and covers the skeletal system and contributes to the body's morphological physique.
Digestive system	Coordinates the activities of the body by responding to stimuli received by sense organs.
Circulatory system	Filters, purifies and eliminates waste products from the body; maintains body's normal composition.
Nervous system	Breaks down food into wastes and nutrients. Nutrients are absorbed and waste eliminated.
Endocrine system	Breathing enables oxygen to be collected from the lungs and, via the blood, reach the body's cells. Carbon dioxide is exhaled as a waste product.
Urinary system	Transports materials around the body via blood and blood vessels.
Integumentary system	Provides immunity and drains fluid from tissues that can include microbes, toxins and other wastes.
Reproductive system	Provides a protective covering for the body.
Respiratory system	Coordinates and regulates processes in the body by means of chemicals (called hormones) released by endocrine glands into the bloodstream.
Lymphatic system	In the context of beauty therapy, the gonads (ovaries and testes) serve an endocrine function, and control the development of the body's ability to reproduce over a lifetime. The fluctuations in hormone levels affect the skin.

2 Blood carries oxygen and nutrients to the cells of the skin from the digestive system. Identify the body system mainly responsible for:

a blood transporting nutrients

b oxygen entering the bloodstream via the lungs

_____

_____

Main organs and functions of digestive and excretory system

Knowledge
Evidence

3 Draw a line to match the three macronutrients with the nutrients the body can use.

MACRONUTRIENT
Proteins
Carbohydrates
Fats

NUTRIENT THE BODY CAN USE
Fatty acids
Amino acids
Monosaccharides

4 What is the name of the enzyme that breaks down carbohydrates in the mouth when we chew food?

_____

5 Stress can affect gut health in a number of ways, including interrupting the normal functioning of peristalsis. What is peristalsis?

_____

_____

6 What is the stomach's main defense against harmful microbes?

_____

7 Describe the three enzymes that are secreted from the pancreas:

a trypsin

_____

_____

b lipase

_____

c amylase

_____

8 What alkaline chemical does the gall bladder store, that emulsifies lipids and adds a yellowish colour to the food?

_____

9 What are the structures that create a very high surface area, allowing more nutrients to be absorbed through the columnar cells in the small intestine walls?

_____

10 What eats the soluble 'dietary' fibre in the large intestine?

_____

11 True or false: Ureters are long thin tubes that transport filtered urine from the kidney to the bladder for urination.

_____

12 True or false: The endocrine system sends hormones to control the action of the bladder.

_____

13 True or false: The liver is responsible for storing vitamins A and D.

_____

LO23.3 PROVIDING ADVICE ON DIETARY GUIDELINES

Performance Evidence

Knowledge Evidence

Assess and evaluate nutritional requirements

1   Assess and evaluate nutritional requirements and dietary health problems for the following three clients with different life stages and nutritional needs. Use the section 'Providing advice on dietary guidelines' in Chapter 23 of your textbook and do some research into beneficial products and treatments for specific skincare requirements in other chapters and online.

CLIENT 1: FEMALE, PREGNANT 20 WEEKS

Skin requirements: puffy, sluggish skin

Skincare priority: *superficial lymph drainage*

• Beauty treatment recommendations (discuss with your teacher or research Chapter 23, other chapters and online):

_____

_____

_____

• Beauty product and homecare recommendations (research Chapter 23, other chapters and online to suggest product ingredients and massage techniques or equipment):

_____

_____

• Nutritional advice (appropriate for the professional to recommend):

_____

_____

• Lifestyle and aftercare advice (use Figure 23.12 in your textbook and think of some points of your own):

_____

_____

_____

_____

• Referral?

_____

_____

CLIENT 2: FEMALE, MENOPAUSE

Skincare requirements: erythmatotelangiectatic rosacea (redness present, no papules present – mild enough to treat in the salon)

Skincare priority: *calm the skin*

• Beauty treatment recommendations (discuss with your teacher or research Chapter 23, other chapters and online):

_____

_____

- Beauty product and homecare recommendations (research Chapter 23, other chapters and online to suggest product ingredients and massage techniques or equipment):

- Nutritional advice (appropriate for the professional to recommend):

- Lifestyle and aftercare advice (Use Figure 23.12 in your textbook and think of some points of your own):

- Referral?

CLIENT 3: MALE, ADOLESCENT

Skincare requirements: acne vulgaris, grade 2 (safe to treat in-salon)

Skincare requirement: *calm the skin, reduce oil flow*

- Beauty treatment recommendations (discuss with your teacher or research Chapter 23, other chapters and online):

- Beauty product and homecare recommendations (research Chapter 23, other chapters and online to suggest product ingredients and massage techniques or equipment):

- Nutritional advice (appropriate for the professional to recommend):

• Lifestyle and aftercare advice (Use Figure 23.12 in your textbook and think of some points of your own):

_____

_____

_____

• Referral?

_____

2 True or false: The beauty therapist is responsible for identifying general nutritional needs in a beauty therapy context.

_____

3 True or false: The beauty therapist is permitted to give specific nutritional advice in a beauty therapy context.

_____

4 Refer to Appendix B Referrals to professionals and use knowledge learnt in Chapter 23 of your textbook to draw a line to match the situation with the most appropriate person for referral.

SITUATION
A client with acne vulgaris grade 4 (severe) wants a referral to a dermatologist
A client with menopause wishing to balance hormones naturally, without medication
A client wants advice about how to move to a vegan diet with balanced nutrition
A client with obesity wishes to lose weight healthily
A client has an eating disorder and wants to talk to someone, like a hotline service, urgently

APPROPRIATE PROFESSIONAL
Dietician
Nutritionist
Relevant government body
Medical practitioner
Complementary therapist

Workplace policies and procedures in regard to beauty therapy treatments

Knowledge Evidence

5 When updating the client card with changes to the client's treatment plan, you need to take into account nutritional factors that impact on treatment outcomes. What are some things you might need to update? Select all that apply.

A Contraindications and restrictions to treatment

B Treatment objectives

C Nutritional advice

D Aftercare and lifestyle advice

E All of the above

_____

6 Your salon policies and procedures always comply with federal work health and safety (WHS) legislation. What does the *WHS Act 2011* say about giving nutritional advice to clients?

_____

_____

_____

CHAPTER 24: BODY STRUCTURES AND SYSTEMS

LEARNING OBJECTIVES

After completing this chapter, you should be able to:

LO24.1 identify the structural levels of organisation in the human body and use this knowledge to discuss presenting conditions and treatment priorities

LO24.2 use knowledge of the musculoskeletal system and its relationship with the human body to determine treatment requirements

LO24.3 understand the relationship the nervous system has to homeostasis and other functions of the human body, and use this to discuss presenting conditions and client priorities

LO24.4 understand the relationship the circulatory system has with skin functions, thermoregulation and homeostasis

LO24.5 use knowledge of the relationship the endocrine system has with hormonal influences on skin and body functions to identify treatment priorities and possible adverse reactions

LO24.6 understand the interdependence of organ systems and the relationship they have with a healthy body.

INTRODUCTION

The chapters in this section have been written differently to those proceeding them. The previous chapters throughout this workbook have been designed around both the knowledge and performance evidences from the training package. This section, however, has separated them: Chapters 24 and 25 are based only on the knowledge evidences of their respective units of competency, while Chapter 26 uses the performance evidences from both units. This is to promote your understanding of the science that underpins beauty therapy and then to demonstrate how you will apply this theory in practice.

The learning activities in this chapter allow you to demonstrate your knowledge and understanding of the unit of competency *SHBBSSC002 Incorporate knowledge of body structures and functions into beauty therapy*, using the knowledge evidence.

LO24.1 STRUCTURAL LEVELS OF ORGANISATION

Knowledge Evidence

1 Fill in the missing parts of the following table to demonstrate your knowledge of the structural organisation of the body.

LEVEL OF STRUCTURAL ORGANISATION OF THE BODY					
CHEMICAL	CELL	TISSUE	ORGAN	SYSTEM	ORGANISM
Collagen/elastin	Fibroblast	Connective tissue		Integumentary	Human
Actin/myosin	Muscle fibre		Skeletal muscle	Muscular	Human
Calcium	Osteocyte	Compact	Bone		Human

Cells

2   Which part of the cell is for energy production?

_____

3   What is the process by which cells reproduce?

_____

4   What is the name of the process by which cells undergo a change in shape and function?

_____

5   What is cell metabolism?

_____

_____

_____

Tissues

6   Watch the YouTube video **https://www.youtube.com/watch?v=lUe_RI_m-Vg** to understand more about epithelial tissues of the body. Answer the following questions.

   a   What is the purpose of the epithelial tissue?

   _____

   b   What part of the skin has epithelial tissue?

   _____

   _____

   _____

   c   Does epithelial tissue have a blood supply?

   _____

   d   What three shapes are the cells of the epidermis?

   _____

   _____

   _____

7   Watch this YouTube video to understand more about connective tissues of the body: **https://www.youtube.com/watch?v=D-SzmURNBH0**. Answer the following questions.

   a   There are four types of connective tissue. One is 'proper', which is found in skin, tendons and ligaments. What are the three other types?

   _____

   b   What does the ground substance of connective tissue do?

   _____

   c   Running through the ground substance gel is a network of _____ known as _____ and elastin.

   d   What are two key chemical proteins that are found in all muscle tissue?

   _____

e What are the three types of muscle tissue called? Note which ones are voluntary and which are involuntary.

_____

_____

_____

f Draw a sketch of smooth muscle tissue and skeletal muscle tissue so the differences between them can be compared.

SMOOTH MUSCLE TISSUE	SKELETAL MUSCLE TISSUE

g What specific type of involuntary tissue does the arrector pili muscle contain?

_____

h What type of muscle tissue is in your quadriceps?

_____

8 What are the functions of blood?

_____

_____

_____

_____

_____

9 How long does it take for a wound to heal?

_____

_____

LO24.2 THE MUSCULOSKELETAL SYSTEM

1 Watch the YouTube video **https://www.youtube.com/watch?v=i5tR3csCWYo**, and then draw a line to match the following cells with the type of tissue listed in the video.

CELL TYPE
Osteocytes
Skeletal muscle fibres
Sensory neurones
Fibroblasts
Adipose cells

TISSUE TYPE
Nervous
Connective
Connective
Connective
Muscle

Knowledge
Evidence

2 In the table below, describe the function of each organ system and label the diagrams.

FUNCTIONS	DIAGRAM
SKELETAL SYSTEM	
	Key structures of the skeletal system 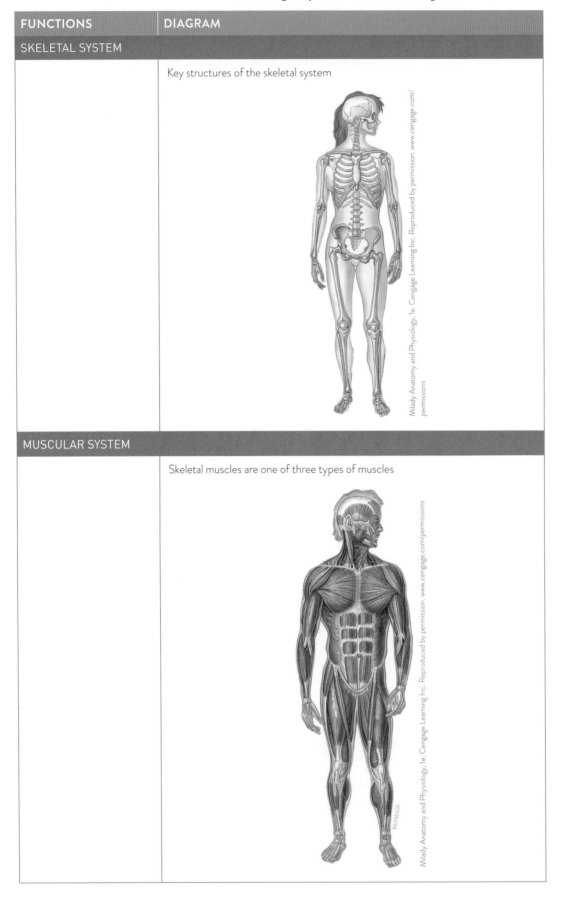
MUSCULAR SYSTEM	
	Skeletal muscles are one of three types of muscles

Milady Anatomy and Physiology, 1e. Cengage Learning Inc. Reproduced by permission. www.cengage.com/permissions

Milady Anatomy and Physiology, 1e. Cengage Learning Inc. Reproduced by permission. www.cengage.com/permissions

The skeletal system

3 What are the key functions of the skeletal system?

_____

_____

_____

_____

_____

_____

MAJOR BONES IN THE BODY

4 Name which body parts contain the following bones:

a tibia

_____

b carpals

_____

c vertebra

_____

d sternum

_____

e scapula

_____

f humerus

_____

g clavicle

_____

h femur

_____

i phalange

_____

j cranium

_____

5 How is a tendon different from a ligament?

_____

STRUCTURE OF BONES

6   Identify a skeletal bone that can be classified as each of the following bone shapes.

   a   Flat bone

   _____

   b   Irregular bone

   _____

   c   Short bone

   _____

   d   Long bone

   _____

7   Which important cells belonging to the circulatory system are produced in the spongy part of the long bone?

   _____

8   Compact bones are composed of structural units called osteons: long and hollow with layers of dense bone matrix, tiny canals and osteocytes (bone cells). What is at the centre of an osteon?

   _____

BONES OF THE FACE

9   Which skeletal bone is your 'cheekbone'?

   _____

10   What do the parietal bone, occipital bone and cranium all have in common? What is their main purpose?

   _____

11   What is the common name for the mandible?

   _____

BONES OF THE LOWER ARMS AND HANDS

12   Use the website **https://www.innerbody.com/image_skelfov/skel20_new.html** to answer the following questions on the bones of the lower arms and hands.

   a   The forearm has two bones that run the length from the elbow to the wrist, the ulna and the radius. Which one is being described below?

   •   It is the longer of the two bones.

   •   It joins the hand on the same side as the pinky finger.

   •   At its proximal end, it forms the hinge of the elbow with the humerus.

   _____

   b   The distal end of the _____ (bone) rotates with the thumb.

c How many bones does one hand have?

_____

d What are the three types of bones in the hands?

_____

_____

BONES OF THE LOWER LEGS AND FEET

13 Use the website **https://www.innerbody.com/anatomy/skeletal/leg-foot** to answer the following questions on the bones of the lower arms and hands.

    a The bones of the legs and feet are very similar in layout to those of the arms and hands. How would you describe the layout of the bones of a leg?

_____

_____

_____

    b What is the technical name for the kneecap? What is its main purpose?

_____

    c Why would the phalanges need to flex or extend?

_____

    d When you toggle between the skeletal system and the muscular system, you are able to see the Achilles tendon (technically called the calcaneal tendon). What does this tendon connect to? Why is it so prone to injury or inflammation?

_____

_____

    e A common injury experienced by sports people where the game requires sudden stops or changes in direction (e.g. soccer, basketball, downhill skiing) is a torn anterior cruciate ligament (ACL). What two bones does this ligament attach to?

_____

JOINTS

14 Draw a line to match the part of the joint (articulation) affected to the relevant disease or disorder. You may need to refer to Appendix A Contraindications and restrictions in your textbook.

PART OF THE JOINT (ARTICULATION) AFFECTED
Bone
Articular cartilage
Synovial membrane and synovial fluid

DISEASE/DISORDER
Osteoarthritis
Rheumatoid arthritis
Fracture

15 Complete the following table by labelling the specific type of synovial joint and how it fits with the bones that make the body movements listed. You can use this YouTube video to help to learn the body movements: **https://www.youtube.com/watch?v=5YcNAPzDxDg**.

POSITION	TYPE OF JOINT	MOVEMENT
Hip		Abduction, adduction, extension, flexion, rotation
Shoulder		As for hip Arm circling involves movement of this joint together with that of the shoulder (pectoral girdle)
Knee		Flexion, extension
Elbow		As for knee
Forearm		Supination (turning the hand palm up), pronation (turning the hand palm down)
Ankle		Dorsiflexion (foot pulled up towards knee), plantarflexion (pointing the foot)
Wrist		Flexion, extension, abduction, adduction
Foot		Inversion, eversion
Hand		Flexion (clenching), extension (stretching)
Toe		Flexion, extension (the joint at the base of the toes allows abduction and adduction)
Finger		As for toe
Thumb		Allows the same movement as the condyloid joint of the wrist

16 Name the type of joint found:

    a in the cranium

    b at the hip

    c at the elbow

17 At a synovial joint, what are the functions of the synovial fluid, ligaments and articular cartilage?

The muscular system

18 Research the location of the muscles listed in the table below by light palpation and describe their location.

Knowledge Evidence

MUSCLE	LOCATION	ACTION
Rectus abdominis		Flexes the spine, compresses the abdomen, tilts the pelvis upwards
Obliques		Both compress the abdomen and twist the trunk, the left internal oblique working with the right external oblique

19 Research and define the following anatomical terms.

    a Rectus

    b Oblique

20 Research the location of the arm and chest muscles listed in the table below. Identify the location by light palpation and describe where you found them.

MUSCLE	LOCATION	ACTION
Pectoralis major		Used in throwing and climbing to adduct the arm, drawing it forwards and rotating it medially
Pectoralis minor		Draws the shoulder downwards and forwards

MUSCLE	LOCATION	ACTION
Deltoid		Abducts the arm to a horizontal position; aids in further abduction and in drawing the arm backwards and forwards
Biceps		Flex the elbow; supinate the forearm and hand
Triceps		Extend the elbow
Brachialis		Flexes the elbow
Brachioradialis		Flexes the elbow
Flexors		Flex and bend the wrist, drawing it towards the forearm
Extensors		Extend and straighten the wrist and hand
Thenar muscles		Flex the thumb and move it outwards and inwards
Hypothenar muscles		Flex the little finger and move it outwards

21 Research the location of the leg muscles listed in the table below. Identify the location by light palpation and describe where you found them.

MUSCLE	LOCATION	ACTION
Gluteals		Abduct and rotate the femur; used in walking and running and to raise the body to an upright position
Hamstrings		Flex the knee; extend the thigh; used in walking and jumping
Gastrocnemius		Flexes the knee; plantarflexes foot
Soleus		Plantarflexes foot; both calf muscles are used to push off (when walking and running)
Quadriceps extensor		Extends the knee; used in kicking. The rectus fibres from the pelvis help to flex the hip
Sartorius		Flexes the knee and hip; abducts and rotates the femur; used to sit cross-legged
Adductors		Adduct the hip; flex and rotate the femur
Tibialis anterior		Inverts the foot; dorsiflexes the foot; rotates foot outwards

22 Research the location of the back muscles listed in the table below. Identify the location by light palpation and describe where you found them.

MUSCLE	LOCATION	ACTION
Trapezius		Moves the scapula up, down and back, to extend the neck

MUSCLE	LOCATION	ACTION
Latissimus dorsi		Used in rowing and climbing; it adducts the shoulder downwards and pulls it backwards
Erector spinae		Extends spine; keeps the body in an upright position

23 Describe the structure of voluntary muscle tissue.

24 What is meant by muscle tone and how can tone be lost or improved?

25 What do the following points describe: a fascia, ligament or tendon?

- It is connective tissue sheath that surrounds a muscle.
- It is slippery to allow muscle/muscle and bone/muscle to slide against one another.
- It attaches skin to muscle.

26 Refer to Chapter 18 Provide body massages in your textbook. Suggest one massage method to stimulate and one method to relax a muscle at the neuromuscular junction.

27 What is the difference between a trigger point, a motor point and the neuromuscular junction?

28 Identify a disorder of the muscular system that might restrict or prevent you from performing a massage treatment. Suggest a medical or complementary therapist who might be able to assist the client.

29 With a partner, identify the superficial muscles by light palpation. Draw a sketch of the body (anterior and posterior) below and list the muscles you have identified.

MUSCLES OF THE FACE

30 To help you learn about the basic muscles of facial expression, you can view this YouTube video: **https://www.youtube.com/watch? v=zGqfKY1rjkM**. (Warning: Aboriginal and Torres Strait Islanders are warned that the video contains an image of a deceased person.)

   a Visit **http://www.innerbody.com/anatomy/muscular/head-neck/ facial-expression-muscles**. Learn the muscles below and label them in the diagram shown:

      i   frontalis

      ii   zygomaticus

      iii buccinator

      iv nasalis

      v   orbicularis oculi

      vi orbicularis oris

      vii mentalis

      viii platysma

   b Answer the following questions:

      i   What part of the muscle connects with bone?

      ii What type of connective tissue is it made of?

      iii What connects bone to bone?

      iv On the face, muscle can be connected to muscle. What connects muscle to muscle?

Shutterstock.com/Sofia Santos

31 Complete the table with the correct information on the muscles of facial expression.

MUSCLE	EXPRESSION	LOCATION	ACTION
Procerus		Covers the bridge of the nose	Wrinkling of the skin over the bridge of the nose
Zygomaticus, major and minor	Smiling, laughing		
	Frown	Between the eyebrows	Draws the eyebrows together
Mentalis			Raises the lower lip, causing the chin to wrinkle

MUSCLES OF THE LOWER ARMS AND HANDS

32 Use the information found at **https://www.innerbody.com/anatomy/muscular/arm-hand** to describe the location and purpose of each of these muscles:

a abductor digiti minimi muscle of hand

b flexor digiti minimi brevis muscle of hand

c lumbrical muscles of the hand

MUSCLES OF THE LOWER LEGS AND FEET

33 Describe what a plantar flexion and a dorsiflexion are.

34 The calcaneous (Achilles) tendon and the plantaris tendon both connect very closely together on the heel bone. Describe the difference between them.

35 Give the location of the following muscles:

a extensor hallucis brevis muscle

b flexor digitorum brevis muscle

c abductor hallucis muscle

INSPIRATION AND EXPIRATION

As air is drawn in, the diaphragm contracts under the control of the autonomic nervous system, creating a suction that allows the thoracic cavity and the lungs to expand. The external intercostal muscles also help to stretch the ribcage for deep inspiration.

Upon expiration, the diaphragm relaxes and the thoracic cavity contracts in response. Internal intercostals pull down the ribcage, forcing air out of the lungs.

36 Research the location of the muscles used in breathing, as listed in the table below. Where possible, identify the location by light palpation and describe where you found them. You may also refer to Figure 24.29 in your textbook.

MUSCLE	LOCATION	ACTION
External intercostals		Used in breathing movements to draw the ribs upwards and outwards when breathing in
Internal intercostals		Draw the ribs downwards and inwards when breathing out
Diaphragm		Contraction of this muscle increases the volume of the thorax

LO24.3 THE NERVOUS SYSTEM

Knowledge
Evidence

1 How do nerve impulses pass along nerve fibres?

2 Where do nerves stimulate muscles to contract?

3 What is the autonomic nervous system?

4 Which autonomic nervous system works favourably during skin treatments?

5 Where are the olfactory bulbs situated and what is their function?

6 How does the limbic system coordinate information received by the senses during aromatherapy?

7 Identify a disorder of the nervous system that might restrict or prevent you from performing a beauty treatment. Suggest a medical or complementary therapist who might be able to assist the client.

8 What is a reflex arc? Explain how the blink reflex works when you feel a hair in your eye.

9 When massage is applied to the skin, the sensory neuron at the skin's surface causes blood vessels to dilate – is this a reflex arc? Explain how this might occur.

10 Which of the following types of sensory receptors do you have in your skin? Consider all of the five senses you feel through your fingertips.

    A Mechanoreceptors

    B Thermoreceptors

    C Photoreceptors

    D Chemoreceptors

11 How do nerve signals travel from one nerve to the other nerve?

12 When the nerve signal travels to the arrector pili muscle (when goosebumps occur), the electrical signal jumps via a chemical known as a n_____.

13 When a client prefers a lighter pressure in massage, is it more likely that they are fragile or that they have more of a certain type of sensory nerve?

14 When performing a facial treatment, it is useful to have an understanding of the dermatome map of the body. What are the four main peripheral or cranial nerves that are relevant to a facial and where are they?

15 What is homeostasis? How does the nervous system contribute to homeostasis?

LO24.4 THE CIRCULATORY SYSTEM

Knowledge
Evidence

1 True or false: Capillaries are one cell-layer thick.

2 True or false: Capillaries have smooth muscle around them in parts.

3 True or false: The strength and elasticity of the capillary walls can be damaged by extreme pressure, heat or UV radiation.

4 Vasodilation and vasoconstriction happen in the body for reasons other than temperature regulation. We can initiate the reflex arc to cause erythema. Which massage movements are best at that? Refer to Chapter 18 Provide body massages in your textbook.

5 Name the three main types of blood vessels and list their distinguishing anatomical structure.

6 Which type of blood vessel carries blood away from the heart?

7 Why do veins have valves?

8 Why do all cells in the body need to be near, or have access to, a blood supply?

9 Which cells in the skin have no direct blood supply?

10 Which cells of the skeletal system have the smallest blood supply?

11 Identify a disorder of the cardiovascular system that might restrict or prevent you from performing a beauty treatment. Suggest a medical or complementary therapist who might be able to assist the client.

The lymphatic system

12 Use the information found at **https://www.visiblebody.com/blog/the-lymphatic-system-innate-and-adaptive-immunity** to complete the below table that describes white blood cells by sketching each cell listed and describing its primary function(s).

WHITE BLOOD CELL	SKETCH	FUNCTION
Neutrophils (small macrophages in pus)		
B-lymphocyte		
T-lymphocyte		
Dendritic cells (such as Langerhan's cells of the epidermis)		

13 When performing beauty treatments, massage movements are best applied towards the lymphatic system structure that drains directly into the bloodstream to ensure lymph is flowing in the right direction. Which structure is that?

A Left subclavian vein

B Thoracic duct

C Cysterna chyli

D Inguinal nodes

14 Where are the popliteal lymph nodes located?

15 Where does lymph come from and where does it drain to?

16 What is distinctive about lymphatic vessels and lymph flow as compared with blood vessels?

_____

_____

_____

_____

17 What are white blood cells?

_____

18 What is the function of the immune system?

_____

19 Identify a disorder of the lymphatic system that might restrict or prevent you from performing a beauty treatment. Suggest a medical or complementary therapist who might be able to assist the client.

_____

_____

_____

IMMUNITY

20 What is immunity?

Knowledge
Evidence

_____

_____

21 How do the following provide immunity in humans?

a   White blood cells

_____

_____

b   Inflammation

_____

_____

LO24.5 THE ENDOCRINE SYSTEM

Knowledge
Evidence

1   Which hormone is responsible for acne vulgaris?

_____

2   Glucocorticoids are often found in medications for rheumatoid arthritis. What do glucocorticoids do?

_____

3   If a person has hyperthyroidism, which hormone is out of balance? What does the hormone usually do?

_____

Note: An exocrine gland, unlike an endocrine gland, secretes a substance via a duct or tube to somewhere external to itself, such as sweat glands, sebaceous glands or the pancreatic digestive glands.

4   What are the five general functions of the endocrine system?

_____

_____

_____

_____

_____

5   Label the glands featured on the diagram of the endocrine system below.

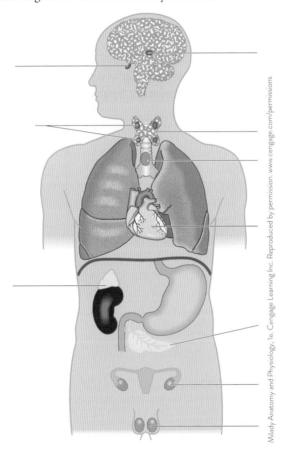

Milady Anatomy and Physiology, 1e. Cengage Learning Inc. Reproduced by permission. www.cengage.com/permissions

6   Complete the following table by naming the hormones produced by each of the glands listed, including the effect each hormone has on the hair, skin and body.

ENDOCRINE GLAND	HORMONE SECRETED	EFFECT ON HAIR, SKIN AND BODY
Thyroid		
Pineal gland		
Adrenal gland		
Adrenal gland		

7   Identify a disorder of the endocrine system that might restrict or prevent you from performing a beauty treatment. Suggest a medical or complementary therapist who might be able to assist the client.

_____

_____

_____

HORMONE ACTION ON SKIN

8 Refer to Figure 24.57 in your textbook and locate the position of the following glands. The major skin-influencing hormones they produce are listed below. Then refer to Figure 24.56 and identify the actions of those hormones on the skin.

    a Pituitary:

        i melanocyte stimulating hormone (MSH)

        ii human growth hormone (GH)

    b Thyroid: thyroxine

    c Parathyroid: parathyroid hormone (PH)

    d Adrenal: corticosteroids

    e Ovary: oestrogens and progesterones

    f Testes: testosterone

9 Which hormone is responsible for causing pigmentation, such as melasma, in pregnancy?

10 Does the following list describe skin that is chronologically aged or photo-aged (dermatoheliosis)?

- Thick epidermis
- Irregular texture to touch
- Surface pattern irregularities
- Dehydrated
- Tough, like leather

11 Male pattern baldness (alopecia) is a result of what hormone?

12 What is telogen effluvium, and what are the hormones involved in it?

LO24.6 ORGAN SYSTEMS AND THEIR RELATIONSHIP TO A HEALTHY BODY

Knowledge
Evidence

The respiratory system

1 List the main organs of the respiratory system.

2 What does the respiratory system need to do efficiently that is vital for the health of the skin?

3 Identify a disorder of the respiratory system that might restrict or prevent you from performing a beauty treatment. Suggest a medical or complementary therapist who might be able to assist the client.

The reproductive system

4 What disorders or hormone imbalances of the reproductive system can affect:

    a hair growth

    b skin growth

5 Which of the following structures most directly relates to *lifting* and *firming* of the breast tissue?

    A Fat (adipose tissue)

    B Cooper's ligaments

    C Lobules of milk-producing glandular tissue

    D Nipple

The integumentary system

Read through the information provided in the section 'Skin sensitivity' in Chapter 25 of your textbook for the following question.

6   For body massage, it is important to know where the skin is more sensitive, so you can be more aware of how the client might be feeling. List these from the most sensitive skin to the least:

- rest of body

- fingers

- sole of foot

- toes

- palm of hand

Disorders of the body

Knowledge Evidence

7   Some fairly common disorders that you will need to recognise during a beauty consultation are listed in the table below. Research and describe the characteristic features of each condition.

CONDITION	CHARACTERISTIC FEATURES
SKELETAL AND ARTICULAR	
Osteoporosis	
Curvature of the spine	
Arthritis	
Fractures	

CONDITION	CHARACTERISTIC FEATURES
Dislocation	
Sprain	
MUSCULAR SYSTEM	
Muscle strains	
Muscle spasms	
Muscle atrophy	
Muscular dystrophy	
Tendonitis	

CONDITION	CHARACTERISTIC FEATURES
Tendinosis	
Fibromyalgia syndrome	

NERVOUS SYSTEM	
Multiple sclerosis	
Dystonia	
Sciatica	
Epilepsy	
Carpal tunnel syndrome	

CONDITION	CHARACTERISTIC FEATURES
RESPIRATORY SYSTEM	
Asthma	
Common cold and influenza	
Sinusitis	
CIRCULATORY SYSTEM	
Varicose veins	
High blood pressure (hypertension)	
Low blood pressure	
Thrombosis or 'deep vein thrombosis' (DVT)	
CONDITION	CHARACTERISTIC FEATURES

CONDITION	CHARACTERISTIC FEATURES
LYMPHATIC SYSTEM	
Oedema, also known as lymphoedema	
Autoimmune disorders	
Glandular fever	
Tonsillitis	
ENDOCRINE AND REPRODUCTIVE SYSTEM	
Amenorrhoea	
Dysmenorrhoea	
Ovarian cysts	

CONDITION	CHARACTERISTIC FEATURES
Pre-menstrual tension	
Breast tumours	

8   Research the inflammatory and autoimmune disease called ankylosing spondylitis. Find out the following points:

- What is an autoimmune disease?

- How does ankylosing spondylitis affect a person's spine?

- What is the effect on posture?

- What is the effect on the respiratory system?

Interdependence of organ systems

The concept of interdependence is that two things require each other to function properly. Review the organ systems of the body to answer the following questions in relation to body massage.

Knowledge Evidence

9   Look at the skeletal system and the muscular system. If we know that tendons attach muscle to bone, what could we say about how the tendon facilitates the interdependence of body systems? Which body system do tendons belong to?

10 How could you describe the interdependence of the nervous system with the integumentary system in providing skin sensations? Give an example.

11 Interstitial fluid helps move oxygen and nutrients to cells from capillaries and brings waste from cells back to the bloodstream. Sometimes, some interstitial fluid is left behind and is later gathered up lymphatic vessels (the interstitial fluid at this point is now referred to as lymph) and returned to the blood stream. How can body massage encourage the interdependence of the lymphatic and circulatory systems?

INTERDEPENDENCE OF THE SKELETAL, MUSCULAR AND NERVOUS SYSTEMS

Answer the following questions to check your knowledge of the interrelationship between the skeletal system, the muscular system and the nervous systems for skin and body function.

Knowledge
Evidence

12 What do the muscle tendons attach to that make movement of facial muscles necessary?

_____

13 If muscles have good tone, how does it make the skin and body appear?

_____

14 If muscles are not connected with the skin very well, how does it make the skin appear?

_____

15 What is the neuromuscular junction?

_____

16 What sets off a muscle contraction?

_____

17 What is flaccid paralysis? Can you think of a reason why flaccid paralysis could occur on the face?

_____

_____

INTERDEPENDENCE OF THE LYMPHATIC, DIGESTIVE, RESPIRATORY AND CIRCULATORY SYSTEMS

Knowledge
Evidence

18 Why is posture important for good lymph drainage?

_____

19 How does deep breathing improve lymph flow?

_____

20 What does a 'clean' digestive system mean to you in the context of skincare? Refer to Figure 21.4 in Chapter 21 of your textbook.

_____

21 What is vasodilation? How can it benefit the skin?

_____

_____

_____

_____

22 Name one autoimmune disorder. What implications can it have for facial treatments? Refer to Appendix A Contraindications and restrictions.

_____

23 What is the difference between homeostasis and thermoregulation? What role does the respiratory system play in thermoregulation? This website will help: **https://biologydictionary.net/how-does-the-respiratory-system-maintain-homeostasis**.

_____

_____

_____

_____

INTERDEPENDENCE OF THE ENDOCRINE AND REPRODUCTIVE SYSTEMS

24 Complete the following table of the effects of hormones on hair, skin and body shape by listing the name(s) of the hormone produced and its effects for each gland listed.

Knowledge Evidence

GLAND	NAME OF HORMONE	EFFECT
Pituitary		
Thyroid		
Parathyroid		
Adrenal		
Ovary		
Testes		

25 In the following table the hormones that control sebaceous gland secretions are provided. State the endocrine gland that secretes the hormone, determine if it increases or decreases sebum, and list one disorder that can affect that endocrine gland.

HORMONE	ENDOCRINE GLAND/ ORGAN THAT SECRETES THE HORMONE	HOW IT AFFECTS SEBUM SECRETION	DISORDER
Androgen (testosterone)			
Thyroxine			
Oestrogen			
Melanocyte stimulating hormone			
Growth hormone			

26 What can menopause do to the skin? Which endocrine organ stops working?

27 At what age does the menopause happen?

28 At what age does puberty happen?

29 What are the characteristics of skin and hair at puberty for both males and females?

30 Adult acne is different to acne vulgaris when a person is younger because the skin takes longer to heal. Why do some people get acne when they are older, even if they have never had acne before?

31 What is sebum mostly composed of?

_____

_____

_____

_____

32 If the skin around a pustule looks inflamed, does that mean it is close to healing? Should you touch it?

_____

_____

33 Can hormonal imbalance cause hair loss? Why is this so, and what are some examples of conditions that are associated with hair loss?

_____

_____

_____

CHAPTER 25: SKIN SCIENCE

LEARNING OBJECTIVES

After completing this chapter, you should be able to:

LO25.1 identify the different structures of the skin and use this knowledge to identify presenting conditions and requirements, and discuss client treatment priorities

LO25.2 understand the function of the various glands of the skin and identify client treatment requirements, identifying possible adverse reactions

LO25.3 use an understanding of the appearance and different types of skin to identify treatment requirements

LO25.4 understand the growth, development, ageing and healing of human skin

LO25.5 understand the scope of practice for a beauty therapist and review client medical history, referring client to an appropriate professional where required.

INTRODUCTION

It is important to have a sound understanding of skin science so that you can clearly explain to clients how beauty therapy treatments and skincare advice will impact on the body's health and wellbeing.

The learning activities in this chapter allow you to demonstrate your knowledge and understanding of the unit of competency *SHBBSSC001 Incorporate knowledge of skin structure and functions into beauty therapy*, using the knowledge evidence.

LO25.1 SKIN STRUCTURE

Knowledge Evidence

The integumentary system

1 What is the integumentary system composed of?

2 Which nerve endings detect touch?

3 What does a dull, aching pain indicate?

4 Explain why a person's absolute threshold to touch affects the massage pressure they might prefer.

Lipids and the skin

Knowledge Evidence

The skin is one of the more active lipid-synthesising tissues of the body. The two major sites of skin lipid synthesis (lipogenesis) are the:

1 sebaceous glands – 95 per cent of surface lipids. These lipids make no contribution to the barrier properties of the skin

2 epidermal keratinocytes – 5 per cent of surface lipids (100% of stratum corneum lipids). Keratinocytes are the producer of ceramides.

While there appears to be only a small contribution from epidermal lipids to the surface sludge, their major function is in the construction of the epidermal barrier and the consequential control of trans-epidermal water loss (TEWL) and percutaneous absorption.

The epidermal lipids within the epidermal keratinocytes form a multi-laminate liquid crystalline lipid membrane structure when they are outside of the cell.

TYPES OF SKIN LIPIDS

The range of chemicals grouped as *lipids* is staggeringly large and complex. Lipids are insoluble in water but soluble in organic solvents such as chloroform, mineral turpentine, benzene, and so on (a case of like dissolving like). Lipids can mix with water provided a third type of chemical is involved. These chemicals are called 'surface active agents' ('surfactants') and include soaps and detergents. They can bind water at one end of the molecule and the lipid at the other; they have 'head' (hydrophilic) and 'tail' (lipophilic) ends. All lipids can be split into two groups according to whether they react with strong alkali:

1   saponifiable – can be processed into soaps by powerful alkalis (e.g. sodium hydroxide, potassium hydroxide)
2   non-saponifiable – cannot be processed into soaps, so they are resistant to alkali attack.

Saponifiable skin lipids

There are two main groups:

1   simple lipids
2   complex lipids.

Simple lipids

The simplest and most abundant lipids are the triglycerides, which include most common fats, oils and waxes.

A triglyceride molecule consists of a backbone of glycerol with three fatty acids tagged onto it. A fatty acid molecule has a long hydrocarbon chain with an organic acid group at one end. The commonest length of the hydrocarbon chain (often just called 'the carbon chain') for skin lipids is 16–18 carbon atoms. Some fatty acids contain all single carbon-to-carbon bonds (saturated), while others contain one or more carbon-to-carbon double bonds (unsaturated).

Simple and complex lipids

The chemical structure of a lipid may be simple or complex. Simple lipids are made of one or two functional units (fatty acids), and complex lipids may contain three or more functional units (fatty acids). Refer to Figure 25.1 for examples of complex and simple lipids in cosmetic chemistry.

Figure 25.1   Common lipids in cosmetic ingredients and the skin

EXAMPLES USING COMMON NAME*	PRIMARY FUNCTION
COMPLEX LIPIDS	
Phospholipids	
Phosphatidycholine	Is broken down to lecithin from seeds; the primary constituent of the cell membrane
Sphingolipids	
Sphingomyelin	Secreted by lamellar bodies along the cell membrane of keratinocytes; reacts with an enzyme to produce a ceramide that contains high levels of linoleic acid
SIMPLE LIPIDS	
Triglycerides (can become fatty acids)	
Glycerol with more than three fatty acids per glycerol unit	May be stored in adipose tissue; broken down into three fatty acids in the cell's mitochondria and converted into adenosine triphosphate (ATP) for cellular metabolism

EXAMPLES USING COMMON NAME*	PRIMARY FUNCTION
Steroids	
Cholesterol	Emollient; maintains the fluidity of an otherwise relatively solid cell membrane
Phytosterols: sitosterol (from chestnuts, rice bran or wheat germ), stigmasterol (from soybean)	Plant-derived; used in cosmetics to mimic the function of cholesterol in the skin
Hormones: androgens, oestrogens, corticosteroids and adrenal hormones	Produced in the gonads and adrenal glands respectively; have an impact on hair and skin growth
Waxes (energy stores, lubrication, waterproofing)	
Carnauba wax	Used as a solidifier in lipstick and other make-up products
Paraffin wax	Used in the paraffin bath; when melted at 37 °C it locks in moisture and infuses the emollient product; it is a non-comedogenic occludent when in a moisturiser
Beeswax	Used in depilatory wax, combined with a sticky resin
Tocopherols	
Vitamin E	Antioxidant found in vegetable oils; stabilised ingredients include tocopherol acetate, tocopherol palmitate
Fatty acids (energy stores, cellular messengers)	
Saturated fat	
Acetic acid	Used to buffer, or control pH of a product
Butyric acid	Has an unpleasant odour, but in cosmetics its esters are used for synthetic 'fruity' fragrances
Caproic acid	Extracted from coconut; is emulsifying and a fragrance fixative
Caprylic and capric acids	Extracted from coconut; act as a thickener, provide a film barrier to prevent TEWL
Lauric acid	Found in coconut and palm kernel oil; is antimicrobial; a common derivative is sodium lauryl sulphate to saponify fats
Myristic and stearic acids	Found in vegetable oils, such as palm and coconut oils; surfactants and cleansing agents
Palmitic acid	From degraded linoleic acid; desaturates to become sapienic acid in sebum; the most common fatty acid in the body
Arachidic acid	Emollient derived from peanut oil
Behenic acid	Thickener, cleanser and surfactant
Monounsaturated fat	
Palmitoleic acid	Found on the skin's adipose tissue
Sapienic acid	Found in sebum; antimicrobial
Petroselenic acid	Extracted from parsley and coriander seeds (linalool); a perfume and antimicrobial
Oleic acid	Nature's most abundant monounsaturated fatty acid; usually found in fish
Erucic acid	Extracted from rapeseed and mustard seed; an emollient
Polyunsaturated fat	
Sebaleic acid	Found in relatively high amounts in sebum as a derivative of sapienic acid
Linoleic acid**	A derivative of ceramides (sphingolipids) found in almost all food and vegetable oils; the most common fatty acid used in cosmetics

EXAMPLES USING COMMON NAME*	PRIMARY FUNCTION
Linolenic acid**	Found in vegetable oils, particularly evening primrose oil
Arachidonic acid**	Derived from linoleic acid; important to form phospholipids in cell membranes

*as indicated on product labels rather than by their chemical name
**EFA – essential fatty acids. Our cells cannot make EFAs and they therefore must be obtained in the diet. The importance of EFAs (especially linoleic acid and linolenic acid) and other fatty acids to the skin is that they are needed as raw materials for more complex forms of lipids; for example, triglycerides, phospholipids, prostaglandins and ceramides.

Complex lipids

Most of these are waxy lipids and are variations on the theme of simple lipids – a small backbone with attached fatty acid side chains. The most important types in skin are *phospholipids*, particularly in the membranes of living cells, both in the dermis and epidermis, and the *sphingolipid ceramides*:

- Phospholipids:
  - have a 'backbone' of glycerol and are important as part of the cell membranes of all living cells. Some have good emulsifying properties and are often used in cosmetic products, for example in making liposomes (e.g. lecithin).

- Sphingolipid ceramides:
  - have a 'backbone' of sphingosine. They are the most abundant and the most important lipids in the stratum corneum. There are at least six different types. Each type is present in the skin in different proportions. Exactly how this 'fine tunes' the barrier properties of the skin is not fully understood. Ceramide composition of the stratum corneum is known to change with age
  - are based on sphingosine fatty acid and a long saturated or monounsaturated carbon chain of 18–26 carbon atoms. Plants are based on a slightly different backbone of phytosphingosine and a fatty acid.

Non-saponifiable lipids

These lipids cannot be broken down by strong alkali. They are usually complex ring structures (of five or six carbon atoms each), with single and double bonds and complex side chains and branches. There are two major subdivisions of the non-saponifiable lipids:

1 steroids and sterols
2 terpenes and terpenols.

Steroids and sterols

Steroids include many important body chemicals and are based on a four-ring molecular structure. Examples include cortisone, vitamin D, testosterones, progesterones and cholesterol. Cholesterol is also based on the steroid molecular structure and is a precursor (raw material) for the steroids of the body and an important part of all cell membranes and the stratum corneum. In fact, the epidermis is one of the most active cholesterol-synthesising tissues in the body.

Terpenes and terpenols

Lipids in the skin that may be terpenes and terpenols include:

- vitamin A (retinol)
- vitamin E (alpha-tocopherol)
- prostaglandins
- beta-carotene.

LIPID STRUCTURE AND THE STRATUM CORNEUM

The stratum corneum is regarded as a two-compartment structure: layers of corneocytes embedded in a cement of lipids. These lipids are synthesised by the keratinocytes (lipogenesis) and expelled as granules (membrane bodies) into the intercellular space, where they begin to unite in the lower stratum corneum.

The corneocytes once had cell membranes made of phospholipids. In the stratum corneum this changes until the membrane is only protein. This protein, called *involucrin*, is important for the connection it provides to the intercellular lipids. There are no phospholipids in the stratum corneum.

The epidermal lipids are mainly polar lipids. Polar lipids have a lipophilic tail and a hydrophilic head group and the molecules arrange themselves spontaneously in water in various structures. In all these structures, the hydrophilic head groups aggregate towards water and the lipophilic tails aggregate away from the water. Ceramides show this type of orderly aggregation of head groups and tails but with some special features. They form as a sandwich (bilayers) of lipid molecules and layered structures – multi-laminate liquid crystalline membranes between the layers of corneocytes.

Some ceramide molecules bind covalently to the membrane protein of the corneocytes to form a protein/lipid envelope complex. Between the corneocytes, each with its protein/lipid envelope, a highly structured sequence of layers of lipids is packed.

Electron micrographs show the lipids of the epidermis to be somewhat disrupted around the remnants of desmosomes – especially those desmosomes on the upper and lower surfaces of the corneocytes, which break down much earlier than the desmosomes on the edges of the corneocytes. This accounts for the skin flaking off in plates (squames) where the cells remain attached at their edges but break or fracture between the plates, a process known as desquamation.

These discontinuities also create channels containing pockets of water-soluble substances, proteins and protein breakdown products. These channels may allow small amounts of water and water-soluble material to penetrate the stratum corneum, otherwise the multi-lamellar structure of epidermal lipids forms a very effective water-resistant barrier.

Oils used in beauty therapy are often defined by their basic properties as 'mineral oil' and 'vegetable oil'. The basic functions are summarised in Figure 25.2.

Figure 25.2 Mineral oil versus vegetable oil

MINERAL OIL	VEGETABLE OIL
Relatively inexpensive	Not cheap
Prevents TEWL as a waterproof barrier	Prevents TEWL
Repels water, causing dryness	Has humectant properties
Does not penetrate the epidermis	Penetrates the epidermis
No expiry date – microbes cannot grow in it	Short expiry date
	Can be comedogenic

Adapted from Dylan Webb's work in Victoria University's *Skin Science Student Resource 2015*.

5   What makes a lipid saponifiable?

_____

6   Which of the following skin lipids are saponifiable?

  A   Fatty acids in sebum

  B   Ceramides (found in between keratinocytes)

  C   Phospholipids (keratinocytes cell membrane)

  D   All lipids can be turned into soap

_____

7   List at least three non-saponifiable lipids found in skin, beauty products or foods.

_____

8   What is the name of the lipid that cell membranes are made of?

_____

Proteins and the skin

9 How do amino acids get to our skin before they are assembled into proteins?

_____

10 What are the two protein types? Give two examples of each.

_____

_____

11 Peptides are popular as hydrating ingredients in skincare. What is a polypeptide?

_____

12 What are three key functions of collagen?

_____

13 Which type of collagen is the most predominant in the dermis?

_____

14 Which type of collagen is found only in the basement membrane?

_____

15 What is a collagen fibril made of ? P_____

16 Describe what the final end product of a collagen fibril looks like.

_____

17 What is a collagenase?

_____

_____

_____

_____

18 Do males or females tend to produce more collagen, and why?

_____

19 Where do you find elastin in your skin?

_____

20 Where is elastin produced and by which cell?

_____

_____

21 Which is the enzyme that degrades elastin for repair and renewal?

_____

Skin structure and function

22 The skin is part of the integumentary system. What else makes up that system?

_____

_____

_____

_____

_____

23 Is the skin a cell, organ or tissue?

_____

24 Where is the skin at its thinnest on the face?

_____

25 Is the epidermis the innermost layer of the skin?

_____

26 List and describe four key functions of the skin.

_____

_____

_____

_____

_____

_____

_____

27 What is the skin's role in homeostasis and thermoregulation?

_____

_____

_____

_____

_____

_____

_____

_____

28 How does the skin provide protection?

_____

_____

29 What is the connection between our skin and the external world?

_____

_____

_____

30 State at which layer (epidermis, dermis or hypodermis) of the skin the following cells are located:

  a melanocytes

_____

  b fibroblasts

_____

  c dermal papilla

_____

  d adipocytes

_____

  e apocrine sweat gland

_____

  f Langerhan's cells

_____

  g keratinocytes

_____

  h pilosebaceous unit

_____

31 What does the skin need to produce vitamin D, and what does the body use vitamin D for, primarily?

_____

THE EPIDERMIS

32 What is a desmosome?

Knowledge
Evidence

_____

_____

33 What is TEWL? What prevents TEWL?

_____

_____

34 Draw a rough sketch of the layers of the epidermis. Name the five strata.

_____

_____

_____

_____

35 Add a melanocyte to your sketched diagram. What does a melanocyte produce?

_____

36 Explain what the following terms mean.

a Keratinogenesis

_____

_____

b Lipogenesis

_____

_____

37 List three different epidermal appendages.

_____

_____

_____

38 Which cell type is responsible for producing ceramide oils?

_____

THE DERMIS

39 Where is the papillary dermis found?

Knowledge
Evidence

_____

40 What is a glycosaminoclycan (GAG)?

_____

41 What does hyaluronic acid do in the dermis?

_____

42 What is the main cell in the dermis and what does it produce?

THE HYPODERMIS

43 What happens to the hypodermis (fat) as we age, and what effect does this have on the skin's appearance?

44 What is the function of adipose tissue?

45 Which hormone is closely linked with body fat?

46 The integumentary system is an organ system consisting of the skin, hair, nails and exocrine gland. List at least three of the main epidermal cells.

47 Which of the following is an epidermal appendage?

A Hair follicle

B Blood vessel

C Eye

D None of the above

48 How does skin help to maintain homeostasis?

49 What does the skin protect us from?

Skin tissues

50 View the following YouTube video about connective tissue: **https://www.youtube.com/watch?v=Jvtb0a2RXaY**. Which of the following connective tissues can be found in the skin? Select all that apply.

A Areolar

B Blood

C Bone

D Adipose

E Reticular

F Cartilage

51 View the following YouTube video about the skin: **https://www.youtube.com/watch?v=Orumw-PyNjw**. Where are blood vessels found in the skin?

52 A plexus is a network of vessels in the skin. Explain how the plexus system works in the dermis.

53 What is a dermal papilla? Where are they located?

54 Where are the nerves located in the skin?

Skin as a sense organ

Refer to the sections 'The nervous system and skin sensations' in Chapter 25 and 'The nervous system' in Chapter 24 of your textbook, and answer the following questions.

Knowledge Evidence

55 How could you explain the process of nerve conduction that happens when you reach out and touch something that is hot?

56 Which division of the autonomic nervous system is responsible for the stress response?

57 Draw a line to match the types of receptors in the skin and what they can sense.

WHAT IT CAN SENSE
Pain
Hot
Cold
Pressure

TYPE OF RECEPTOR
Pacini corpuscles
Kraus end bulbs
Ruffini corpuscles
Free nerve endings

58 During the nerve impulse, what chemical rushes into the neuron's axon membrane to set off the action potential that enables the electrical signal to occur?

59 What is a nerve?

_____

60 Which of the following statements is correct?

A Major nerves tend to be located alongside major blood vessels.

B Major nerves tend to be located inside bones.

_____

61 A _____ transports an electrical signal from the axon of one neuron to the dendrite of another neuron.

62 When consulting for a massage we need to ask the client what pressure they prefer. What is it that makes every person different in what we perceive to feel?

_____

Differences in skin depending on body location

63 Draw a line to match the structure of the parts of the skin to their functions.

Knowledge
Evidence

STRUCTURE
Stratum corneum
Pacinian receptor
Dermal papilla
Langerhan's cells
Fibroblasts
Keratinocytes

FUNCTION
Detects pressure sensation
Hard, dead layer of the epidermis that sloughs off
Signalling cells of the immune system
Supplies nourishment from blood to neighbouring epidermal or hair cells
Principal hard, protective cell of the epidermis
Cells that produce collagen, elastin and the gel matrix

64 When describing the skin, we often discuss it in terms of the layers and their depth. Therefore, we use the terms 'superficial' for layers that are closer to the surface and 'deep' for when they are further away from the surface. Decide whether these statements are true or false:

a The epidermis is deep to the hypodermis.

b The muscle is superficial to the hypodermis.

_____

_____

65 What is your microbiome?

_____

66 What is the difference between normal body flora and opportunistic flora?

_____

_____

_____

67 What do you think could happen when the skin's microbiome is out of balance; for example, when opportunistic flora outnumber the normal body flora?

LO25.2 GLANDS OF THE SKIN

Knowledge
Evidence

Skin glands

Knowledge
Evidence

Refer to the section 'Glands of the skin' in Chapter 25 of your textbook and answer the following questions.

1 Moist skin contains a lot of water at the surface. If a skin is moist, which glands are most likely active?

    A Eccrine glands

    B Apocrine glands

    C Sebaceous glands

---

2 Where are most of the eccrine glands distributed over the body?

---

3 Do we grow more eccrine glands or lose them over the course of our lives (from foetus to adult)?

---

4 Does eccrine sweat production happen all the time? Or only when we can see sweat?

---

5 Which areas on the body produce the most apocrine sweat?

---

6 Sketch a single diagram showing where the sebaceous gland, the eccrine gland and the apocrine sweat gland open onto the skin's surface.

---

7 Where are most of the sebaceous glands distributed on the body?

---

pH AND MICROBIOME

The balance of the sweat and the sebaceous secretions contribute to the skin's pH. If eccrine sweat is predominantly water, the pH is relatively neutral at 7. Sebum is approximately pH 4.5. The balance we aim for is around pH 5.5 for the skin's acid mantle. The slightly acidic pH makes it unfavourable for bad bacteria and favourable for normal body flora.

Refer to the information in the section 'Glands of the skin' in Chapter 25 of your textbook and answer the following questions.

8   Which of the following cleansing techniques and skincare advice will *not* help to maintain the normal body flora?

    A   Use a pH balanced cleanser and toner.

    B   Use strong cleansers and exfoliants.

    C   Cleanse with gentle cleansers and exfoliants to maintain the skin's acid mantle.

    D   Keep cleansing sponges clean.

9   Where is the sebaceous gland typically located?

10   Does cleansing your skin stop the production of sebum?

11   Where do you find apocrine glands?

12   What is different about the sebaceous glands in your eyelashes, the ear canal and around the lip vermillion?

13   What three things is sebum mainly made of (less than 95%)?

14   What is not in sebum that provides protection against TEWL in the epidermis?

15   List three functions of:

    a   sebum

    b   eccrine sweat

    c   apocrine sweat

16 What part of the brain controls thermoregulation to set off the eccrine glands?

_____

17 What five things stimulate sweat production from the eccrine glands?

_____

_____

_____

_____

_____

LO25.3 APPEARANCE OF SKIN

Knowledge
Evidence

Skin types and conditions

Client characteristics are a person's identifying features, such as skin type, skin condition and body type. The client requirements can change considerably after a thorough skin analysis. To learn how to perform a complete skin analysis, refer to 'Step-by-step: Perform a skin analysis' in Chapter 21 of your textbook.

SKIN TYPE

The skin's basic structure does not vary from person to person, but our DNA – our genetic blueprint for physiological functioning and building anatomical structures – is different, and this is why we have different skin types.

Skin reflects general health and responds quickly to any changes. Genetics and the activity of your body systems such as endocrine function, immune function and digestive function influence the skin's appearance and constitution. The key skin types are categorised according to sebaceous activity:

* normal
* dry
* oily
* combination.

Full descriptions with images of the skin types can be found in the section 'Classify the skin' in Chapter 21 of your textbook.

SKIN CONDITIONS

Skin conditions relate to external or internal influences on the skin. Examples include:

* sensitive-allergic
* diffuse red (e.g. couperose)
* pigmented (e.g. photo-aged, melasma)
* damaged (e.g. photo-aged, acne scarring, wrinkles)
* mature
* dehydrated
* congested
* other medical conditions/problems (e.g. acne vulgaris, seborrheic dermatitis, rosacea).

Read about the skin conditions in the section 'Classify the skin' in Chapter 21 of your textbook.

1 What are the different characteristics of diffuse red skin and sensitive-reactive skin?

_____

_____

_____

----

2   Couperose is the genetic predisposition to flushing. Refer to Figure 21.13, 'Skin conditions, diseases and disorders treated in facials and specialist facials' in Chapter 21 of your textbook and answer: What is the best treatment plan for couperose skin?

----

3   Refer to Figure 21.7, 'Touch and visual skin analysis techniques' in Chapter 21 of your textbook.

   a   Describe the appearance and characteristics of:

      i   oily skin

----

      ii   dry skin

----

   b   List and describe four types of pigmentation that you might find on the skin in a visual and touch diagnosis.

----

4   Combination skin:

   A   is a combination of normal and oily skin

   B   is an oily t-zone

   C   is a combination of normal and dry skin

   D   could be any of the above

----

5   Refer to Figure 21.10, 'Skin types and expected outcomes' in Chapter 21 of your textbook and list the 10 characteristics of mature skin.

Keratin

Keratins belong to a class of cellular chemicals that form the cytoplasmic scaffolding of epithelial cells. They work like those flexible fibreglass rods that hold up dome tents – they give the cell shape and form, and are possibly chemically interactive both with the cell and between other cells.

Keratins are only found in epithelial cells. They are tissue- and location-specific. The epithelium of the epidermis has a different keratin composition to the epithelia of eye cornea, ovarian epithelia, blood vessel lining, intestinal lining and female reproductive lining. The components of this architectural scaffolding are:

* actin microfilaments (6 nm diameter)
* intermediated filaments (IF) (7 to 10 nm diameter). There are six subgroups of IF, types I–VI. Keratins belong to type I and type II of this middle (IF) cluster
* microtubules (25 nm diameter).

In the past, the term 'tonofilament' was applied to keratin IF. However, do not use the term 'tonofilament' today, as it is confusing and misleading. There are quite a number of types of keratins. Fifty-four different types have been identified (though not all of these are human keratin). Individual keratin molecules are identified by the letter 'K' or 'H' and a number; for example, K14 (a common epidermal keratin).

The genes for keratin are clustered on only two chromosomes: numbers 12 and 17.

KERATIN STRUCTURE

All keratins are fibrous proteins, i.e. they form insoluble fibres when microfibrils of keratin aggregate in long, parallel strands. Amino acids are their building blocks. In a single keratin intermediate filament (IF) there are many, many polypeptide chains (keratin molecules) that are held together by cross-links rather like a fistful of spaghetti. Many of these cross-links are strong covalent disulphide bonds. These disulphide bonds depend on sulphur-containing amino acids occurring in the polypeptide chains. This sulphur-containing amino acid is called *cystine*.

The molecules have several parts to them. They are all linear rods with four separate helical sections in the mid part – which probably accounts for the stretch that is so obvious in hair when you pull on it. The IF connect up with each other and the desmosomes to form a 3D meshwork. There are two fundamental types of keratins:

1 cytokeratin
2 trichokeratin.

Cytokeratin: soft keratin

The soft type is characterised by its relative suppleness and pliability, low sulphur content (around 1–2%) and a fairly high lipid content. It is ready stainable with routine histological stains. It is found in the epidermis and hair medulla.

Trichokeratin: hard keratin

The hard type is characterised by its toughness and firmness. It has a high sulphur content (around 4–8% with 17% cystine) and a low lipid content. It is not possible to stain with some histological dyes. Hard keratin is relatively permanent; it forms strong covalent sulphur-to-sulphur cross-links. It occurs in the nails, and in the hair cuticle and cortex.

Keratin chemistry

Around 18 different amino acids make up keratin of the epidermis. The simplest of all amino acids, *glycine*, is once again (like in collagen and elastin) the biggest player (accounting for 20% of the keratin in the epidermis) with glutamic acid and serine next in line.

The amino acids that have sulphur atoms as part of the side chain are perhaps the best contributors to keratin durability because they form very strong covalent cross-links to adjacent molecules. This makes the protein 'tough' and resilient. These amino acids are methionine and, most importantly, cystine. These can form sulphur-to-sulphur cross-links between two keratin molecules.

A coil of keratin filaments is a heterodimer of two types of keratin. Heterodimers are matched pairs of acidic and basic (alkaline) molecules of keratin. These pairings cross-link with other heterodimers to form large cross-linked polymers that turn into microfilaments.

Figure 25.3 shows the types of keratins that you would expect to find in the hair, skin and nails.

Figure 25.3 Types of keratins in the hair, skin and nails

KERATIN TYPE	ACIDIC/BASIC	MOLECULAR SIZE	KERATIN EXAMPLES
Type I	Acidic	Smaller	K9 to K20
Type II	Basic	Larger	K1 to K8

Keratinogenesis: keratin formation in the epidermis

The cytoplasm of basal cells (stratum germinativum) and of spinous cells (stratum spinosum) contain thin, wavy fibrous microfibrils (2 nm diameter) which are an early stage in the formation of the tough, fibrous protein mass of keratin. To begin with, this is K5 and K14.

In the granular layer (stratum granulosum), dense granules called *keratohyalin* and matrix material (known as *filaggrin*) form along the length of IF bundles. Keratin K1 and K10 have replaced the basal type keratin from the granular layer onwards.

In the stratum corneum, keratinocyte has lost the nucleus and its organelles (autolysis). Keratin fibres increase and eventually flatten into keratin 'plates'. Filaggrin forms a gel-like matrix in which the keratin fibres are embedded. The flattened mat of keratin inside the body bag of involucrin constitutes nine-tenths of the corneocytes' bulk.

Epidermal keratinogenesis

Healthy normal cells with normal differentiation:

- basal keratinocytes – K5 (basic) + K14 (acidic)
- corneocytes – K1 (basic) + K10 (acidic)
- palmoplantar epidermis – K9.
  Cells are activated during wound healing:
- Keratinocytes become hyperactive and mobile.
- K6, K16 and K17 genes are activated.
- K16 is associated with keratinocyte migration.
- Basement membrane contraction via K17 within keratinocytes.
- Once healing has finished, the basal keratinocytes revert to K5 and K14 production.

Keratinogenesis in the hair and nails

The keratins present in the cuticle, cortex and medulla probably have different structures, with varying amounts of sulphur (and therefore different amounts of disulphide cross-links). The cortex contains a mass of cross-connected keratinised cells.

- Exposure to UVB (290 to 320 nm wavelength):
  - K19 and to a lesser extent K6, and K5/K14 increases
  - UVB 'wounds' the skin and a hyper-proliferative response is triggered.

- Exposure to UVA (320 to 400 nm wavelength):
  - mysteriously, only K17 is increased (K17 has been found to increase in basal cell carcinomas).

Cosmetic aspects

A chemical that has the capability of breaking down keratin is referred to as *keratolytic*. For this to happen, that chemical must have a pH equal to or greater than pH 12. Perm solutions alter the hair shaft structure by the use of powerful chemicals (reducing agents); for example, thioglycolic acid, which releases hydrogen sulphide (rotten egg) gas. Thioglycolic acid breaks the disulphide bonds in keratin. Rollers are applied to wrench the molecules of keratin apart and to realign them with new partners. The hair is then 'neutralised' (this has nothing to do with acids or alkali), which rejoins the disulphide bonds in new arrangements that hold the keratin fibres in the new shape.

However, not all disulphide bonds are reformed, so the hair fibres are weaker after the process. Only around 30 per cent of sulphur cross-links find new partners. The rest are irreversibly damaged.

In the bleaching process, the disulphide bonds may also be broken by powerful oxidising agents (e.g. hydrogen peroxide), leading to more brittle hair.

With depilation, the softening actions of reducing agents, such as calcium thioglycolate and calcium hydroxide (which break disulphide bonds), on keratin allows unwanted hair to be scraped away. It releases a smelly hydrogen sulphide gas.

6　What is the difference between a soluble and an insoluble protein? Provide one example of each.

7　Which hormone is responsible for producing stronger collagen?

8　Why does procollagen leave the cell?

9　How many different types of collagen are there?

10　What type of collagen is found predominantly in the dermis?

11　What is the special type of collagen that the basement membrane is made of?

12　Which cells are responsible for keratinogenesis?

13　Which of the following layers of the epidermis is without keratin?

   A　Stratum corneum

   B　Stratum germinativum

   C　Stratum spinosum

   D　Stratum granulosum

14　What key skin proteins do fibroblasts produce?

Skin colour

Knowledge Evidence

15　What are the three main factors that contribute to skin colour? In the table below, describe how you would:

   a　identify the skin condition

b note it down on the client record card during a skin analysis.

FACTORS THAT CONTRIBUTE TO SKIN COLOUR	IDENTIFYING FEATURE	HOW YOU WOULD NOTE IT ON THE CLIENT RECORD CARD
Epidermal thickness		
Circulation		
Pigmentation		

16 True or false: Delayed tanning is a long-lasting tan that is UVB triggered and lasts for weeks to months

17 What happens to melanosomes in the keratinocytes as they pass through the stratum lucidum on their way to the stratum corneum?

18 Melanocytes are said to be 'dendritic'. How would you describe 'dendritic'? (Sketch something that is dendritic if it's easier.) How does the structure help the melanocyte to do its work?

19 Describe the structure and location of the following:

a melanocytes

b keratinocytes

20 In beauty therapy, energy from various parts of the electromagnetic spectrum is used, so it is important to understand its different components. Undertake some research online, then order the following from shortest to longest wavelength: radio waves, visible light, infra-red light, UV light, X-ray.

21 What effect does UV light have on the skin?

22 There are two different types of melanin pigment that have varying tones of brown, which contributes to skin colour. Name the types and the colours they reflect.

23 Other than melanin, a number of other things contribute to skin colour. What are they? (Note also the colour they contribute to the skin.)

24 Does sunscreen act the same as sunblock? If not, what is the difference?

25 Explain, as you would to a client, the importance of wearing sunscreen to protect the skin.

26 What could you recommend to a client who does not like wearing sunscreen?

The Fitzpatrick scale

Knowledge
Evidence

27 What does the Fitzpatrick scale measure?

_____

_____

28 We need to know about the Fitzpatrick scale for specialised facial treatments to be aware of the risk of adverse reactions. What are the distinctive risks associated with the following skin types; for example, when performing extractions?

a   Type I–II

_____

b   Types IV–VI

_____

29 UV radiation is both the main cause of skin cancer and our main source of vitamin D. Dark skin absorbs much of the UV. Read content on the following weblink and answer the question below: **http://www.sunsmart.com.au/ skin-cancer/risk-factors#skin-types**.

a   Can Fitzpatrick type VI skin get away without sunscreen, as it has a higher MED? Explain your answer.

_____

_____

_____

_____

_____

_____

_____

_____

Light and the skin

Knowledge
Evidence

30 Identify the skin parts that are responsible for the following optical properties:

a   absorption

_____

b   reflection

_____

c   transmission

_____

d   scattering

_____

31 Which Fitzpatrick skin type can have the highest MED?

_____

SUNSCREENS

Sunscreens prevent the penetration of UVA, UVB and UVC radiation to the skin. Sunscreens that protect the skin from all types of rays should be labelled 'broad spectrum'. Ingredients that are sunscreens work by two mechanisms – chemical blocks and physical blocks:

- Chemical blocks absorb the sunlight and release the energy as heat; they are made from organic chemicals. Some are 'broad spectrum', though most predominantly block UVB radiation. They become deactivated by UV light, which is why sunscreens must be re-applied every 2–3 hours.

- Physical blocks are not absorbed by the skin and act by reflecting light away from the skin. They are made from inorganic chemicals (mineral ingredients) that sit on the skin's surface. They provide 'broad spectrum' protection (from UVA and UVB radiation). Zinc oxide and titanium dioxide are examples of physical blocking agents.

Sunscreens are available in oil, foam, spray, mousse, lotion, stick, milk, gel and cream preparations.

Figure 25.4   Sunscreen ingredients for specific UV radiation and skin characteristics

Type of radiation	Positive effects on skin	Negative effects on skin	Sunscreen ingredients that prevent absorption of radiation
UVA	• May be used medically to treat psoriasis • Stimulates Vitamin D production • Prevents rickets, eczema and jaundice	• Ageing (especially wrinkling) • Skin cancers • Photodermatoses (photosensitivity causing a rash) • Pigmentation	• *Organic sunscreen ingredients* that prevent UVA radiation: oxybenzone, ecamsule (Mexoryl SX®) dioxybenzone, and avobenzone • *Inorganic sunscreen ingredients*: zinc oxide, titanium dioxide
UVB	Delayed (lasting) tanning – protects the skin against UV by absorption by melanin pigment (if regularly using correct sun protection, this should not be necessary)	• Skin cancers • Ageing • Burning • Tanning	• *Organic sunscreen ingredients* that block UVB radiation: cinnamates (e.g. octinoxate, Cinoxate), PABA (para aminobenzoic acid; although now less common as its acidity can be irritating), padimate O (a PABA derivative), salicylates (e.g. octisalate salicylate trolamine salicylate) • *Inorganic sunscreen ingredients*: zinc oxide, titanium dioxide
UVC	UVC radiation is completely filtered by the Earth's ozone layer. It does not reach the skin	However, if it were to reach the skin, it would be the most damaging form of UV radiation	N/A

32 Which type of UV radiation penetrates the deepest?

_____

33 Which is the most dangerous for ageing?

_____

34 Which is the most dangerous for burning and SSC/BCC skin cancers?

_____

35 Which type of radiation is in the solarium?

_____

36 When UVA and UVB penetrate the atmosphere, what are the defensive responses enacted by the keratinocytes and melanocytes?

_____

37 Other than melanisation, what are some other effects the sun can have on the skin?

_____

_____

_____

_____

_____

38 Why (in your own words) should 30+ SPF sunscreen be used on *all* Fitzpatrick skin types?

_____

_____

Age-related changes

Knowledge
Evidence

39 To demonstrate your understanding of the effect of hormones on skin, draw a line to match the following hormones to the effect on the skin.

NAME OF HORMONE	EFFECT
Parathormone (parathyroid hormone)	Excess (in adults) causes coarsening of skin, increased hair growth and more muscular appearance.
GH (growth hormone)	Increases melanin pigment production; increases sebum production.
Melanocyte stimulating hormone (MSH)	Lack of hormone not only affects the bones but causes abnormal production of keratin, affecting hair, skin and nails.
Oestrogen and progesterone	Excess causes Cushing's syndrome, characterised by a redistribution of fat producing a 'moon face', 'buffalo hump', large abdomen and thin limbs; purple stretch marks and bruises may appear on the skin; deficiency causes Addison's disease with weight loss and darkening of the skin.
Aldosterone	Excess can cause oedema.
Testosterone	Excess of androgens causes virilism in women (deepening of the voice, growth of facial and body hair, muscle development and sometimes male pattern baldness); excess of oestrogens causes feminisation in men – breasts will enlarge.
Glucocorticoids	Keep skin and hair in good condition; control distribution of body hair at puberty and influence the typical 'female' shape by causing fat to be stored in breasts, hips and thighs.
Corticosteroids (sex hormones)	Causes growth of facial and body hair at puberty; causes muscular development influencing 'male' body shape; encourages fat to be deposited around the waist and abdomen.

40 What does oestrogen do to sebum production?

_____

41 A lack of sebum in skin makes it:

    A dehydrated

    B rough

    C seborrhoeic

    D both B and C are correct

42 What does excess circulating testosterone do to women experiencing polycystic ovarian syndrome (PCOS)?

43 Are women experiencing menopause likely to have a higher or lower sebum secretion?

44 Are older women and men likely to sweat more or less?

45 What do we know about our hormones that explains why the skin follows a diurnal rhythm and grows and heals better when we are less stressed?

46 Which hormones stimulate hair growth?

47 What is the most common cause of hair loss known as alopecia areata?

48 What effect do these three hormones have on hair loss:

    a thyroid deficiency

    b corticosteroids

    c oestrogens

49 What is in the HRT medicine?

50 What is the effect of HRT on the skin?

51 What initiates the production of skin pigment?

52 Which two systems of the body are responsible for sweat gland secretion?

    A Circulatory system and endocrine system

    B Nervous system and endocrine system

    C Muscular system and circulatory system

    D Endocrine system and reproductive system

53 Read the following article and answer the following questions about menstruation, hormones and the effect on skin: **https://www.sinclairdermatology.com.au/period-skincare-yes-you-can-hack-your-skincare-to-match-your-menstrual-cycle**.

    a Which four hormones are responsible for the hormonal fluctuations in the menstrual cycle?

    b What menstrual event is occurring when most of these hormones, except progesterone, are at their highest levels in your blood ('peaking' when looking at the graph)?

        A Menstruation (bleeding)

        B Ovulation (egg leaves the ovary)

        C Nothing

    c When does your skin get dry and sensitive in your menstrual cycle?

    d When are the two times acne can flare during the menstrual cycle?

    e Why is it suggested that women track their periods?

LO25.4 GROWTH, DEVELOPMENT, AGEING AND HEALING OF HUMAN SKIN

Growth and development of human skin

Knowledge
Evidence

1 For how long in the cell cycle is mitosis happening? Which is the longest phase in the cell cycle?

2 Which hormones affect a person's cell division?

3 What types of medications can affect cell division of the epidermis?

4 In your own words, define 'differentiation' as it applies to cell type changes.

5 In your own words, define 'desquamation'.

Ageing

6 Explain the difference between chronological ageing and photo-ageing.

Knowledge
Evidence

7 What is the other word for photo-ageing, that means sun-damaged?

8 What are the three repair systems we naturally have to fight UV damage?

9 List three free radicals that cause damage to DNA in our skin.

10 List three chemicals in the skin that fight against free radical damage.

11 Refer to Figure 25.29 in your textbook, 'Comparison of chronological ageing and photo-ageing' and identify which of the following is photo-ageing and which is chronological ageing (mark with a 'P' for photo-ageing and 'C' for chronological ageing).

- Skin cancers common
- Reduced TEWL
- Pigmentation often very blotchy
- Increase in 'junk' elastin
- Regular thinning

- Collagen decreases

- Capillary loops reduced with massive twisting and thickening or remaining vessels

- Skin cancers not common

- Rough, often leathery skin

- Epidermis thickens

- Increase in cell numbers (more hypercellular)

Wound healing stages

Knowledge
Evidence

The four wound healing stages are:

1 inflammation
2 contraction
3 collagen formation
4 epithelialisation.

1. INFLAMMATION

Immediately following a cut, a blood clot forms. There follows a short delay before the events of inflammation begin:

- dilation of small blood vessels (vasodilation)
- increased permeability of small blood vessels
- slowing of blood flow (vasoconstriction)
- walling off of injured area
- migration and aggregation of leukocytes at site of injury
- phagocytosis by leucocytes and macrophage.

The events of inflammation are:

- Tissue damage with or without contaminating micro-organisms. Damage may be caused by physical stress, such as abrasions or cuts, chemical irritation, micro-organisms or burns.
- Immediately, the inflammatory response is triggered.
- Immediately after the injury, bleeding begins and mast cells in the surrounding connective tissue respond. Mast cells release chemicals (histamine and heparin). These chemicals affect blood vessels near the injury.
- Blood vessels dilate to increase blood flow to the area. This brings nutrients, oxygen and white blood cells into the area, and removes dissolved wastes and toxic chemicals.
- Dilation of blood vessels makes the skin in the area appear reddish in colour and warm to the touch, typical symptoms of inflammation.
- Blood vessel walls in the area become more permeable, and fluid containing dissolved materials 'leaks' into the surrounding injured tissue. The area becomes swollen (oedema).
- The pressure of this extra tissue fluid as well as chemicals released by injured cells stimulates nerve endings in the area to produce a sensation of pain.
- Blood begins to clot almost immediately with the formation of insoluble fibres of fibrin.
- Clot formation walls off the inflamed region, slowing the spread of cellular debris or bacteria into surrounding tissues. Eventually, a scab forms from the clot after several hours and temporarily restores the surface of the skin and restricts the entry of more micro-organisms as well as loss of tissue fluid.
- Phagocytes are attracted to the area where they squeeze through the capillary walls to join the attack. These phagocytic cells remove debris and more phagocytes arrive with the enhanced circulation. Actively phagocytic cells are short-lived, most surviving for just a few hours before dying and disintegrating. Pus is the mixture of living and dead cells and tissue debris that develops within an area of inflammation. Antibodies may be produced, which help destroy or inactivate invading micro-organisms or foreign toxins.

2. CONTRACTION

Normal contraction:

- Inward migration of myofibroblasts causes tension to draw the wound margins together.
- Contractile protein – actin.
- The extent of contraction may result in diminished movement, especially at joints.
- Contraction may be lessened by the direct use of:
  - corticosteroids
  - vitamin A
  - a pressure bandage.

How the above three actually work is largely unknown.

Assisted contraction: suturing, stitching, taping, gluing.

3. COLLAGEN FORMATION

Collagen formation usually follows this pattern:

1   accelerated synthesis of collagen fibres by activated fibroblasts
2   deposition of fibres within the boundaries of the wound
3   degradation and removal of excess collagen used as temporary packing (usually achieved through the action of collagenases).

Collagen is manufactured in the cells (mainly the fibroblasts). Cross-linking does not occur until collagen is in the matrix outside the cell. In normal skin there is an equilibrium of manufacture and degradation of collagen.

Types of collagen in skin

- Type I: 'Adult collagen': this type of fibre is tougher and stronger but more 'expensive' to make.
- Type III: 'Juvenile collagen': this is weaker, easier to make and is the first laid down as 'quick fix' temporary binding and packaging.

4. EPITHELIALISATION

Epithelialisation is the final growth and differentiation of cells over established granulated dermis with its effective blood supply. If this process starts before the dermis is restored, it will not work, because the dermis will not have the strength to hold the wound together.

FACTORS THAT AFFECT WOUND HEALING

- Good blood supply
- Local infection
- Delayed primary closure
- State of health, both physiological and psychological
- Age
- Diet: nutritional status, protein, vitamin C, vitamin A, zinc
- Drugs: corticosteroids, immunosuppressants
- Genetic haemophilia, Reynaud's Syndrome
- Occupational conditions; for example, repeat injury
- Fictitious wounds self-inflicted, psycho-social conditions

NORMAL SCAR TISSUE

Scar tissue is normally an unruly jumble of collagen fibres with intertwining blood vessels that grew into the wound site in order to supply the repairing fibroblasts and macrophages with sufficient resources and also to speed up the removal of waste products.

A healing wound is a very busy area.

As the healing progresses over the next 12–18 months, the vessels withdraw. This changes the scar colour from pinky-red to paler. It never is invisible mending. Around three months after an injury, a scar has achieved approximately 90 per cent of the healed strength. The remaining 10 per cent takes another nine to 12 months. Nevertheless, it will always be weaker than the surrounding undamaged normal skin.

ABNORMAL SCAR TISSUE

An interruption in the normal wound healing process, such as by infection, disease (e.g. diabetes) and skin trauma (e.g. acne cysts), can result in abnormal scar tissue. Common issues are hypotrophic scarring (pitting in the skin), hypertrophic scarring (thickened scars) and keloids, as described below.

Hypertrophic scar tissue

Hypertrophic scars occur when the active phase of scar formation is increased and the scar becomes thick, red and itchy. Even in severe cases, however, the size of the scar does not increase after 12 weeks although the irritation may persist for 3–6 months. Hypertrophic scars are always confined to the area of the original wound. They are more common in women than men. Strain across a wound caused by poor handling during early stages of healing may be an important factor in the development of hypertrophic scars.

Keloid scarring

Keloid scarring is not confined to the original wound but spreads beyond the wound margins sometimes producing huge cauliflower like growths in the skin. The normal regulation of the repair process does not operate and there appears to be an over production of collagen. Keloid scars are uncommon but not rare in people with pale skin. Paler-skinned women keloid more than paler-skinned men and the regions most affected are the chest and back, and head and neck. They may be contraindicated for piercings. Post-pubescent and premenopausal years are the years when people are most affected. Keloid scarring seems to be hormonal as well as racial and gender directed. Adults with compromised immune systems do not keloid – adding to the mystery.

Keloids are more common in people with the darker Fitzpatrick skin types.

It seems that somehow the activity of collagenase is repressed but exactly how is unknown.

More recent evidence has demonstrated that keloids are associated with the under production of type VI collagen, which seems to play an important role in demarcating the wound area much like fences demarcate fields and houses. Surgical or laser treatment is generally unsuccessful. Suppression by injecting corticosteroids works but has to be repeated at 6- to 12-month intervals. The pain of the injection is described as 'excruciating'.

Adapted from Dylan Webb's work in Victoria University's *Skin Disorders Resource 2015.*

12 Place the four phases of wound healing in order.

- Contraction

- Inflammation

- Epithelialisation

- Collagen formation

----

13 List the six things that happen during inflammation.

----

14 What happens in the 'epithelialisation' phase of wound healing?

15 When you are treating the skin of a client prone to acne vulgaris, you notice a follicle that has inflammation around it. What would be the danger of performing extractions?

16 How would you identify that a scar is keloid rather than hypertrophic based on the way that it grows?

17 Can you perform extractions with a needle on a person prone to keloid or hypertrophic scarring?

18 If a person is having a recurring comedone in the one follicle, it is most likely because:

   A   they are constantly touching the one spot

   B   permanent scar tissue has formed around that follicle (which is why we don't pick)

   C   of hormones

19 True or false: Abnormal scar tissue is due to some type of interruption in the normal wound healing process.

The genetics of skin disorders

Knowledge
Evidence

Inheritance is the term given to the passage of hereditary traits from one generation to another. The science dealing with inheritance is called genetics.

THE GENE THEORY

The gene theory of inheritance allows us to make predictions about the probability that the offspring of two given parents will have a particular characteristic. These generalisations are based on Mendel's Laws. Gregor Mendel, an Austrian monk, developed his theories in 1866.

Mendel's Laws

Mendel's first law states that genes exist in pairs. When gametes form, these pairs separate so that each gamete contains only one of each kind of gene.

Mendel's second law states that segregation of each pair of genes is random. Therefore, great numbers of characteristics can be inherited simultaneously and offspring can resemble one parent in certain traits and the other in different traits.

Note: Mendel's theories have been modified since the discovery of chromosomes. We now know that genes are transmitted as components of chromosomes, each carrying many different genes.

Genes

Genes are sequences of DNA arranged along a chromosome. Each cell, except gametes, contains 23 pairs of chromosomes, the diploid number.

Twenty-two of these pairs of chromosomes are called autosomes and one pair forms the sex chromosomes. One in each pair is from the mother and the other from the father.

Homologue is the name given to a pair of chromosomes. Each homologue contains the information for the same traits and these are found in the same location on both homologous chromosomes. These pairs of genetic information are called alleles.

Mutation

Mutation refers to a permanent change to a gene that causes it to have a different effect than it previously had. Mutations in genes cause diseases and disorders to occur throughout our lives.

Genetically inherited skin disorders

- Acne vulgaris
- Alopecia areata
- Androgenic alopecia
- Atopic dermatitis: also linked with hay fever, asthma and very dry skin
- Hypertrichosis
- Pigmentation disorders such as vitiligo and albinism
- Psoriasis
- Seborrhoeic dermatitis

20 When a condition is genetic in origin, are you always born with it?

_____

21 What is a genetic mutation?

_____

22 How can a genetic mutation cause a skin disorder? Research a skin disease of your choice, and explain the genetic mutation and what it does to the skin cell process.

_____

_____

_____

_____

23 Circle which of the following conditions are congenital (genetic in origin or with a genetic link).

- Psoriasis
- Atopic dermatitis
- Contact irritant dermatitis

LO25.5 SCOPE OF PRACTICE

1 What are the three things that define your scope of practice?

_____

_____

_____

2 When would you need to refer a client for the following procedures? Who would you refer them to?

   a Extract milia

_____

_____

   b Distinguish cysts and nodules

_____

3 Who is responsible if you work outside your scope of practice?

_____

Skin conditions

4 Refer to Appendix A Contraindications and restrictions and answer the following questions.

   a What is a normal skin response to irritation and trauma for the first couple of hours?

      A Inflammation

      B Oedema

      C Rash

      D Hives

   b What is erythema?

_____

5 Refer to Appendix A and source information from anatomy, physiology and skin science publications to complete the following table.

CONDITION	APPEARANCE	GENETIC FACTORS	POSSIBLE MEDICAL TREATMENTS	LIMITATIONS OF FACIAL TREATMENTS ON SKIN CONDITIONS
Acne vulgaris				
Eczema and atopic dermatitis				

Knowledge Evidence

CONDITION	APPEARANCE	GENETIC FACTORS	POSSIBLE MEDICAL TREATMENTS	LIMITATIONS OF FACIAL TREATMENTS ON SKIN CONDITIONS
Hirsutism				
Vitiligo				
Telangiectasia				
Seborrheic dermatitis				

Nutrition and healthy skin

Knowledge
Evidence

6   Refer to Chapter 23 of your textbook to answer the following questions.

   a   Refer to the section 'Vitamins'. What does vitamin E do for the skin when it comes from your food?

_____

_____

_____

b Refer to the section 'Water'. Why is water important to all of the tissues of the body?

_____

_____

c Refer to the section 'Weight management'. When someone is on a diet to manage weight – a calorie reduction diet – what are five adverse effects it might have on their skin?

_____

_____

_____

_____

_____

_____

_____

CHAPTER 26: INCORPORATE KNOWLEDGE OF ANATOMY AND PHYSIOLOGY INTO BEAUTY THERAPY

LEARNING OBJECTIVES

After completing this chapter, you should be able to:

LO26.1 use knowledge of body and skin structures and systems to design treatment by integrating knowledge to meet the client requirements, design the treatment and develop the treatment plan

LO26.2 advise the client by using correct anatomical and physiological terminology to discuss benefits, effects and post-treatment advice

LO26.3 record relevant data by integrating and using accurate anatomical and physiological terminology to record and update information relating to the skin, skin conditions, body parts, functions and treatment effects and by communicating with colleagues and medical practitioners

LO26.4 maintain knowledge of anatomy and physiology by identifying and using opportunities to source, update and expand knowledge, share current information and compile and maintain credible sources of information about anatomy and physiology.

INTRODUCTION

Communicate client conditions with the client, with team members and when liaising with medical practitioners using correct anatomical and physiological terminology to a professional and ethical standard.

The learning activities in this chapter allow you to demonstrate your knowledge and understanding of the units of competency *SHBBSSC001 Incorporate knowledge of skin structure and functions into beauty therapy* and *SHBBSSC002 Incorporate knowledge of body structures and functions into beauty therapy*, using the performance evidences.

LO26.1 USING KNOWLEDGE OF BODY AND SKIN STRUCTURES AND SYSTEMS TO DETERMINE CLIENT NEEDS AND DESIGN THE TREATMENT

It is important to have an understanding of anatomy and physiology so that you can communicate with clients effectively to determine their needs and perform the treatment effectively.

Performance Evidence

Body massage

Refer to Chapter 18 in your textbook to help answer the following questions.

1 Which massage movement helps to release nerve tension from muscles?

A Effleurage

B Petrissage

   C Tapotement

   D Vibration

2 Which massage movement compresses muscles to increase circulation?

   A Effleurage

   B Petrissage

   C Tapotement

   D Vibration

3 Which massage movement tones the nerves and stimulates circulation?

   A Effleurage

   B Petrissage

   C Tapotement

   D Vibration

4 How should you vary the repetition, rhythm and variation of massage movements to adapt to the following clients?

   a Elderly client with fine dry skin

   b Athlete with muscle tension and lactic acid build-up

Facial massage

Refer to Chapter 21 in your textbook to help answer the following questions.

5 If a client has a lack of skin sensation in a particular area, what parts of the facial do you need to omit, and where should you avoid?

6 What are the benefits of facial massage to the bones?

7 Are bones connected to the muscular and nervous systems? What effect does massage have on the bones?

The nervous system and treatment modification

Any nervous system condition is a contraindication to service, and you need to know about the anatomy of the body to be able to modify the treatment.

Let's use a client who has had wisdom tooth dental surgery as an example.

8 Which nerve should you avoid if a person has had wisdom tooth surgery in the past six months?

_____

9 Refer to the image below. If a client had surgery on the cheek near the facial nerve, which areas would you have to avoid when using electrical machines?

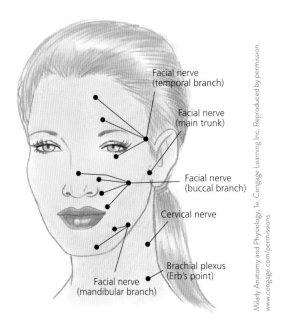

Milady Anatomy and Physiology, 1e. Cengage Learning Inc. Reproduced by permission. www.cengage.com/permissions

10 What types of unexplained pain are symptoms for someone experiencing a dysfunction with their nerve sensation?

_____

11 Refer to the image above. If a client has had an operation with nerve damage to the mandibular extension of the trigeminal nerve, which area of the face do you need to avoid with electrical machines?

_____

_____

Design and develop the treatment

12 Complete the following table of skin anatomy and skin function affected, and the range of treatment options for the client requirements shown. Use the two rows at the bottom of the table to come up with different skin requirements a client might have.

CLIENT REQUIREMENTS	SKIN ANATOMY AFFECTED	SKIN FUNCTION AFFECTED	RANGE OF SKIN TREATMENT OPTIONS
Reduce visible pores			

CLIENT REQUIREMENTS	SKIN ANATOMY AFFECTED	SKIN FUNCTION AFFECTED	RANGE OF SKIN TREATMENT OPTIONS
Pigmentation			
Excess dark hair on lower leg			

13 When recommending and designing a series of treatments, clients who have budgetary or time constraints might not be happy with too many treatments scheduled. Choose one of the client requirements you have discussed in the table in Question 12 above and specify how you would justify the frequency of the treatments based on accurate scientific reasoning.

_____

_____

_____

14 Develop a treatment plan for the following clients:

a William has been shaving his chest hairs for many years, and is wondering what you would advise to be a better way of removing hair.

_____

_____

_____

b Elena has visible pores, and wants to try to improve that.

_____

_____

_____

LO26.2 ADVISING THE CLIENT
Knowledge of anatomy and physiology

Performance
Evidence

1 Write the opposite direction for each of the following:

a proximal

_____

b superior

_____

c posterior

_____

d medial

_____

e superficial

_____

2 Draw a line to match the following body region names to their area on the body.

BODY REGION
Axillary
Inguinal
Cervical
Occipital
Lumbar
Popliteal

AREA
Lower back
Back of knee
Lower aspect of back of head
Neck
Under the arm
Groin

3 Describe the location of the following body parts in relation to each other by circling the anatomical direction.

a The throat is _____ to the neck. (anterior/posterior)

b The navel is _____to the chest. (superior/inferior)

c The wrist is _____ to the fingers. (proximal/distal)

d The ear is _____to the nose. (medial/lateral)

4 Use a medical dictionary to help you to complete the following table. Divide the word according to its word root (prefix or suffix). Include the definitions. 'Conjunctivitis' has been completed as an example.

WORD	PREFIX	WORD ROOT	SUFFIX
Conjunctiv-itis		Conjunctiv- membrane lining the eyelid	-itis inflammation of
Mast-ectomy			
Hyper-tension			
Melano-cyte			
Epi-dermis			
Ven-ule			
Pilo-sebaceous			

Type of client

5 The following clients come to your salon and start discussing treatments with you. Assess what type of client each is, and determine what you would do in each situation, along with contraindications you would identify and previous adverse reactions.

a Marie comes to the salon regularly. She has always had fake eyelashes put on for special occasions, but she likes them so much that she has been getting the treatment done more often than needed, which you can see reflected in the dates she has had treatments done.

----

b John comes to the salon regularly for body massage treatments, but he has decided he needs a facial to help with his oily complexion. He starts asking you about the products you use and the massage you perform, and what their effects are on sebum production.

----

c Elizabeth has been to the salon sporadically over the years. She says she is keen to have her eyebrows waxed, and maybe some permanent hair removal, as well as to relax, and possibly a demonstration of make-up application.

d Alex has come to the salon for the first time. He is getting married in two weeks and has been getting acne over his lower face, chin and neck.

e Tania is feeling like she needs help. She's not sure exactly what she needs, but she knows that the skin on her legs and arms has been dry. When you discuss different body treatments you could provide, she says she's unsure, and asks for your opinion.

f Georgia has called up, saying that she's read about a specialised facial treatment called desincrustation and how it can help with pore refining, which she feels will fix her skin instantly. Georgia is already a client, and you can see that she is over 60 years old.

6 What are the steps to reviewing the client medical history?

Skin treatment

7 Look at the relevant chapters to fill out the table below. Fill out at least one benefit and one effect of each treatment.

TREATMENT	BENEFITS	POSSIBLE ADVERSE EFFECTS
Facial		
Specialised facial		
Lash and brow tinting		
Lifestyle change		
Product selection and use		

8 Use the correct skin anatomy and physiology terminology to accurately explain why a client should always wear sunscreen when outside. What treatments would you recommend to remediate existing damage?

_____

_____

_____

_____

_____

_____

_____

_____

_____

_____

Hair treatment

9 Complete the table below with the effects on hair of each treatment, and one piece of post-treatment advice for each, using correct skin anatomy and physiology terminology.

TREATMENT	POST-TREATMENT ADVICE
Lash and brow tinting	
Shaving	
Tweezing	
Waxing	

Body treatment

10 For each of the following beauty therapy treatments or client concerns, state how you would advise a client on the effects and benefits on their health and appearance. You might have to review the relevant chapter for each one.

BEAUTY TREATMENT	EFFECT	BENEFIT
Body massage		
Body treatments		
Aromatherapy massage		
Lifestyle changes		
Product selection and use		

Lifestyle advice

11 Which muscle is responsible for causing wrinkles at the bridge of the nose? Which expression should your client try not to make?

12 What can cause muscle tension (and wrinkles), and what lifestyle advice can you give your client?

LO26.3 RECORDING RELEVANT DATA

1 What are the three main reasons you need to be clear and accurate when recording relevant client information?

2 Explain why the following communication points are heeded when filling out client record cards:

   a   correct for the right client

   b   timely

   c   avoid humour and personal commentary

3 When communicating your consultation or treatment notes with a medical practitioner, why is accurate skin anatomy and physiology terminology and identification so important?

4 Complete the table below with accurate skin anatomy and physiology terminology for skin, skin conditions, functions and treatment effects. Use the two rows at the bottom of the table to come up different client requirements for other body parts.

CLIENT REQUIREMENTS	SKIN	SKIN CONDITIONS	FUNCTIONS	TREATMENT EFFECTS
Reduce visible pores				

CLIENT REQUIREMENTS	SKIN	SKIN CONDITIONS	FUNCTIONS	TREATMENT EFFECTS

LO26.4 MAINTAINING KNOWLEDGE OF ANATOMY AND PHYSIOLOGY

Performance Evidence

1   One great way to gain and maintain information on the beauty industry, including anatomy and physiology, is to use an industry mentor. Find and interview an industry mentor, ideally someone with at least three years' experience in the beauty industry. Ask them the following questions:

   a   How do you incorporate scientific information in your consultation in a way that the client can understand?

   b   Do you recommend every aftercare instruction (listed on the bottle) for product safety or are you specific to what each client requires?

   c   Can you give an example of a situation in which you've had to cope with a sales decline/rejection?

   d   What types of lifestyle advice to you recommend (e.g. face yoga, yoga)? Are you specific with diet and exercises or refer clients to a professional?

   e   When future treatment planning:

      i   How many treatments do you plan ahead?

      ii   Do you ever refer to allied professionals such as dermal clinicians, nutritionists, etc. for combined therapies?

      iii   What percentage of skincare would you consider to be professional/in-salon and what percentage to be retail/homecare?

   f   How do you make sure clients are happy with your advice?

   g   How does your salon conduct online consultations?

      i   Do the staff have training support to conduct online consultations according to salon protocol?

      ii   What resources does the salon have to access their clients online?

      iii   Are the clients willing to use skincare at home as much as you think they should?

      iv   Are clients purchasing the same types of products that they used to (pre-COVID-19 pandemic)?

Sharing and accessing information

2   What is one way you can share current information from trusted sources about skin anatomy and physiology with:

a   colleagues?

b   clients?

3   You have started work at a new salon, and as the manager is giving you an induction to the salon, she points out a cupboard that sits below the computer in the reception area and says this is the 'reference library'. Based on your understanding of how credible sources are compiled, what would you expect to find in there?

CROSS-SECTOR APPENDIX: USING SOCIAL MEDIA AND ONLINE PLATFORMS

After reading this chapter, you should be able to:

LOCSA.1 source information on the general impacts of social media

LOCSA.2 comply with industry and organisational ethical and professional codes of conduct for online activities

LOCSA.3 maintain a personal online presence consistent with organisational standards

LOCSA.4 protect customer privacy and maintain confidentiality of organisational information

LOCSA.5 source information on copyright

LOCSA.6 engage professionally with customers online.

INTRODUCTION

More people are using social media to research and discover new brands, and to view content, influencer endorsements and current client recommendations. Having an online presence, both in social media and a website, has now become an essential part of the beauty business.

Effective performance when utilising social media and online platforms for diverse purposes is an underpinning skill for salon ownership, freelance and working as an employee. You will be able to source information on, and work according to a range of ethical and professional standards when using social media and online platforms for business purposes. It can be applied in any type of beauty business that utilises social media and online tools for customer engagement.

The learning activities in this cross-sector appendix allow you to demonstrate your knowledge and understanding of the unit of competency *SIRXOSM002 Maintain ethical and professional standards when using social media and online platforms*, using the performance and knowledge evidences.

LOCSA.1 SOURCE INFORMATION ON THE GENERAL IMPACTS OF SOCIAL MEDIA

Performance Evidence

Knowledge Evidence

Terms of service

Locate the 'terms of use' of social media platform Instagram and answer the following questions to identify and evaluate some key features.

1 When you use Instagram, you give 'permission to use your username, profile picture ...' in what capacity?

_____

_____

_____

_____

2 How is the Instagram service funded?

_____

_____

_____

_____

COPYRIGHT

3 With reference to copyright, what do Instagram's terms mean by the statement: 'We do not claim ownership of your content, but you grant us a license to use it'?

_____

_____

_____

_____

_____

_____

_____

_____

_____

_____

_____

4 What do you think are the positive aspects of Instagram's service that allow them to operate within privacy and copyright legislation in every country?

_____

_____

SHARING USER INFORMATION

5 From information you have collected and read, write an overview evaluation of the terms of service for the most common social media platforms, including how they share user information and how this may impact either positively or negatively on users.

_____

_____

_____

_____

_____

_____

_____

_____

_____

The general impacts of social media on organisations and users

Knowledge Evidence

6 What are the three types of information mined from social media sites with the intention of selling it for advertising?

_____

_____

_____

7 From three different sources, identify both the positive and negative impacts of social media, on:

- the salon/business' reputation
- social impacts on users.

In your own words, complete the following table with your findings.

	POSITIVE IMPACTS	NEGATIVE IMPACTS
Impacts on the salon/business' reputation		

Social impacts on users		

8 From information you have collected and read, write a half-page overview evaluation of the general impacts of social media on salon/businesses and social media users, both positive and negative.

_____

_____

_____

_____

_____

_____

_____

_____

_____

_____

_____

_____

_____

Cyberbullying and its impacts

Performance Evidence   Knowledge Evidence

9 What are some of the effects of cyberbullying on people – how can it make people feel? You can find the answer at **https://www.healthdirect.gov.au/cyberbullying**.

_____

_____

_____

_____

_____

_____

10 Read section 4 of the report *Cyberbullying and Indigenous Australians: A review of the literature*, and answer the following questions: **https://research-management.mq.edu.au/ws/portalfiles/portal/92634728/MQU_Cyberbullying_Report_Carlson_Frazer.pdf**.

a How can cyberbullying impact young people in general; in particular, the victim?

b How can cyberbullying impact young people in general; in particular, the cyberbully?

c Read this summary of the findings of the report: **https://apo.org.au/node/260076**. What do the three key findings of this report tell us about the cyberbullying issues facing Indigenous Australian communities?

11 In your own words, complete the below table using your knowledge of image-based abuse, doxing, catfishing and reputation attacks.

	WHAT IS IT?	SUGGEST AT LEAST ONE SOCIAL MEDIA PLATFORM THAT MAY BE USED
Image-based abuse		

	WHAT IS IT?	SUGGEST AT LEAST ONE SOCIAL MEDIA PLATFORM THAT MAY BE USED
Doxing		
Catfishing		
Reputation attacks		

12 Cyberbullying can escalate much more quickly compared to bullying in-person. The following factors have been known to contribute to the severity of the problem:

- being anonymous

- having a large audience

- having a range of attack methods; e.g. apps, SMS and online forums

- being unable to communicate face to face

- having the ability to contact the victim 24 hours a day.

   Reflect on an incident of cyberbullying you have experienced or have been a bystander to and identify which of the above factors was relevant to the experience.

ORGANISATIONAL IMPACTS

13 Cybercrime is an illegal action that can have negative impacts, not just to the victim, but to a salon too, in terms of money lost. ReportCyber is part of the Australian Cyber Security Centre (ACSC), the government's cybercrime reporting system that directly channels your situation to the relevant authority, such as the police, for assessment. Visit their website at **https://www.cyber.gov.au/acsc/report** and list in the table below the types of cybercrime you can report and not report to the ACSC via ReportCyber.

CYBERCRIME YOU CAN REPORT	WHAT YOU CANNOT REPORT

14 Visit this website and answer the following questions: **https://www.betterhealth.vic.gov.au/health/ HealthyLiving/Cyberbullying**.

    a Read the section 'What does cyberbullying look like?' and draw a line to match the following situations set in a beauty industry context with the ways a staff member can be cyberbullied because of salon connections and/or via the salon's social media sites.

SITUATION IN A BEAUTY INDUSTRY CONTEXT	WHAT CAN CYBERBULLYING LOOK LIKE?
You have started a local community beauty group on Facebook to draw attention to your salon, and so you think it's best not to 'friend' the staff at local salons.	You are being trolled or stalked online
You find that there's another person with your name doing something very similar but is posting some defamatory content on Instagram that could affect yours and the salon's reputation.	People are trying to stop you communicating with others
Someone keeps commenting on your Instagram posts to say that the products or treatments are bad.	You are being left out of social forums
You are in a product knowledge webinar and the host is continually muting you, though it is clear that other participants are having the opportunity to speak.	Someone sets up and uses fake profiles pretending to be you

    b Read the section 'Are you a cyberbully?'. List three reasons why you may have cyberbullied someone.

15 In the cross-sector appendix, review the section 'Evaluate ways to avoid negative impacts on users of organisational social media sites'. List eight processes your salon can do to respond to incidents, and also avoid incidents, as a duty of care to protect individuals from cyberbullying.

_____

_____

_____

_____

_____

16 From information you have collected and read, write a half-page overview evaluation of cyberbullying and its impacts.

_____

_____

_____

_____

_____

_____

_____

_____

_____

_____

_____

_____

LOCSA.2 COMPLY WITH INDUSTRY AND ORGANISATIONAL ETHICAL AND PROFESSIONAL CODES OF CONDUCT FOR ONLINE ACTIVITIES

Performance Evidence

There are two relevant codes of conduct for online activities for the beauty industry:

- Australian Code of Practice for Disinformation and Misinformation developed by DIGI
- the Code of Ethics published by the Australian Association of National Advertisers (AANA).

Read more about these codes of conduct in the cross-sector appendix in the section 'Identify industry codes of conduct, objectives and scope, and seek advice from relevant personnel on applicability to organisation'.

Organisational code of conduct in the beauty industry

Knowledge Evidence

1 Download a copy of the Australian Code of Practice for Disinformation and Misinformation from **https://digi.org.au/disinformation-code**.

  a Read section 4.1. What two online activities does the Code apply to?

_____

_____

b Read section 4.2 and list one of the services and products that are *excluded* from the Code. List an example of an online platform that they are referring to.

_____

_____

_____

c Read Objective 1. What type of 'Harm' can come from publicly broadcasting misinformation about COVID-19 workplace health and safety procedures in the salon? Who would you refer to for advice?

_____

_____

2 What does the Code of Ethics published by the AANA apply to?

_____

_____

3 Complete Activity CSA.8 in the cross-sector appendix.

Find the Code of Ethics published by the AANA at **https://aana.com.au/self-regulation/codes-guidelines/code-of-ethics** and answer the following questions.

a List the attributes of people who could potentially be discriminated against or vilified on social media or online platforms.

_____

b What are the two ways you can be 'exploitative'?

_____

_____

_____

_____

4 From information you have collected and read, write a half-page overview evaluation of the industry's Codes and their importance in guiding ethical and professional behaviours.

_____

_____

_____

_____

_____

_____

_____

_____

_____

Policies and procedures

In the cross-sector appendix, review the sections 'Identify organisational online code of conduct, objectives and scope, and determine applicability to own job role' and 'Evaluate contents of codes of conduct and significance to own personal and work-based online behaviours and communications'.

Review the following list of expected communication behaviours.

CLIENT AND PUBLIC RELATIONS

1   Are you permitted to talk about the salon at all in your personal posts?
2   Who is expected to post on the salon's social media platforms, and how?
3   How frequent are the posts?
4   Can staff befriend clients or the salon only?
5   Is the content monitored, or collected by particular team member(s)?
6   Who is responsible for emailing?
7   Do they require special training for posting and emailing?

SECURITY PROTOCOLS

8   Training: does everyone know how to use the system?
9   How often are passwords changed?
10 What is the information back-up system? Is it a secure hard drive?
11 Who maintains the security system and who is the main contact?
12 What is the plan of action in the case of a security breach?

PRIVACY AND CONFIDENTIALITY

13 Ensure staff are careful to keep client information secure, for example: Are the devices they are using protected with firewalls? When the client record card is open, is the screen hidden from view of the general public?
14 Keep all salon information confidential. For example, your reply to a client should not contain information that is irrelevant to their post.

PROFESSIONAL ETIQUETTE AND PROMOTION

15 Staff should only upload industry-standard images.
16 Recommend staff create an industry profile to prevent client access to personal images.
17 Avoid any language or content that is or can be seen as defamatory, discriminatory, threatening or abusive.

COPYRIGHT

18 If you did not create an image, and you wish to use one you have found, you need to identify when you can use it, how you can use it or whether you simply cannot use it.
19 Credits and cross-promotion: when the image is not at the business, hashtag or credit that business; for example, a beauty expo. When you have the logo of a professional brand in the background, hashtag that brand as well.

5 Using the above list of expected communication behaviours, complete the table below by listing one or more communication behaviours relevant to each online activity.

ONLINE ACTIVITY	SALON PROCEDURE/CODE OF CONDUCT	EXPECTED COMMUNICATION BEHAVIOURS
Video: vlogging and tutorials	Demonstrate new and exciting products and treatments Clients can also enter vlog	
Community awareness	Use of relatable 'memes' and others to inspire clients Reaching out on a personal level to various client types; e.g. mothers, diverse groups	
User-generated content	Hashtagging Tagging followers Re-posting content	
Link socials	Sending emails to link to socials	
Speaking the brand	Voice salon values	
Online consultations	Socially-distanced option Convenient consultation option	
Emailing	Follow-up to consultation Marketing promotions	

6 From information you have collected and read, write a half-page overview evaluation of the salon's policies and procedures for online activities relevant to your current or prospective job.

_____

_____

_____

_____

_____

_____

_____

_____

_____

_____

_____

_____

_____

Roles and permissions

Performance Evidence

When using the salon's software system, linked to the website, the roles are determined for all staff as users and any required permissions for release need to be obtained by the salon manager or owner. The following definitions apply.

- _Role_ (or 'team space roles'): are used to grant permissions for a group of users to access and perform tasks relevant to their job, set in a hierarchy. An example of roles you may find on your salon's website are:
    - Admin – able to do everything in the space; allocates users and their roles to the space
    - Editor – can approve and publish content
    - Author – can create content but not publish it
    - User – can access the content, use the system but not make changes.
- _Permissions_: refers to the ability of a user to able to create, update and delete any role within the space. For example, for the online booking system, the salon owner may be Admin, the salon manager may be Editor and the staff may be User.
- _Release_: refers to content published. That may be a post to social media, sending an email, editing content on the website or making an online booking.

7 In the table below, list which team space role you expect each salon job description to have when it comes to the amount and type of financial information accessed.

SALON JOB DESCRIPTION	TEAM SPACE ROLE
Salon owner	
Salon manager	
Staff	

8 Brainstorm with some colleagues a list of benefits to the business of having the hierarchy of team space role levels.

_____

_____

_____

_____

_____

_____

_____

LOCSA.3 MAINTAIN PERSONAL ONLINE PRESENCE CONSISTENT WITH ORGANISATIONAL STANDARDS

Knowledge Evidence

1 It is important to ensure your personal posts are not assumed to be representing your salon. Review 'Step-by-step: Build a professional–personal presence' and suggest the methods or steps that can ensure your personal social media posts are not seen by other users as representing your salon.

_____

_____

_____

_____

_____

_____

_____

_____

_____

_____

_____

_____

_____

_____

_____

_____

_____

---

2   What are the four things to consider when you are posting to your professional account?

---

3   In the table below, list who is affected by your ability to maintain a professional online presence. Use the following terms to complete the table:

Skincare company (supplier); Industry reputation; HR management (staff efficiency); Client engagement

	WHO IS AFFECTED?	HOW?
Knowing the audience		Management will try to determine what type of posts, likes and comments gain more positive attention, and know the types of clients that are listening. They will also compare the audience on every online platform.
Setting goals and objectives		The salon communicates to the audience with an agenda in mind before posting. For example, a sales strategy to promote a new product.
Building processes to support objectives		Create content to post and scheduling when to post it. The salon will inevitably be posting high volumes of content, and will need to develop strategies to manage it, such as outsourcing.
Deliver the social media objectives.		Posting and maintaining the online presence in a timely and professional manner.

LOCSA.4 PROTECT CUSTOMER PRIVACY AND MAINTAIN CONFIDENTIALITY OF ORGANISATIONAL INFORMATION

In order to understand and comply with the *Privacy Act 1988*, you need to be familiar with the Australian Privacy Principles (APPs), as well as your state/territory privacy legislation. The APPs are technology-neutral, so they can be adapted to all activities on social media and online platforms.

Performance Evidence

1   Search the OAIC website at **https://oaic.gov.au** for the 13 Privacy Principles and in your own words write a brief description of each principle.

State and territory privacy legislation

Knowledge Evidence

Every state or territory has its own legislation for data protection. You will be able to access an online guide that clearly explains what is required of you and what you can do to ensure digital data privacy protection.

2   Access the data protection legislation from your state or territory and locate their 'how-to' guide that makes the legislation easier to interpret and follow.

   a   List one privacy risk in the guide.

   b   List two things, recommended in the guide, that you can do to protect a client's privacy.

3   Using information you have collected and read, write a half-page overview evaluation of the APPs and any applicable local, state or territory privacy regulations relevant to your current or prospective job as a beauty therapist.

_____

_____

_____

_____

_____

_____

_____

_____

_____

_____

_____

The content of organisational privacy policies and procedures

4 Complete the table below to apply salon privacy policies and procedures to a beauty industry context and to show the importance of guiding ethical and professional behaviours. The guiding ethical and professional behaviours are listed below and can be found in the cross-sector appendix in the section 'Collect, use, maintain and protect information provided by customers and others, according to privacy regulations and organisational policies and procedures'.

- Observe local traditions and restrictions.
- Gain verbal or written consent.
- Prevent harm.
- Respect the person.
- Protect children.
- Make no profit.

To complete the table, write the guiding ethical and professional behaviour for each sample situation, and add some specific detail of how you would remedy or behave in the sample situation.

APP TITLE AND NUMBER	APPLICATIONS TO SOCIAL MEDIA AND ONLINE PLATFORMS IN POLICIES AND PROCEDURES	SAMPLE SITUATION	GUIDING ETHICAL AND PROFESSIONAL BEHAVIOURS
APP 1. Open and transparent management of personal information	Salons should have a privacy policy that relates to client information and one for social media and online platforms. The privacy policy should be made available to clients as well as staff.	Client has recently been a victim of domestic violence and wants to be sure that her ex-partner cannot obtain her contact information.	
APP 2. Anonymity and pseudonymity	Anonymity means the person is dealing with you without any identifiable characteristics or personal information. Pseudonymity means the person you are dealing with has another name as their username. There is no need for you to know their correct name online, but your client record cards that are signed should have their legal name.	Anonymous posts: when an Instagram user with no profile photo and a username 'MsX' posts a comment.	

APP TITLE AND NUMBER	APPLICATIONS TO SOCIAL MEDIA AND ONLINE PLATFORMS IN POLICIES AND PROCEDURES	SAMPLE SITUATION	GUIDING ETHICAL AND PROFESSIONAL BEHAVIOURS
APP 3. Collection of solicited personal information	Beauty therapists collect personal information for online consultations, such as age bracket (skincare ranges), medical conditions (contraindications), address (for marketing). Sensitive information online generally requires client consent.	A client booking online has told you she is Muslim and you are recommending a facial treatment. You need to confirm whether she is happy to be treated by a male beauty therapist.	
APP 4. Dealing with unsolicited personal information	If you receive information about a client that was not 'solicited' or given to you by the client, you are required by law to destroy it or 'deidentify' it.	A client has accidentally copy-pasted a section of personal information in a DM to you, and it was irrelevant to you and the salon.	
APP 5. Notification of the collection of personal information	You should explain why you are asking for information, such as medical conditions.	Your client wants to know why you are asking about all of 'these' medical conditions.	
APP 6. Use or disclosure of personal information	An example of an exception to this rule is when you refer to an appropriate professional. It is easier to navigate this legislation if you ask the client to contact the professional themselves.	You want to store the information on the client record card for other salons and online consultations (same salon business).	
APP 7. Direct marketing	Newsletters, promotions that are directly sent to the client who has provided their details directly to the salon, and if it's reasonable to expect the salon to use their information.	The salon wishes to send a newsletter to clients after getting the personal details from their client record card.	
APP 8. Cross-border disclosure of personal information	The salon must take steps to protect personal information before it is disclosed overseas.	Your salon website is being used by people overseas.	

APP TITLE AND NUMBER	APPLICATIONS TO SOCIAL MEDIA AND ONLINE PLATFORMS IN POLICIES AND PROCEDURES	SAMPLE SITUATION	GUIDING ETHICAL AND PROFESSIONAL BEHAVIOURS
APP 9. Adoption, use or disclosure of government related identifiers	There should be no need for you to have a person's drivers' licence number, passport number, Medicare card number, etc. for a beauty salon-related transaction online.	N/A	
APP 10. Quality of personal information	Personal information must be accurate, up-to-date and complete.	The salon must take steps to ensure that online consultations are updated at every visit.	
APP 11. Security of personal information	Protect personal information it holds from misuse, interference and loss, and from unauthorised access, modification or disclosure.	The client is not confident that your online transaction was secure, that there was security in the client's payment card information.	
APP 12. Access to personal information	The client has the right to access their personal information you have stored but not commercial information.	Client wishes to access all personal records you have of them, within 30 days.	
APP 13. Correction of personal information	You must take reasonable steps to ensure your client's personal information is up-to-date and correct.	You know a client that is transgender has changed their legal name.	

Sensitive information

5   In the cross-sector appendix, review the section 'Information in the beauty business that is sensitive'. Describe briefly, or list in bullet points, the three main types of information generally considered to be sensitive and deemed not for public release.

Knowledge Evidence

6   Complete the table below by listing which type of sensitive information is described in the situation on social media or online platform.

TYPES OF SENSITIVE INFORMATION	SITUATION ON SOCIAL MEDIA OR ONLINE PLATFORM
	Social media: you notice the salon's product promotion of skin colours that don't cover all the Fitzpatrick range.
	Social media: the salon is showing alliance to the 'me too' political movement.
	Social media: your boss is celebrating Christmas and does not post for Eid, Hannukah and every other major religious and philosophical event.
	Social media: a colleague was sharing his opinion of the various beauty industry associations.
	Social media: you are posting the rainbow flag in support of LGBTQIA+ rights.
	Online activities: You have seen on Reddit that a client has been involved in a burglary, and they are scheduled to come in next week.
	Online activities: the online consultation form for facials asks about body shape.
	Social media: your client's face was recognised on your post by facial recognition.

7   For each of the types of sensitive information listed in Figure CSA.8, 'Sensitive information', suggest ways you can maintain confidentiality as part of your salon policy when using social media or other online platforms.

Consent

Knowledge Evidence

Answer the following questions to demonstrate your understanding of consents used when publishing information and/or images of others.

8   Which of the following is *express* consent in a beauty industry context?

    A   The person has implied that they are happy to have the service by their actions, by sitting down for a manicure service.

    B   A paper-based or e-signed form has been provided for the online consultation.

    C   The salon team have expressed that they are happy to have the Christmas photos posted on Instagram.

9   Which of the following is *implied* consent in a beauty industry context?

    A   The person has implied that they are happy to have the service by their actions, by sitting down for a manicure service.

    B   A paper-based or e-signed form has been provided for the online consultation.

    C   The salon team have expressed that they are happy to have the Christmas photos posted on Instagram.

10   Give one example of a situation where you or your salon should acquire written consent via the template shown in Figure CSA.7, 'Written consent template'.

LOCSA.5 SOURCE INFORMATION ON COPYRIGHT

What copyright protects and does not protect

Knowledge Evidence

1   When is material that you have created considered 'copyrighted'?

Refer to the Figure CSA.9, 'Material that is protected and unprotected by copyright' to answer the following five questions.

Performance Evidence

2   How long does copyright last for?

3   List the three 'moral rights' of the copyright owner.

4 In the case of 'fair dealing', how much can you copy of someone's content?

_____

5 If you post something to the salon page, who owns the copyright?

_____

6 In relation to 'Definition of material', how could a facial treatment method possibly be protected by copyright?

_____

7 Describe how to:

a obtain permission to use a photo you have found on Instagram and credit the person in your post.

_____

_____

_____

_____

_____

_____

_____

b repost a photo using the Regram app.

_____

_____

_____

_____

Exclusive rights of owners of copyright

Knowledge Evidence

8 What does it meant to have 'exclusive rights' to your material as the copyright owner – 'exclusive rights' to what?

_____

_____

9 There are four ways you can identify copyright:

1 Legal license of copyright material.

2 Apply a statement of copyright.

3 Brand an image.

4 Apply a software lock.

List the best way to identify your copyright in the following instances:

a When you are posting images of your make-up portfolio on Instagram or Pinterest.

_____

b When you do not want people sharing your brand logo with cut/paste.

c When you want to add an image to Creative Commons so that it can be shared globally for sharing, reusing and remixing.

d When you do not want someone to reproduce any of your material.

10 Using information you have collected and read, write a half-page overview evaluation of copyright in Australia – what it protects and does not protect, exclusive rights of owners and the requirements for permissions.

LOCSA.6 ENGAGE PROFESSIONALLY WITH CUSTOMERS ONLINE

Performance
Evidence

Defamation via social media

1 What is the difference between slander and libel in the definition of defamation?

Performance
Evidence

Knowledge
Evidence

2 Research and list three different sources of information about defamation via social media in Australia. Include in your research some specific information from your state or territory.

3 List at least three things you can do that can increase your risk of defaming someone when using social media.

4 When would it be appropriate for a person to 'sue' you for defamation?

5 What should you do if someone accuses you or the salon of defamation? Does insurance cover it financially?

6 Which of the following activities can be classed as defamation?

   A a deleted defamatory post

   B a 'liked' defamatory post

   C a post you didn't write

   4 all of the above

7 Using information you have collected and read, write a half-page overview evaluation of defamation via social media.

Salon criticism via social media

Performance
Evidence

Research information from three sources about criticism of a business, either from clients or staff, via social media.

Read the article links provided below about what can be done when employees complain on social media, and answer the following questions.

https://www.hrinasia.com/retention/what-to-do-to-employees-who-complain-on-social-media

https://www.socialmediatoday.com/content/7-tips-responding-negative-comments-social-media

8 What can employers do when staff complain on social media?

_____

9 List seven tips for responding to negative comments on social media.

_____

_____

_____

_____

_____

_____

_____

10 Refer to the following link, or research another social media policy for salon owners: **https:// salonmarketingexperts.com/social-media-policy**. What do employers and contractors who use social media on their behalf have to do in order to minimise risk?

_____

11 Read this article about customer complaints about a business via social media, and answer the following questions: **https://www.theguardian.com/media-network/2015/may/21/customer-complaints-social-media-rise**

   a What per cent of the customers surveyed had made complaints on social media?

_____

   b What are the two things that customers also do online that can be of great benefit to the business?

_____

_____

   c Describe the two points of advice for businesses to support employees on social media in a way that is meaningful to the salon environment.

_____

_____

_____

12 From information collected, provide a half-page overview evaluation of client or staff-based criticism of the salon via social media.

_____

_____

_____

_____

_____

_____

_____

_____

_____

_____

Avoiding risk to the salon's reputation

Performance Evidence

In the cross-sector appendix, review the section 'Review and ensure personal online posts and activities do not damage reputation of organisation and those associated with it'.

13 For the following three situations that present a risk to the salon's reputation, identify a way you can avoid the risk or respond to the risk. Write your answers for each as step-by-step guidelines that the salon can use in their policies and procedures.

a You are interviewing someone in an online vlog, and when a competitor is mentioned, the person attempts an exaggerated, dramatic way of saying things that you know is intended to manipulate the audience. What can you do?

_____

_____

_____

_____

b A friend told you they saw someone blogging some hateful comments about your salon on their personal Twitter account. You ask them to take it down and they won't.

_____

_____

_____

_____

_____

_____

c On your salon's Instagram account, one user is continually commenting on every post featuring your products asking questions about the product packaging. They are querying about how sustainable the packaging is. You have responded with the details, but they have continued to post similar comments to three other product promotional posts.

_____

_____

_____

_____

_____

_____

Principles of positive online communication and language

Knowledge
Evidence

14 Suggest two ways you can connect with clients in a way that builds rapport.

_____

_____

_____

_____

15 What is the recommended response timeframe to a person who has:

a emailed you?

_____

b posted a comment to your Facebook page?

_____

16 In the cross-sector appendix, review the section 'Types of positive language'. In the table below, write a positive comment in response to the client comments that follows positive communication principles.

CLIENT COMMENTS ON YOUR POST OR THREAD:	POSITIVE COMMUNICATION	WHAT CAN YOU SAY?
Your salon doesn't have anyone that does threading.	Suggest what can be done.	
I can't afford your facial treatments.	Provide options as solutions.	
That moisturiser (pictured) is too heavy for me – free to a good home!	Be encouraging and helpful.	
The last time I had waxing done in this salon I got ingrown hairs.	Focus on positive outcomes due to positive actions.	

17 Follow the steps outlined in Figure CSA.12, 'How to respond positively to negative online behaviours on social media' for the following two client complaints.

CLIENT COMPLAINT	STEP 1 RESPOND QUICKLY	STEP 2 ACKNOWLEDGE ANY MISTAKES IN THE PUBLIC SPHERE	STEP 3 TAKE THE CONVERSATION OFFLINE	STEP 4 CHECK THAT THERE IS NO AUTO-REPLY (AS APPROPRIATE)	STEP 5 IDENTIFY THE PROBLEM AND SOLUTION, ACKNOWLEDGING THE CLIENT'S SITUATION
I had a treatment at that salon, and when I went to pay they didn't honour the gift voucher, even though there was no expiry date on it.					
I noticed the staff like to talk about each other behind their backs at the manicure tables.					

INDEX

A

abdomen
    massage technique 273
    muscular system 395
    xerosis (dry skin) 329
ABIC 3–4
abnormal scar tissue 448–9
absorption pathways 332–3
Accident Report book 74
accidents 74
acetone 78
acid mantle 322
acids and bases 321–3
acne and congestions
    facial treatments and skincare 310
    massage movements recommended 267
acne vulgaris 112, 310
    nutritional factors 373, 385–6
adhesives, eyelash extensions 204
advanced nail art 164–5
    cleaning service area 166–7
    range and variety 165
    reviewing 165–6
adverse reactions
    aromatherapy massages 288–9
    aromatic plant oil blends 303–4
    aromatic plant oils 301
    body massages 274–5
    client feedback 344–5
    cosmetic tanning 253–4
    lash and brow services 192–3
    manicure and pedicure services 153
    specialised facial treatments 365
    See also contraindications
advertising treatments and services 56
advice, beauty products and services 32–42
aftercare advice
    aromatherapy massages 288–290
    body massages 274–7
    cosmetic tanning 252–3
    eyelash extensions 206–7
    facial treatments and post-treatment
    skincare 345–6
    lash and brows 195
    make-up 224–5
    manicure and pedicure services 156
    remedial camouflage make-up 234
    specialised facial treatments 366–7
    waxing services 181
Afterpay transactions 64, 68
ageing
    men and women 326
    nutritional factors 373
age-related skin changes 442–4, 445–6
allergic dermatitis 129–30
allergies and sensitivities to ingredients 41–2,
    309, 344, 349
alopecias 174, 408
alpha tocopherol (vitamin E) 340, 375, 422
alpha-hydroxy acids 334
amino acids 435, 436
amphoteric surfactants 337

anagen 170, 184
analogous colours 214
anatomy and physiology
    data in client records 463–4
    effects and benefits of body massage
    257–60
    knowledge 458–9
    maintaining knowledge 464–5
    using correct terminology 268
animal carriers of disease 95
animal parasites 118–19
anionic surfactants 336
ankylosing spondylitis 414
antibiotics 112
anti-discrimination and equal employment
    opportunity (EEO), state and territory
    government boards and commissions 5
anti-discrimination and equal employment
    opportunity (EEO) laws 8
antifungal treatments 114
antioxidants 340–1
APAN 3–4
appointments, receiving clients and making 23
arms and chest, muscular system 395–6
arms and hands 400
    anatomy and physiology 145–7
    bones 392–3
    massage technique 273
aromatherapy massages 278–92
    cleaning the treatment area 290–2
    client medical history and medications 280
    client records 281, 288
    contraindications and scope of practice
    279, 293–4
    designing and recommending treatment
    281–6
    effects and benefits of aromatherapy oils
    281–4
    establishing client priorities 278–81
    massage routine 285
    preparing for 286–7
    providing 287
    providing advice of effects and benefits to
    client 462
    reviewing service and aftercare advice
    288–90
    suitability for client needs 287
    treatment, duration, frequency 286
    work health and safety 287
aromatic plant oil blends 293–305
    adverse reactions 303–4
    application methods 296–7, 303
    to avoid 294
    botanical names 300
    carrier oils 295–6
    characteristics and properties 297–300
    cleaning treatment area 305
    client records 301, 304
    designing 295–300
    establishing client priorities 293–5
    organic chemistry 298–300
    organisational policies and procedures 301

    preparing 302
    profiles and plant information 299
    providing treatment using 302–3
    reviewing treatment and post-treatment
    advice 303–4
    setting up for blending 301–2
    suitability for client needs 294–5
arteries
    hand and forearm 146, 147
    legs and feet 149, 150
aseptic procedure 90
AS/NZS 4815 *Reprocessing of reusable*
    medical devices in health service
    organisations 124
AS/NZS 4817 *Office-based health care*
    facilities - Reprocessing of reusable medical
    and surgical instruments and equipment,
    and maintenance of the associated
    environment 124
atoms 317–19
attenuation 358
Australian Competition and Consumer
    Commission (ACCC) 338
Australian Certified Organic Standard 2013
    324
Australian Competition & Consumer
    Commission (ACCC) 34, 57
Australian Consumer Law 56
Australian Dietary Guidelines 378
Australian Guidelines for the Prevention and
    Control of Infection in Healthcare (2010) 104
Australian Standards 124
autoimmune diseases 414
autonomic nervous system 401
awards *See* industrial awards

B

back muscular system 397–8
bacteria 91, 106, 109
    gram negative and positive bacteria 111
    products and services suitability 110
    reproduction 110
    simplified classification according to
    shapes 109–10
    skin diseases 92
    transmission in salons 110–13
beauty industry
    career opportunities 59–60
    expectations of staff 54
    mentors 464
    relationship with other related industries
    55–6
    research and application of information
    50–61
    scope of practice 131–2
    working conditions 54–5
beauty products 32–42
    cosmetic chemistry 335–8
    demonstration 39–42
    product demonstration preparation 38
    product knowledge development 32–6
    salon services recommendations 36–8

beauty professionals
  providing advice on dietary guidelines 384–6
  scope of practice 364, 450–3
beauty professionals' organisational requirements compliance 3–18, 7–8, 354
  effective work habits 16–18
  personal presentation maintenance 15–16
  supporting work teams 8–15
beauty therapy treatments
  applying to body systems 382–3
  assessing and evaluating client needs 459–60
  designing and developing 456–8
  work health and safety policies and procedures 386
bioavailability 334–5
biofilms 108, 110, 112
biological activity 335
birthmarks 227
blood supply
  hand and forearm 146–7
  legs and feet 149–50
body disorders 409–14
body language 20, 44
body massages 257–77, 454–5
  analysis for 265–6
  areas to treat and expected outcomes 269
  cleaning treatment area 277
  contraindications and scope of practice 260–4
  designing treatment 266–9
  effects and benefits of massage movement 266–8
  effects and benefits on anatomy and physiology 257–60
  establishing client priorities 257–66
  legal requirements 272
  lymphatic system 404
  preparing for treatments 270–2
  providing 272–4
  providing advice of effects and benefits to client 462
  reviewing and aftercare advice 274–7
  suitability for client needs 268
  treatment, duration, frequency and cost 269, 275–6
body piercing 37
body system disorders
  body massage restrictions 260–4
  modifying aromatherapy treatment 279
body systems 144–5
  applying knowledge of beauty therapy treatments 382–3
  effect of aromatherapy treatments 284–5
  first aid for acids and bases 323
body systems and structures 387–418
  circulatory system 403–5, 412, 415–16
  digestive system 415–16
  endocrine system 405–8, 413–14, 416–18
  lymphatic system 404, 413, 415–16
  muscular system 410–11, 415
  musculoskeletal system 389–401
  nervous system 401–2, 411, 415
  organ systems and relationship to healthy body 408–18
  reproductive system 416–18
  respiratory system 412, 415–16
  skeletal system 409–10, 415

  structural levels of organisation in 387–9
  using knowledge in designing treatment to suit clients 454–8
body treatment advice 462
body's defences 97
bonds 319–20
  breaking peptide 342
  saturated and unsaturated 324–5
  water molecules and hydrogen 327–8
bones
  arms and hands 392–3
  facial 392
  hand and forearm 145
  joints 393–5
  legs and feet 147–8, 393
  major 391
  structure 392
botanical name of plants 300
bound water 328
brittle nails 136
buffering 322
bullying behaviour 77, 84

C

callus 152
Candida albicans 113
capillary naevus 228
carbohydrates 373
career opportunities 59–60
carer's leave 12
carrier oils 283–4, 295–6
case studies
  eyelash tint procedure 189
  risk management processes 103
cash registers 63
cash sales
  counting 64
  ensuring security 66–7
  GST 62–3
  managing float 68
  recording takings 69
catagen 170, 184
cationic surfactants 337
cells 388
ceramides 423
chain of command, in salons 14–15
chemical blocking agents 441
chemical change 321
chemical facial peels 342–3
chemical spills, risk assessment and control 78
chemistry, keratin 436
cholesterol 377–8, 422
chronological ageing 445–6
circuit breakers 86
circulatory system 144–5, 403–5, 415–16
  disorders 412
circumducting 273
cleaning
  aromatherapy treatment area 290–2
  aromatic plant oil blends treatment area 305
  body massage treatment area 277
  cosmetic tanning 255–6
  eyelash extensions treatment area 208
  facial treatment area 346–7
  lash and brow service area 195–6
  make-up services 225
  manicure and pedicure treatment area 155–6
  nail art service areas 166–7
  photographic make-up tools 241

  remedial camouflage treatment area 234
  specialised facial treatment area 368
  waxing treatment area 182–3
cleaning rosters 90
cleansers 338
cleansing skin
  deep and superficial 314
  for extractions 314–15
  product demonstration preparation 38
  using ultrasonic machines 355–60
client appointment scheduling 27–8
client complaints responses 28–30
  eyelash extensions 202
  remedial action resolution 28–9
client feedback
  aromatherapy massages 288
  eyelash extensions 207
  facial treatments 344–5
  lash and brow services 193–4
  make-up services 223
  remedial camouflage make-up 233
  specialised facial treatments 364
  waxing services 181
client feedback forms 153–5, 194
client image and occasion, make-up 213–14
client needs 213, 268, 287, 294–5, 312, 459–60
client priorities 128–40, 157–61, 169–76, 184–8, 197–200, 210–13, 226–30, 244–6, 278–81, 293–5, 348–51
client records
  access and maintenance 26
  aromatherapy massages 281, 288
  aromatic plant oil blends 301, 304
  cosmetic tanning 246
  eyelash extensions 198, 207
  facial treatments and skincare 314
  photographic make-up 238–9
  privacy 26
  recording relevant data 463–4
  specialised facial electrical treatments 352
  specialised facial treatments 353
clients
  ability to follow homecare advice 40
  dealing with difficult and abusive 31
  receiving and making appointments 23
client's briefs, interpreting 239–40
clients with special needs 30–1
  effective communication with 45
  manicure and pedicure services 128–9
  nail art considerations 129, 158
closed questions 44
codes of practice, infection control 104–5
collagen 326, 330, 331, 333, 340, 447 *See also* elastin
colour analysis and design 214–16
  remedial camouflage make-up 230–1
colour harmony and colour context 214
colour wheels 214
colours
  in cosmetics 215–16
  ensuring even tanning application 251
  nail art and advanced nail art 160
  tanning shades 246–7
  warm versus cool 214–15
Comcare 83
comedones 315
communication
  face-to-face 22
  retail customer sales 44–5
  verbal and non-verbal 19–21

complementary colours 214
complex lipids 420, 422
compounds 324
conductive gel 357
consent for treatments 232, 272
consumer law 43
    products and services 46–7
    refunds and exchanges 68
consumer protection and trade practices 28, 34, 57–8
contact dermatitis 129–30
    cosmetic irritants 349
    treatments 130
contagious skin disorders 107–8
contamination 97–8
contingencies 24
    electricity safety 84–8
    infection control procedures 72–3
contract urticaria (hives) 130
contraction of skin 447
contracts *See* employment terms and conditions
contraindications 41
    aromatherapy massages 279
    aromatic plant oil blends 293–4
    body massages 260–4
    cosmetic tanning 244–5
    eyelash extensions 199–200
    facial treatments and skincare 309–10
    infection control action 94
    lash and brow services 187–8
    make-up services 210–12
    manicure and pedicure 129–30
    nail art treatment 158–9
    photography make-up 237–8
    remedial camouflage services 229–30
    specialised facial treatments 348–9
    waxing services 175–6
Coppertone® 244
corrective make-up
    facial shape and 216–18
    false eyelashes 223–4
cosmetic chemistry 317–23
    in beauty products 335–8
cosmetic emulsions 335
cosmetic ingredients 349–50
    regulations for 338–41
cosmetic tanning 244–56
    applying product with spray gun 250–1
    cleaning treatment area 255–6
    client records 246
    contraindications and scope of practice 244–5
    establishing client priorities 244–6
    hygiene 249–50
    preparing to apply 246–50
    procedures followed and products used 248–9
    reviewing and aftercare advice 252–5
counselling and discipline 12–13
couperose skin 433, 434
COVID-19 116
    impact on salons 58
credit card/eftpos transactions 64–5
cultural considerations, body massage treatments 272
cultural safety 45
curls, eyelash extensions 204
customer and industry expectations 47

customer service principles, website resource 43
cutaneous absorption 332–3
cutaneous adsorption 333
cuticle 135
    care 152
cytokeratin (soft keratin) 435

D

damaged, chipped broken nail art 164
damaged eyelashes 198
data security 66–7
deep cleanse 314
deep ridges in nails 136
demodex folliculorum 119
depilatory waxing 169
dermatomycoses 114
dermatophytes 114
dermis 427–8
detergents 336, 337, 338, 339
dietary fibre 377
dietary guidelines 384–6
diffuse red skin 350–1, 433 *See also* couperose skin
digestive system 383, 415–16
Dihydroxyacetone (DHA) 249
discolouration of nails 136, 145
discrimination and harassment 13
diseases 92–3
    distinguishing between organisms and 106
    fungal 114
    infectious 107–8
    transmission 95–6
    transmission by blood 96
disinfection of sterilising equipment 121–2

E

early anagen 184
earth leakage circuit breakers (ELCBs) 87
eating disorders 381
ectoparasites 118
effleurage technique 266, 272, 287, 316
elastin 325–6, 330
electric shocks 85–6
electrical circuits 86
electrocution 85–8
electrostatic discharge 353
email
    appointment scheduling 27–8
    correct communication techniques 22
emergency procedures 84–5
    following 98–9
emollients 339–40
employees *See* staff
employer associations 3–4 *See also* ABIC; APAN; HBIA
employer workplace rights and responsibilities 8–10
employment laws basics 5–6
employment rights and responsibilities 3–8
employment terms and conditions 13–14
emulsifiers 336
emulsions 337
    cosmetic 335
endocrine system 405–8, 416–18
    disorders 413–14
    hormone action on skin 406–8, 417–18
endoparasites 118

environmental issues and requirements 53
epidermal keratinocytes 419–20, 422, 435, 436
epidermis 132, 426–7
    absorption pathways 333
epithelial tissues 435
epitheliasation 447–9
eponychium 135
equal employment opportunities 13
Equal Opportunity Act 2010 13
equipment and tools
    body massage treatments 270–1
    eyelash extensions maintenance 201
    make-up 219–20
    nail art and advanced nail art 162
    risk assessment for faulty 78–9
    safety checks 313
    specialised facial treatments 351–2, 366–7
erythema 309, 310, 403
Erythrulose 249
estimated energy requirements (EER) 379–80
ethical considerations
    products and services suitability 42
    professional behaviour 30
    research 53–9
exchanges *See* refunds and exchanges
excretory system, 383
exfoliants 342
exfoliation
    cosmetic tanning 250
    facial treatments 314, 333, 342, 360
    feet 152
    video 360
exocrine gland 405
extractions, 310, 314–15
eye cleansers 338
eyelash extensions 197–208
    applying and selecting types 203–4
    benefits and risks 200
    cleaning treatment area 208
    client records 198, 207
    contraindications 199–200
    establishing client priorities 197–200
    preparing treatment area, staff and client 200–3
    removing 202–3
    reviewing and providing post-service advice 206–7
    to suit eye shape and facial features 204–5
eyelash lift 191
eyelash mites 119
eyes, corrective make-up 216–18

F

face-to-face communication 22
    client appointment scheduling 28
facial massages 316–17, 455
facial muscles 399–400
facial peels
    basic 361–2
    chemical 342–3
facial products, chemical formulations 341–3
facial shape and corrective make-up 216–18
facial shape and photographic make-up 238
facial treatments and skincare
    recommendations 307–47
        applying specialised products 317–43
        cleaning treatment area 346–7

cleansing and exfoliating and performing extractions 314–15
client feedback and adverse effects 344–5
client records 314
contraindications 309–10
cosmetic chemistry 317–23
designing and recommending 311–12
effect and benefits 307–8
explaining treatments 312
glycosaminoglycans (GAGs) 330–1
organic chemistry 323–5
percutaneous penetration of cosmetic ingredients 332–5
preparing for 313–14
providing facial massage 316–17
recommending post-treatment skincare regimen 345–6
reviewing 343–5
scope of practice 310
suitability for client needs 312
water and skin 326–30
Fair Work Act 2009 5
Fair Work Commission 4
Fair Work Ombudsman 4, 12
farewelling techniques 24
fats (lipids) 373–4
feet 132–5
  exfoliation 152
fibroblasts 326, 330
financial transactions 62–70
  completing POS transactions 64–7
  completing refunds 67–8
  operating POS equipment 62–4
  payment arrangements 24
  reconciling takings 70
  removing takings from register/terminal 68–70
fingernails 133–4
first aid, acids and bases 323
Fitzpatrick scale 440
fleas 119
Folliculitis 108, 112, 309
follow-up advice 254–5
food intolerance 381
formaldehyde 344
free edge 134, 135
free radicals 340
free water 328
frictions technique 266, 272, 287
full-set applications 197
functional groups 324
fungal diseases 92
  antifungal treatments 114
fungi 91, 113–14
  skin diseases 92
  See also moulds; yeasts

G

genetics of skin disorders 449–50
glands, skin 431–3
glycosaminoglycans (GAGs) 330–1
golden ratio 228
Golden staph *See Staphylococcus aureus*
goods and services tax (GST) 62
gram positive and negative bacteria 111, 112
greyscale 215
grievances 13
grooming and personal presentation 23

H

hair
  keratinogenesis 436
  using anatomy and physiology knowledge to explain treatment 461
hair analysis 173–4
Hair and Beauty Industry Award 6
hair disorders 174–5
hair growth 169–72
  cycles 170–2
  excessive 174
  eyelashes and brows 184–6, 200
  factors affecting 173
  patterns 170
  variations 170
hair removal, alternative methods 172–3
handbooks, staff 7
hands and feet, skin on 132–3 *See also* arms and hands; legs and feet
handwashing hygiene 74
harassment provisions 9–10
hazards, near misses and incidents 74–7
  risk assessment and control 78–81
HBIA 3–4
head lice 119
health effects, from electric shocks 85
henna 187, 190–1
Herpes simplex viruses I and II 116
Herpes varicella zoster virus (HVZ) 117
hives 130
hormone action on skin 406–8, 417–18
hot waxing 179
human carriers of disease 95, 109
humectants 330, 339
hydrocarbons 323–4, 344
hydrocolloid emulsions 336, 342, 357
hydrophilia 333, 336, 337, 420, 423
hygiene 58, 89–90, 301–2
  body massage treatments 271
  cosmetic tanning 249–50
  eyelash extensions requirements 202
  food and nutrition legislation 378
  handwashing 74
  make-up 222
  monitoring hygiene of premises 105
  nail art and advanced nail art 162–3
  remedial camouflage make-up 231–2
  See also work health and safety
hypertrophic scars 448
hypodermis 428
hyponychium 135

I

immunisation 97
immunity 405
incidents 74–7
  case study 189
  risk assessment and control 78–81
industrial awards *See* Hair and Beauty Industry Award
industrial relations issues 53–4
industry associations 50–1
industry awards 6
infection control guidelines 104
infection control procedures
  contingencies 72–3
  contraindications and required action 94
  following 90–8
  industry codes of practice 104–5

infection control standards maintenance 102–25
  awareness of clinic design for control of infection risks 124–5
  complying with legal obligations 102–5
  monitoring hygiene of premises 105
  skin penetration treatments 106–21
  sterilising equipment and maintaining steriliser 121–4
infectious diseases 107–8
in-fill services 197
inflammation of skin 446
influenza 109
information sources 7–8
  for beauty industry 50–2
  product 32
  sharing and accessing 61, 465
ingredients
  cosmetics 241–2
  facial products 341–3
  lipids in cosmetic 420
  penetration of cosmetic 332–5
  regulations for cosmetic 338–41
  remedial camouflage products 231
  sunscreens 441
  tanning products 249
  understanding cosmetics effects on skin 317–19
insects 119
insurance, public indemnity 57–8
integumentary system 409, 419
interstitial fluid 414
ionic surfactants 336
iontophoresis treatment 358, 361
irritant dermatitis 129

J

jewellery for nail art 160
joints 393–5

K

keloid scars 448
keratin 435–7
keratinogenesis 436
keratolytics 437

L

labelling
  allergens 344
  cosmetics 338
  organic 324
lanolin 340, 344
l-ascorbic acid (vitamin C) 340, 375
lash and brow services 184–96
  aftercare advice and post-treatment recommendations 195
  chemically treating 190–1
  cleaning treatment area 195–6
  contraindications and scope of practice 187–8
  establishing client priorities 184–8
  incidents case study 189
  preparing for 188–9
  reviewing and providing post-service advice 192–5
  shaping eyebrows 186–7, 191–2
leave, personal and carer's 12
legal issues, research 53–9

legal requirements
    beauty products demonstration 39–40
    body massage treatments 272
    infection control 102–5
    nail product safety 142
    skin penetration guidelines 90
    specialised facial treatments 354–5
    work health and safety 82–3
legs and feet 400
    anatomy and physiology 147–50
    bones 393
    massage technique 273
    muscular system 396–7
letters, correct communication techniques 22
lice 119
licensing and registration 103
lifestyle advice 462
    skin health 330, 346, 365–6
lifting heavy boxes 83
lighting situations
    photographic make-up 236, 237, 239
    remedial camouflage make-up 228, 231
lipids, skin structure 419–23
lipophilia 332, 333, 337, 340, 420, 423
liposomes 334, 338
lips, corrective make-up 218
listening techniques, clients with special
  needs 31
lunula 134, 135
lymphatic system 404, 415–16
    disorders 413

M

magazines, trade 51
make-up *See* photographic make-up; remedial
  camouflage make-up
make-up
    applying false eyelashes 223–4
    cleaning service area 225
    contraindications 210–12
    designing make-up plan and application
    techniques 213–22
    establishing client priorities 210–13
    hygiene regulations 222
    products that may cause reactions 220–1
    providing post-service advice 224–5
    providing services 222–3
make-up tools 219–20, 221
males, massage movements recommended
  268
manicure and pedicure services 128–56
    areas requiring special attention and
    needs 128–9
    cleaning treatment area 155–6
    common contraindications 129–31
    establishing client priorities 128–40
    nail analysis 132–5
    preparing for nail service 141–4
    products used in 142–4
    providing nail services 144–52
    reviewing 152–5
manufacturer's instructions 123
marine extracts in skincare ingredients 309
masks 341–2
    foot 152
measles 109
mechanical exfoliants 342
medical history and medications
    considerations for waxing services 176
    modifying aromatherapy treatment 280

melanocytes 438
Mendel's Laws 449
menopausal clients
    nutritional requirements 385
    sebum production 442–3
mentors 464
merit-based employment decisions 10
microbes 90–3, 95–6
    harmless 108–9
    naming organisms and and diseases 106
    scientific and common names 106
microbiome 430–1, 432–3
microcurrent 362
microemulsions 335
microparticles 334
*Milady's Aesthetician's Series: Advanced Hair
  Removal* (Hill & Blackmore) 169
milia 315
mineral oils 423
minerals 376
moisturising treatments 327
moisturisers 339
moulds 113–14
mucopolysaccharides *See* glycosaminoglycans
  (GAGs)
muscles
    hand and forearm 145–6
    legs and feet 148–9
muscular system 144, 395–401, 415
    disorders 410–11
    inspiration and expiration of air 401
musculoskeletal system 389–401
    muscular system 395–401
    skeletal system 391–5
mutations in genes 450
multiphase emulsions 335–6

N

nail analysis 133–5, 158
    practice questions 136–40
nail art 163–4
    areas requiring special treatment 158
    cleaning service area 166–7
    contraindications 158
    design 159–61
    establishing client priorities 157–61
    organisational policies and regulations
    161–2
    preparing service area 161–3
    range and variety 160–1
    reviewing 165–6
nail bed 134
nail buffing 151
nail disorders 129
nail grooves 133, 135
nail matrix (nail root) 134, 135
nail plate 133, 134, 135, 136
nail polish 152
    ingredients used in 142–3
    matching colours to skin tone 140
nail service 141–4
    providing 144–52
nail shapes 135, 151
    nail art considerations 160
nail walls 135
nails
    effect of health and disease on 135–6
    keratinogenesis 436
    minimising damage during treatment 141
    See also fingernails; toenails

nanoparticles 334
National Employment Standards (NES) 5
National Health and Medical Research
  Council (NHMRC) 104
natural moisturising factor (NMF) 328, 338–9
near miss 74
neck and chest massage technique 273
needlestick injuries 73
    infection control standards 120
    risk assessment and control 78
nervous system 401–2, 415
    disorders 411
    modifying treatments to avoid
    contraindication 456
    skin structure 429–30
neutralisation 322
New Zealand, EEO government boards and
  commissions 5
NHMRC Australian Guide to Healthy Eating
  378
non-cash sales 64–5
    ensuring security 66–7
    recording takings 69
non-ionic surfactants 336, 337
non-saponifiable lipids 420, 422
non-verbal communication 19–20, 44
normal body flora 108, 113, 432
normal scar tissue 447–8
NSW Fair Trading 47
nutrients
    guidelines 379
    needs across lifespan 380
    role in weight management 379–80
nutrition
    common diet-related health problems
    381
    healthy skin 452–3
    legal requirements 378
    skin health 346, 363–4, 365–6
Nutrition Australia Healthy Eating Pyramid
  378
nutrition principles 371–81
    nutritional composition of common
    foods 372–7
    in regards to treatment procedures 371–2
NYSTATIN 114

O

obesity and weight management 381
occludents 339
oedema, massage movements recommended
  267
Office of the Australian Information
  Commissioner (OAIC) 69
oil
    emulsions 335–6, 337
    mineral vs vegetable 423
olfactory sense 297–8
online platforms 466–92
onycholysis 129
open questions 44
open wounds 74
opportunistic flora 108, 112
ordering procedures 34
organ systems and relationship to healthy
  body 408–18
    disorders 409–14
    integumentary system 409
    interdependence 414–18
    respiratory and reproductive systems 408

organic chemistry 323–5
    aromatic plant oil blends 298–300
organic labelling 324
organisational policies and procedures
    accepting, declining and amending roster
        hours 11–12
    aromatic plant oil blends 301
    eyelash extensions 200
    photographic make-up 241
overloads, electricity 87–8

P

parasites 91
    animal 118–19
    infection in salons 92–3
paronychia 129, 135
paraffin 339–40
patch testing 198
    recognising skin reaction 245
pathogenic flora 108–9
payment arrangements 24
Payment Card Industry (PCI) Data Security
    Standards 66–7
payment handling procedures 63–4
    cash sales 62–3
    determining change and denominations
        65–6
    non-cash sales 64–5
peels
    basic 361–2
    chemical 342–3
penetration enhancers 334
percutaneous penetration of cosmetic
    ingredients 332–5
periodic table of elements 318
perionychium 135
person conducting a business or undertaking
    (PCBU) 82, 83
personal and carer's leave 12
personal presentation maintenance 302
    body massages 271
    grooming 23
personal protective equipment (PPE) 72–3,
    250
personal service environments, organisational
    requirements compliance 3–18
petrissage technique 266, 273, 287, 316
pH levels
    cosmetics 337
    effect on health 322–3
    microbiome 432–3
pH scale 321–2
photo-ageing 329, 445–6
photo-allergic dermatitis 130
photographic make-up 235–42
    analysing photography context 235–7
    cleaning tools and equipment 241
    client records 238–9
    contraindications and scope of practice
        237–8
    designing make-up plan 238–40
    in different conditions 240–1
    establishing requirements 237–8
physical blocking agents 441
physical change 321
pigmentation disorders 245
pinworm 118
plant oils 282–3
    carrier oils 284
    storing 291

point of sale transactions
    completing 64–7
    operating 62–4
polar lipids 423
polymer structures 330–1
popliteal lymph nodes 404
post-treatment lifestyle advice
    aromatherapy massages 290
    aromatic plant oil blends 304
    basic facial peel 362
    body massages 276–7
powder 239
pregnancy
    massage movements recommended 268
    nutritional requirements 384
pricing strategies, beauty products 56–7
primary colours 214
print context, photographic make-up
    principles 236–7
prions 117
privacy 54
    client records 26
Privacy Act 1988 46, 54, 69, 378
product
    nail safety 141–2
    providing advice of effects and benefits
        462
    reviewing effect on skin 344–5
product knowledge development 32–6
    clients who will benefit from treatments
        34–6
    features and benefits 33–4
    lash and brow 191
    sourcing information 32
    specialised 40
    updating 59–61
product recommendations 57
product returns 25
products
    cosmetic tanning 248–9
    cosmetics that may cause reactions 220–1
    effects and benefits of aromatherapy oils
        281–4
    eyelash extensions 205–6
    ingredients in waxing post-treatment 179
    lash and brow 190
    lash and brow post-treatment 191
    manicure and pedicure contraindications
        129–30
    manicure and pedicure services 142–4
    services and special packages 26
    shelf-life and use-by-date 34–5
    in workplace range 34
products, techniques and services, industry
    expectations 56–7
products and services suitability 40–2
    ethical considerations 42
products demonstration, legal requirements
    39–40
Propionibacterium acnes 108
propylene glycol 334, 357
proteins 372
    and skin 424
protozoans 118–19
pubic lice 119
public indemnity insurance 57–8

Q

Quick Tanning Lotion 244

R

recommended dietary intakes 379
record keeping
    client 26, 157
    eyelash extensions 198
    hazards, near misses and incidents 74–7
    lash and brow services 189
    manicure and pedicure services 140
    nail art 157
    remedial camouflage make-up 230
    signed consent 232
    waxing performed 182
reflex arc 402
refunds and exchanges 67–8
remedial camouflage make-up 226–34
    applying 232–3
    cleaning treatment area 234
    contraindications 229–30
    designing make-up plan 230–2
    establishing client priorities 226–30
    record keeping 230
    reviewing 233–4
remedial camouflage products 231, 233–4
removal services, eyelash extensions 198
reproductive system 408, 416–18
    disorders 413–14
reprovisions 10
rescheduling
    cosmetic tanning 255
    eyelash extensions 207
research and application of information 50–61
    information sources for beauty industry
        50–2
    legal and ethical issues 53–9
    updating knowledge of industry and
        products 59–61
residual current circuit breaker (RCCB) 87
residual current device (RCD) 87
respiratory system 408, 415–16
    disorders 412
retail customer sales 43–9
    establish customer needs 43–5
    facilitate sale of products 48–9
    provide advice on products and services
        46–8
retinoids (vitamin A derivatives) 340, 376, 422
retinol 333
retouching/repair
    advanced nail art 164–5
    nail art 163–4
ringworm 114
risk assessment and control 77–82
    case study 103
    faulty equipment 78–9
    infection control 102
    specialised facial treatment machines
        353–4
    ultrasonic/galvanic machines 355–6
    waxing 79–80
Risk Evaluation Form 82
Risk Management sheets 80–2
roster hours for salons, considerations 11–12
roundworm 119 *See also* pinworm
rubella 108–9

S

Safe Work Australia 82–3, 142
safe work practices 82–4 *See also* work health
    and safety

safety data sheets (SDSs) 78, 80–2, 110, 313
sales *See* retail customer sales
sales techniques and principles 48–9
sales transactions *See* cash sales; non-cash sales
salon services 19–31
    ability to identify client needs and constraints 36–8
    client appointment scheduling 27–8
    client complaints responses 28–30
    clients with special needs responses 30–1
    customer service 19–24
    presentation of features and benefits 37–8
    rapport, contingencies, product promotion, process sale, returns and refunds 24–6
    selection of suitable products and treatments 37
    types of infections in 90–1
salons
    bacterial transmissions 110–13
    impact of COVID-19 58
    sustainability practices 161
sales and refunds 25
saponification process 336, 420
saturated and unsaturated bonds 324–5
scars 447–9
sealants 163, 164
sebaceous glands 419
sebum 442–3
secondary colours 214
sensitive/ticklish areas, massage movements 268
serums, specialised facial treatments 360–2
sexual harassment 84
shapes, eyebrows 191–2
sharps and needles, infection control standards 120, 121
shea butter 340
short circuit 86
signed consent 232, 272
simple lipids 420–2
skeletal system 144, 391–5, 415
    disorders 409–10
skin
    aromatherapy effect on 285
    bacterial diseases 92
    cleansing using ultrasonic/galvanic machines 355–60
    colour choice for tanning 246–7
    common types and contraindications to photographic make-up 237–8
    conditions requiring remedial camouflage 227–8
    contraction 447
    designing specialised facials 352
    fungal diseases 92
    hands and feet 132–3
    hormone action on 406–8, 417–18
    hydration 326–30, 332–5
    lifestyle advice 330
    measuring rate of ingredients penetration 334–5
    measuring water content in 328
    nutrition and lifestyle factors 346, 363–4, 365–6
    understanding cosmetic chemistry effects 317–19
    using anatomy and physiology knowledge to explain treatment 461

viral diseases 92
    xerosis (dry skin) 329–30
skin barrier 332, 357–8
skin blemishes 360–2
skin cleansers 338
skin conditions 451–2
skin diseases/disorders 129
    aromatherapy massage restrictions 280
    aromatic plant oil considerations 294
    body massage restrictions 264–5
    fungal 114
    genetics 449–50
skin glands 431–3
skin penetration
    guidelines 90
    treatments and infection control 106–21
skin science 419–53
    appearance of skin 433–44
    glands of the skin 431–3
    growth, development, ageing and healing 444–50
    scope of practice 450–3
    skin structure 419–31
skin structure 419–31
    differences depending on body location 430–1
    function 425–8
    integumentary system 419
    proteins 424
    skin as sense organ 429–30
    skin lipids 419–23
    skin tissue 428–9
skin tone, matching nail polish to 140
skin types and conditions 212–13, 433–5
    age-related changes 442–4
    Fitzpatrick scale 440
    keratin 435–7
    light and skin 440–2
    skin colour 437–9
social media 466–92
sodium hypochlorite (bleach disinfectant) 78
sodium laureth sulphate (SLES) 339
sodium lauryl sulphate (SLS) 339
software systems 63
sonophoresis treatment 358, 361
sound energy (ultrasound waves) 356–8
    skin barrier properties 357–8
    thermal and non-thermal effects and benefits 358–60
specialised facial treatments 348–68
    adverse reactions 365
    benefits and effects of using machines 351–2
    cleaning treatment area 368
    client feedback 364
    completing 363
    contraindications 348–9
    designing and recommending 351–3, 367
    diffuse red skin 350–1
    establishing client priorities 348–51
    formulation, function and action of cosmetic products 349–50
    preparing for 353–5
    providing post-treatment advice 363–7
    relevant medical history and medications 350
    removing minor skin blemishes and infusing serums 360–2
    suitability of electrical treatments and recording outcomes 352

SPF and effect in photographic make-up 237
spider naevus 228
spills, infection control standards maintenance 122–3
spray guns 250–1
    associated risks 255–6
    cleaning 255
staff
    effective team membership 17
    general role boundaries 14–15
    industry expectations of 54
    personal presentation maintenance 15–16
    training 57
    work activities planning and organisation 17–18
staff handbooks 7
Staphylococcus aureus 106, 110, 309
state and territory governments
    consumer law 47
    EEO 5
    role in complaints management 9
    work health and safety 83
static electricity 88
steam treatments 308, 315, 321, 334
sterilisation process 110
steroids 422
sterols 422
stock availability 36
storage requirements 36
    plant oils 291
stratum corneum 132, 332, 333
    lipid structure 422–3
    water content 327, 328–9
stratum germinativum 329, 330
stratum lucidum 132, 340
strawberry naevus 228
strip waxing 179
sugar waxing 179
sunscreens 441–2
superficial cleanse 314
surfactants 336–8, 339, 350, 420
surge protectors 87
sustainability practices 161
    aromatherapy massages 292
    aromatic plant oil blends 305
    cosmetic tanning 256
    photographic make-up tools 241
'Swedish' massage 257

T
tanning *See* cosmetic tanning
tanning enhancers 254–5
'Tap and go' payments 65
tapotement technique 267, 272, 273, 316
tattoos 228, 230–1
telephone communication 20–1
    client appointment scheduling 27
telogen 171, 184, 408
terms and conditions of employment *See* employment terms and conditions
terpenes 422
terpenols 422
Therapeutic Goods Act 1989 301
thermal effects and benefits 358–60
threading 169
tineal infections 114, 129
tint and henna 187, 190–1
tissues 388–9
    skin 428–9

toenails 133
tonofilament 435
trade magazines 51
Trade Practices (Consumer Product
    Information Standards) (Cosmetics)
    Regulations 1991 338
trade unions 7–8
training, staff 57
trans-epidermal water loss (TEWL) 328–9,
    332, 335, 339, 420
transgender clients 280
trapezius lift 273
trichokeratin (hard keratin) 435
triglycerides 420
tweezers 206
tweezing 169
T-zones 311

U

ultrasonic/galvanic machines 355–60
    modes 358
ultrasound transducer 356
underage clients 246
underperformance management 12–13
unsafe work practices 77
urea 334
urticaria 309
UV radiation 326, 329, 440, 441

V

vaporisation 304
vegetable oils 423
veins 389
    hand and forearm 146, 147
    legs and feet 149, 150
ventilation for aromatic oil blending 302
verbal communication 19–20, 44
vibrations technique 267, 273, 316
videos
    body movement 394
    bones 392, 393
    emergency procedures 98
    exfoliation 360
    facial muscles 399
    short circuit 86
    size of virus 116
    sound energy transmission 357
    static electricity 88
    tissues 388
    tissues and veins 389
virinos 117

viruses 91, 115–18
    hosts 116
    infections 116–17
    life cycle 116
    skin diseases 92
    structure 115–16
    transmission 117–18
vitamin A (retinoids) 340, 376, 422
vitamin C (l-ascorbic acid) 340, 375
vitamin E (alpha tocopherol) 340, 375, 422
vitamins, role and function of essential 374–5
voice tonality and volume 20

W

warranties 36
water 376
    emulsions 335–6, 337
    role in skin function 326–30
waxing services
    adverse outcomes 180–1
    aftercare advice 181
    applying and removing wax 177–9
    cleaning treatment area 182–3
    considerations for people under 16 years
    37
    contraindications and scope of practice
    175–6
    establishing client priorities 169–76
    preparing for 176–7
    relevant medical history and medications
    176
    reviewing and providing post-service
    advice 179–82
    risk assessment and control 79–80
weak nails 136
websites
    ACCC 57
    ACCC labelling standards 338
    acne scarring 310
    anti-discrimination complaints 9, 30
    biodegradability of items 156
    cleaning waxing pots 182
    Comcare 83
    communication with clients with special
    needs 45
    customer service principles 43
    data security 66
    dietary guidelines 378, 379
    EEO laws 5, 8
    eye make-up effects 223
    Fair Work laws 4, 12

GST information 62
    industry awards 6
    life cycle of virus 116
    lymphatic system 404
    menstruation, hormones and skin 444
    muscular system 400
    NHMRC Standard Precautions 104
    nutrition 372
    organic chemistry 299
    payment handling 64
    personal information protection 69
    pricing strategies 56
    refunds and exchanges 68
    skin cancer 440
    states and territories fair trading
    standards 47
    vitamins 376
    work health and safety 80
work ethics 57
work health and safety 29–30, 59, 72–101, 83
    applying safety procedures 74–84
    aromatherapy massages 287
    aromatic plant oil blends 301–2
    cleaning procedures 99–101
    infection control procedures 90–8
    inspections 104
    minimising infection risks 89–90
    nail product safety 141–2
    nutritional advice to clients 386
    procedures for emergencies 98–9
    protecting against infection risks 72–4
    specialised facial treatments 354–5
    standard and additional precautions for
    treatments 77
    using electricity safely 84–8
    website 80
Work Health and Safety Act 2011 59, 378
work performance analysis 52
working conditions 54–5
workplace relations 55
wound healing stages 446–9

X

xerosis (dry skin) 329–30

Y

yeasts 113

Z

ZipPay transactions 64